CW01337406

The Essentials of
Linear State-Space Systems

The Essentials of
Linear State-Space Systems

J. Dwight Aplevich

John Wiley & Sons, Inc.

New York • Chichester • Weinheim • Brisbane • Toronto • Singapore

ACQUISITIONS EDITOR	Bill Zobrist
MARKETING MANAGER	Katherine Hepburn
PRODUCTION EDITOR	Ken Santor
COVER DESIGNER	David Levy

This book was set in Times Roman by the author and printed and bound by Hamilton Printing Company. The cover was printed by Phoenix Color Corp.

This book is printed on acid-free paper.

The paper in this book was manufactured by a mill whose forest management programs include sustained yield harvesting of its timberlands. Sustained yield harvesting principles ensure that the numbers of trees cut each year does not exceed the amount of new growth.

Copyright © 2000 John Wiley & Sons, Inc. All rights reserved.

MATLAB® is a registered trademark of The Math Works, Inc.
MATRIX$_X$® is a registered trademark of Integrated Systems, Inc.

No part of this publication may be reproduced, stored in a retrieval system or transmitted in any form or by any means, electronic, mechanical, photocopying, recording, scanning or otherwise, except as permitted under Sections 107 or 108 of the 1976 United States Copyright Act, without either the prior written permission of the Publisher, or authorization through payment of the appropriate per-copy fee to the Copyright Clearance Center, 222 Rosewood Drive, Danvers, MA 01923, (508) 750-8400, fax (508) 750-4470. Requests to the Publisher for permission should be addressed to the Permissions Department, John Wiley & Sons, Inc., 605 Third Avenue, New York, NY 10158-0012, (212) 850-6011, fax (212) 850-6008, E-Mail: PERMREQ@WILEY.COM.

```
Aplevich, J. D. (J. Dwight), 1943-
   The essentials of linear state-space systems / J.D. Aplevich.
      p.    cm.
   Includes bibliographical references and index.
   ISBN 0-471-24133-4 (alk. paper)
   1. Linear systems.  2. State-space methods.   I. Title.
   QA402.A64  1999
   006'.74--dc21                                    99-13831
                                                       CIP
```

Printed in the United States of America

10 9 8 7 6 5 4 3 2 1

To the memory of my father, who knew the value of the written word, and to Pat, Noelle, Claire, and now Robert.

Preface

This book is intended to be a thorough introduction to the properties of linear, time-invariant models of dynamical systems, as required for further work in feedback control system design, power system design and analysis, communications, signal processing, robotics, and simulation. The state-space model is used throughout, since it is a fundamental conceptual tool, although the background analysis applies to other models. The material is based on a course which has been taught for several years to motivated electrical and computer engineering undergraduates at the University of Waterloo, most of them in the first term of their senior year. Beginning graduate students from the same department, as well as from other departments, often have attended. The course has two goals: first, to give a thorough grounding in linear state-space modeling and, second, to give a mature understanding of some applications of the required background, which often has been learned quickly and used little. The content is presented in thirteen weeks, usually supplemented by a modest computational project.

The required background is working knowledge of linear algebra, differential equations, Laplace transforms, and $\mathcal{Z}$ transforms. The new material is supplemented with many examples, some of which extend the theory. There is also a good dose of proofs, to promote logical thinking and to show the elegance of some of the techniques, but hints about practicalities and floating-point numerical computation are also given, where it is possible without greatly extending the material.

I have attempted to emphasize the algebra common to a diversity of applications rather than specialized analysis, and where theoretical details have been omitted, they usually have been those of lesser application. Consequently, two themes recur: the importance of the Markov sequence in defining external system behavior, and the importance of the normal form of a matrix, and its generalizations, for matrix-related analysis. In setting the context of LTI systems, time-varying and nonlinear systems are mentioned, but not treated in detail.

Many good books containing similar subject material have appeared over the years. However, we have had difficulty adapting them to our requirements, often because the content is intended exclusively as an introduction to theoretical work, because their volume is far in excess of a one-semester course, or because they contain specialized analysis or design that does not fit with the objective of an introductory course applicable in a diversity of disciplines. Consequently, it has been convenient to tailor a package at a suitable level and of suitable size.

How to use this book: Some choices are available in reading or presenting the material in these chapters. Chapters 1 to 4 can be covered in a different order than presented here. However, provided that care is taken to motivate the use of state-space models, the details of their derivation from physical laws and other model forms are covered most economically by using both time-domain solutions and transform methods. Consequently, the book steps directly into solutions in Chapters 2 and 3 before the detailed modeling of Chapter 4. Not all of the material and examples included here can be covered in a typical course of thirteen weeks. Rather, slightly more material is included to allow customization. Not all of the sections of Chapter 4 are essential for understanding later material, and sections of Chapters 8 to 10 can also be omitted without major difficulty. In Chapter 4 for example, only one of the sections on electric circuits and Euler-Lagrange equations might be covered. The second half of Chapter 8 on stability might be omitted in order to cover the algebraic applications in Chapter 10. A selection from the examples can be made, depending on desired emphasis. Some of the material not explicitly covered in class can be used as a source of computational projects.

I recommend that the reader skim most of the examples and problems, since many of them embellish the main theoretical points of the text, although it is not necessary to examine each of them in detail.

Traditional texts present material on matrices and vector spaces either at the beginning or in appendices rather than in the middle, as done here in Chapters 5–7. Because we include the material in class, I have put it in the body and in the location at which we usually cover it, although a different order of presentation could be chosen.

Some readers might be surprised at seeing the B. L. Ho algorithm in an introductory course. In fact this algorithm is at once a natural application of the normal form of a matrix and of the Cayley-Hamilton theorem, an introduction to system identification and the concept of minimality, and an application of the singular-value decomposition. The proof contains moderate detail but no particular difficulty. This material could be covered in conjunction with minimality in Chapter 9 but I find it a motivating real-world application of the normal form, which has just been introduced in Chapter 5.

In the later chapters, it is useful to distinguish between essential theoretical constructions, especially the Jordan form and the Smith form, and more reliable computations. Hints are given in examples and problems.

The material is self-contained, although at the end of each chapter there is a section on further reading. The references are inevitably a personal selection from publications over four decades, but I have attempted to include some classics, under the philosophy that it is better to read the masters than their students, together with a selection of contemporary material.

The theory given here can only be applied easily with the assistance of a computer and software for floating-point computation, although for small systems, exact symbolic computation can be helpful. An excellent computing-laboratory tool for this material is the commercial MATLAB software, which provides easy invocation of professional-quality numerical routines, a programming language, and a graphical model-building environment. I encourage students to learn to use this or comparable software, and to develop a healthy sense of caution about its blind use on inexact floating point data, especially for results that depend on the computed rank of a matrix. I usually assign one or more modest computational projects, such as implementing the Ho algorithm or testing other types of system identification. Many of the examples and problems are easily performed by using MATLAB and its toolboxes. However, the hints provided are intended to apply equally to floating-point arithmetic in embedded software, as well as to design computations conducted on an engineering workstation. A thorough introduction to the concepts of condition and the floating-point stability of algorithms is beyond the scope of this book, even if such an introduction is essential prior to the professional use of floating-point implementations of the theoretical material given here. For this as for other topics, I have attempted to provide the essentials, without attempting to be encyclopedic.

This book was produced by using LATEX and PSTricks. The diagrams were prepared with the m4 macro processor, the dpic interpreter for the pic language, and PSTricks.

Acknowledgements: Many thanks to Rob Gorbet, who used a draft of this book to teach the material, and sent perspicacious comments. Thanks also to other readers who offered comments, particularly Carolyn Beck.

I would like to thank the National Research Council of Canada Innovation Centre, Vancouver, and George Wang in particular, for allowing me to work on this material in addition to other projects during a stay there.

The production of a book of this kind involves many steps, and the people at John Wiley and Sons who made it possible have my gratitude: thanks to editor Bill Zobrist, to designer Madelyn Lesure for advice and for approving an idiosyncratic design, and to Kenneth Santor and the other production people. All have been sources of professional help and advice at the other end of an email connection.

These notes have been greatly influenced by the demanding and perceptive attention of students over several years. I hope that the transparency of exposition approaches the clarity of their questions.

Dwight Aplevich
aplevich@uwaterloo.ca
May 6, 1999.

Contents

| **Appendix** | **Solutions** | **249** |

Introduction

An understanding of dynamical system models is important in diverse disciplines: automatic control, communications, filter design, specialized computer circuit design, power systems, and robotics, as well as in other branches of engineering, and indeed, in many branches of the physical, biological and social sciences. Many of the dynamical system models of interest can be put into a common framework, that of the so-called "state-space" model. Engineering examples will be emphasized here, and the important class of linear, constant models will be analyzed in detail.

State-space models are sets of equations of a particular form, with several useful attributes: first, their form contributes to an intuitive understanding of the behavior of many dynamical systems; second, efficient computational techniques are available for solving them; and third, a large body of theory is available for analyzing them.

A general description of state-space equations will be given, and then several simple examples will be put into state-space form. The modeling consequences of linearity and time-invariance will be given, and then the simplest systems, those that are both linear and time-invariant, will be introduced. Finally, a technique for finding a linear approximation of a nonlinear system near a known solution will be given.

1 The structure of state-space models

State-space models are collections of equations corresponding to the logical structure illustrated by Figure 1.1. As will be seen, this logical structure applies to a variety of physical objects, such as electromechanical systems, digital computers, and digital software processes. Writing the equations is extremely simple for a model corresponding to this diagram, but the starting point may be a model of different structure, in which case the corresponding equations have to be manipulated to rewrite them in state-space form.

1

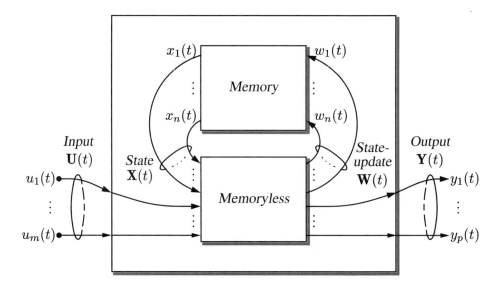

Fig. 1.1 The components of a dynamical system.

All variables in the model are assumed to be a function of an independent variable, usually time, which justifies the adjective "dynamic" for the system. Thus there is a set, typically $\mathbb{T} = \mathbb{R}$, the set of real numbers for analog systems, for which time is continuous, or $\mathbb{T} = \mathbb{Z}$, the integers, for discrete-time systems such as computer circuits.

At each time $t \in \mathbb{T}$, a set of m independent external quantities u_i, $i = 1 \cdots m$ called *inputs* are assumed to affect the system, and at each time the system is assumed to produce a set of p quantities y_i, $i = 1 \cdots p$ called *outputs*, which may affect the external environment.

For modeling purposes, inside the system there are two logically distinct types of components, grouped together in Figure 1.1. There is a set of memoryless components that compute instantaneous functions of the external inputs and of internal variables, and a set of memory elements that store internal quantities. The variables stored in the memory are called *state* variables. Thus, the memoryless part computes the functions

$$y_1(t) = g_1(x_1(t), \cdots x_n(t), u_1(t), \cdots u_m(t), t)$$

(1.1)
$$\vdots$$

$$y_p(t) = g_p(x_1(t), \cdots x_n(t), u_1(t), \cdots u_m(t), t).$$

These functions specify the p system outputs at each time t in terms of the input and state variables at time t, and in terms of t itself. By convention, the above

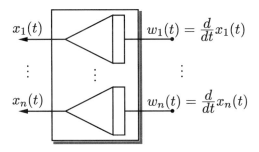

Fig. 1.2 Continuous-time memory.

equations are often written as

(1.2) $\mathbf{Y}(t) = \mathbf{G}(\mathbf{X}(t), \mathbf{U}(t), t).$

In this equation, $\mathbf{Y}(t)$ is a column of variables $y_1(t), \cdots y_p(t)$, and is referred to as a vector of dimension p, or a p-vector, with entries (or elements) $y_i(t)$, $i = 1, \cdots p$. Similarly $\mathbf{X}(t)$ is an n-vector and $\mathbf{U}(t)$ is an m-vector. The entries of vector $\mathbf{G}$ are the quantities $g_i(\cdots)$, $i = 1, \cdots p$, which are functions of the entries $x_i(t)$, $i = 1, \cdots n$ of the state vector $\mathbf{X}(t)$; of the entries $u_i(t)$, $i = 1, \cdots m$ of the input vector $\mathbf{U}(t)$; and of t.

Similarly, the state-update values in Figure 1.1 are given by equations of the form

$$w_1(t) = f_1(x_1(t), \cdots x_n(t), u_1(t), \cdots u_m(t), t)$$

(1.3)

$$\vdots$$

$$w_n(t) = f_n(x_1(t), \cdots x_n(t), u_1(t), \cdots u_m(t), t)$$

and these equations are abbreviated, by convention, as

(1.4) $\mathbf{W}(t) = \mathbf{F}(\mathbf{X}(t), \mathbf{U}(t), t).$

Two possible memory types will be considered. The continuous-time memory shown in Figure 1.2 contains n integrators, for which, using vector notation,

(1.5) $\dfrac{d}{dt}\mathbf{X}(t) = \mathbf{W}(t),$

where the time-derivative of a vector is the vector of derivatives of its entries. The discrete-time memory of Figure 1.3 contains n delays, for which

(1.6) $\mathbf{X}(t+1) = \mathbf{W}(t).$

State-space forms When the equations for the memory and memoryless parts are combined to eliminate $\mathbf{W}(t)$, the continuous-time state-space equations take the form

(1.7a) $\quad \dfrac{d}{dt}\mathbf{X}(t) = \mathbf{F}(\mathbf{X}(t), \mathbf{U}(t), t)$

(1.7b) $\quad \mathbf{Y}(t) = \mathbf{G}(\mathbf{X}(t), \mathbf{U}(t), t),$

and the equations for a discrete-time system have the form

(1.8a) $\quad \mathbf{X}(t+1) = \mathbf{F}(\mathbf{X}(t), \mathbf{U}(t), t)$

(1.8b) $\quad \mathbf{Y}(t) = \mathbf{G}(\mathbf{X}(t), \mathbf{U}(t), t).$

The right-hand sides of (1.7) and (1.8) contain only state variables and system inputs, and the equations explicitly determine $\frac{d}{dt}\mathbf{X}(t)$ in (1.7), $\mathbf{X}(t+1)$ in (1.8), and $\mathbf{Y}(t)$ in both cases. Writing state-space equations from some other starting point requires the above left-hand variables to be solved exclusively in terms of the above right-hand variables.

Often the model inputs and outputs represent real-valued quantities, that is, $\mathbf{U}(t) \in \mathbb{R}^m$ and $\mathbf{Y}(t) \in \mathbb{R}^p$, in which case the state vector $\mathbf{X}(t)$ normally contains real values as well, that is, $\mathbf{X}(t) \in \mathbb{R}^n$; but as will be seen, it may be convenient to allow $\mathbf{X}(t) \in \mathbb{C}^n$, the complex n-vectors.

Other kinds of variables are possible, however. For example, in binary computer circuits, all variables except t take values from the set $\{0, 1\}$, and provided the model is linear, it is possible to write $\mathbf{U}(t) \in \mathbb{Z}_2^m$, $\mathbf{Y}(t) \in \mathbb{Z}_2^p$, $\mathbf{X}(t) \in \mathbb{Z}_2^n$, where $\mathbb{Z}_2$ is the set of integers modulo 2.

In summary, defining a state-space model requires definition of the input, output, and state vectors; the time set $\mathbb{T}$; and the vector functions $\mathbf{F}(\cdots), \mathbf{G}(\cdots)$ in (1.7) or (1.8). Systems described by ordinary differential equations (1.7a) contain integrators in the memory of the corresponding conceptual model. Discrete-time systems described by (1.8a) contain delay elements that store quantities for one time interval. In both circumstances, the state variables are conveniently chosen as the contents of the memory elements.

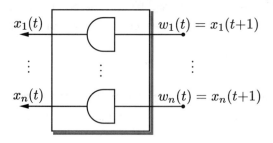

Fig. 1.3 Discrete-time memory.

Example 1
Stored quantity
associated with
state variable

Writing continuous-time state equations from physical laws often requires identifying where physical quantities are accumulated, or stored. The container
shown in Figure 1.4 is an elementary example of such a system.

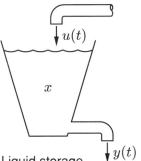

Fig. 1.4   Liquid storage.

Liquid flows in at rate $u(t)$, and flows out at rate $y(t)$. The volume x of
accumulated liquid can be used to obtain a model in the desired form, with
output equation

$$y = g(x)$$

where $g(\cdot)$ gives output flow rate as a function of x and of the physical charac-
teristics of the liquid and the container. The state-update equation is

$$\frac{dx}{dt} = u - g(x).$$

Example 2
Newton's
second law

Some mechanical components have natural state-space models. For example,
Newton's law relating the momentum q of an object and the applied force f is
$dq/dt = f$, and this equation is in the form which describes a memory element
in Figure 1.2. Thus q is a natural choice for a state variable in a model containing
the object. When $q = mv$, where the mass m is constant and v is velocity,
the equation becomes $dv/dt = (1/m)f$, so velocity v is an alternative choice
of state variable. If the object position x is included in the model, then the
additional equation $dx/dt = v$ results, and the two state variables v, x are
associated with this motion of the object.

Example 3
State variables
for capacitors
and inductors

In electric circuits, each capacitor accumulates charge proportional to the integral of current, and for each inductor the magnetic flux is proportional to the
integral of the voltage, so it is to be expected that state variables are associated

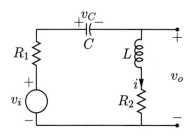

Fig. 1.5 An electric circuit.

with the inductors and capacitors. For the electric circuit shown in Figure 1.5, given input $v_i(t)$ for $t \geq t_0$, it is desired to find output $v_o(t)$ for $t \geq t_0$.

Suitable differential equations are written as follows, using Kirchhoff's laws and equations describing the circuit elements: from Kirchhoff's current law,

$$C\frac{dv_C}{dt} = i,$$

and from Kirchhoff's voltage law,

$$L\frac{di}{dt} = -v_C - (R_1 + R_2)\,i + v_i.$$

The output can be found in terms of the capacitor voltage and inductor current:

$$v_o(t) = L\frac{di}{dt} + R_2 i.$$

As is typical in many situations, the input and output variables are prespecified, but the state variables must be chosen. In the above equations, v_C and i appear differentiated once, so it is natural to choose these variables as state variables. Solving for their derivatives gives

(1.9a)
$$\frac{d}{dt}\begin{bmatrix} v_C \\ i \end{bmatrix} = \begin{bmatrix} (1/C)\,i \\ -(1/L)\,v_C - (R_1/L + R_2/L)i + (1/L)\,v_i \end{bmatrix},$$

which is in the form of (1.7a). Expressing the right-hand side of the output equation in terms of state variables and inputs gives

(1.9b) $$v_o(t) = -v_C - R_1 i + v_i,$$

which is in the form of (1.7b), so that (1.9a) and (1.9b) are in the form of state-space equations. In order to calculate $v_o(t)$ for $t \geq t_0$, not only the input $v_i(t)$ for $t \geq t_0$ must be known, but also the initial state $v_C(t_0)$, $i(t_0)$.

Example 4
Inherently
discrete-time
system

Some processes are inherently discrete-time, since the time variable is incremented by counting. Let $x(t)$ be the amount of money stored in a cashier's cash drawer before the drawer is opened for the t-th time, and let $q(t)$ be the net amount added at the t-th opening. Then the evolution of $x(t)$ as a function of the input $q(t)$ is described by

$$x(t+1) = x(t) + q(t),$$

which is of the form of Equation (1.8a), and is illustrated in Figure 1.6.

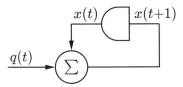

Fig. 1.6   A discrete-time system.

1.1   The concept of state

The importance of Figure 1.1 and the previous discussion is that if the independent variable t signifies time, as it normally will in these chapters, and if it is possible to model a physical system using equations exclusively of the form (1.7) or (1.8), then a state-space model is said to exist for the system, and the vector $\mathbf{X}(t)$ is defined to be the *state* at t.

State-space theory can be developed as a mathematical topic, and several axiomatic definitions, not all equivalent, of dynamical systems and their state can be found in the literature. However the word "state" was chosen because of its intuitive meaning. A convenient working definition of the state of a system is a set of quantities which are (1) independent of each other and (2) independent of present and future inputs, and for which (3) the present state is sufficient for predicting the present and future output given the present and future inputs. Condition (1) means that knowledge of $n-1$ of the entries of $\mathbf{X}$ does not imply knowledge of all n entries. Systems satisfying condition (2) are said to be causal or nonanticipative.

Equations that adequately model a physical system must have solutions. However, during model development it may be important to know that solutions exist for a postulated set of equations. Equations of the form (1.7) or (1.8) can be shown to have solutions under relatively mild technical assumptions on the vector functions $\mathbf{F}(\cdots)$, $\mathbf{G}(\cdots)$, as will be described in Chapter 2.

Consider equations (1.8), for example. Given the vector $\mathbf{X}(t_0)$ at time t_0, and given the values of the input vector from t_0 to t_f, then it is possible, by solving the equations, to predict $\{\mathbf{X}(t_0),\ \mathbf{X}(t_0+1),\ \cdots \mathbf{X}(t_f)\}$, and $\{\mathbf{Y}(t_0),\ \mathbf{Y}(t_0+1),\ \cdots \mathbf{Y}(t_f)\}$. The entire previous history of the system affects only the quantity $\mathbf{X}(t_0)$ in this prediction. Therefore the word "state" is used for $\mathbf{X}$ since the state of an object is intuitively the cumulative result of its past. Furthermore, *any* set of independent values at time t_0 which, together with $\mathbf{U}(t)$ for $t \geq t_0$, is sufficient to predict the system output $\mathbf{Y}(t)$ for $t \geq t_0$, qualifies as a definition of the system state at time t_0. Thus the choice of state variables is often not unique.

For systems completely described by a set of ordinary differential or difference equations with independent initial values as boundary conditions, the variables specified as initial values may often be taken as the state variables.

Example 5
Newton's law of cooling

Newton's law of cooling, written $\dot{T} = -\alpha T$, with initial condition $T(t_0)$, requires one state variable. A composite model will require a state variable for every such equation that is independent.

Example 6
Newton's second law and initial conditions

In mechanical systems, Newton's second law, written $\ddot{y} = (1/m)f(t)$, requires two initial conditions $y(t_0)$, $\dot{y}(t_0)$, for solution, so the choice $x_1 = y$, $x_2 = \dot{y}$ is a suitable set of state variables. In a more complex mechanical system, two state variables will be required for each independent invocation of Newton's law.

Example 7
Stateless programs

Internet server programs are often said to be "stateless," meaning that the response, which is a sequence of characters, to an input sequence does not depend on previous inputs. Then, in principle, a model of the input-response behavior can be given entirely by the algebraic equations of Equation (1.8b), without requiring Equation (1.8a). Stateless servers have distinct advantages in the event of hardware or network failures, since the client program need only retry a request until the server responds.

Example 8
Independent capacitor and inductor variables

In electric circuit models, each capacitor is described by an equation of the form $\dot{q} = i(t)$, requiring a state variable for each capacitor for which the initial value $q(t_0)$ can be specified independently. Similarly, each inductor, described by an equation of the form $\dot{\lambda} = v(t)$ with initial value $\lambda(t_0)$, will require a state variable, provided the initial value is independent.

Example 9
Initial conditions
for Figure 1.6

The discrete-time system of Figure 1.6 has solution $x(\ell) = x(t_0) + \sum_{k=t_0}^{\ell-1} q(k)$, which contains initial condition $x(t_0)$. Consequently, a suitable state variable is $x(t)$.

Example 10
Infinite-
dimension
state

Some systems have infinite-dimensional state vectors. If the state-space model of a system contains n independent differential equations, then only n independent initial conditions are necessary for their solution. However, a perfect analog delay line, with input $u(t)$ and output $y(t)$, for which $y(t+\tau) = u(t)$, cannot be modeled exactly by using a finite-dimensional state vector (see Figure 1.7). The output for $t \geq t_0$ is a function not only of $u(t)$ for $t \geq t_0$ but also, for $t_0 \leq t < \tau$, the output is the exact replica of the waveform $u(t)$ over $-\tau \leq t < t_0$, which contains an infinite number of points. It is intuitively clear that the function $u(t)$

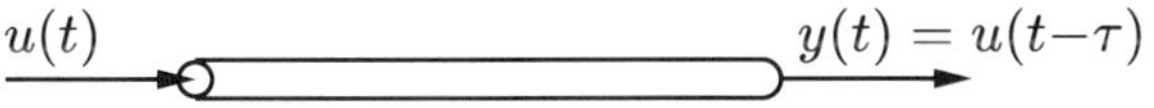

$$u(t) \qquad\qquad\qquad\qquad y(t) = u(t-\tau)$$

Fig. 1.7　Analog delay line.

over $-\tau \leq t < t_0$ satisfies the definition of initial state, since the points on it are independent of each other and of future inputs and are sufficient to predict the output, given the input. Whether this function can properly be called a "vector" will be the subject of problems at the end of Chapter 6. In practice the bandwidth of a delay line may be limited so that only a finite number of data points are required to produce a sufficiently accurate model.

2　Linear models

For a large and important class of models, each of the right-hand sides of (1.7) can be written as a linear combination of the state and input variables, with possibly time-varying coefficients, so (1.7) becomes

(1.10a)

$$\frac{d}{dt} \begin{bmatrix} x_1(t) \\ \vdots \\ x_n(t) \end{bmatrix} =$$

$$\begin{bmatrix} a_{11}(t)x_1(t) + \cdots a_{1n}(t)x_n(t) + b_{11}(t)u_1(t) + \cdots b_{1m}(t)u_m(t) \\ \vdots \\ a_{n1}(t)x_1(t) + \cdots a_{nn}(t)x_n(t) + b_{n1}(t)u_1(t) + \cdots b_{nm}(t)u_m(t) \end{bmatrix}$$

(1.10b)
$$\begin{bmatrix} y_1(t) \\ \vdots \\ y_p(t) \end{bmatrix} =$$

$$\begin{bmatrix} c_{11}(t)x_1(t) + \cdots c_{1n}(t)x_n(t) + d_{11}(t)u_1(t) + \cdots d_{1m}(t)u_m(t) \\ \vdots \\ c_{p1}(t)x_1(t) + \cdots c_{pn}(t)x_n(t) + d_{p1}(t)u_1(t) + \cdots d_{pm}(t)u_m(t) \end{bmatrix},$$

or, rewritten more conveniently using matrices,

(1.11a) $\quad \dfrac{d}{dt}\mathbf{X}(t) = \mathbf{A}(t)\,\mathbf{X}(t) + \mathbf{B}(t)\,\mathbf{U}(t)$

(1.11b) $\quad \mathbf{Y}(t) = \mathbf{C}(t)\,\mathbf{X}(t) + \mathbf{D}(t)\,\mathbf{U}(t).$

State-space equations in this form are said to be *linear*. The matrices in (1.11) are

$$\mathbf{A}(t) = \begin{bmatrix} a_{11}(t) & \cdots & a_{1n}(t) \\ a_{21}(t) & \cdots & \\ \cdots & & \\ a_{n1}(t) & \cdots & a_{nn}(t) \end{bmatrix}, \quad \mathbf{B}(t) = \begin{bmatrix} b_{11}(t) & \cdots & b_{1m}(t) \\ b_{21}(t) & \cdots & \\ \cdots & & \\ b_{n1}(t) & \cdots & b_{nm}(t) \end{bmatrix},$$

$$\mathbf{C}(t) = \begin{bmatrix} c_{11}(t) & \cdots & c_{1n}(t) \\ c_{21}(t) & \cdots & \\ \cdots & & \\ c_{p1}(t) & \cdots & c_{pn}(t) \end{bmatrix}, \quad \mathbf{D}(t) = \begin{bmatrix} d_{11}(t) & \cdots & d_{1m}(t) \\ d_{21}(t) & \cdots & \\ \cdots & & \\ d_{p1}(t) & \cdots & d_{pm}(t) \end{bmatrix}.$$

For computational purposes, the vector $\mathbf{X}$ can be taken to be an $n \times 1$ matrix, and similarly $\mathbf{U}$ and $\mathbf{Y}$ are matrices with one column.

It will be shown later that if the equations are linear, then certain of the system input-output properties, that is, the relations defining the entries of $\mathbf{Y}$ as a function of those of $\mathbf{U}$, are linear.

Analogous results hold for discrete-time equations (1.8), resulting in the linear form

(1.12a) $\quad \mathbf{X}(t+1) = \mathbf{A}(t)\,\mathbf{X}(t) + \mathbf{B}(t)\,\mathbf{U}(t)$

(1.12b) $\quad \mathbf{Y}(t) = \mathbf{C}(t)\,\mathbf{X}(t) + \mathbf{D}(t)\,\mathbf{U}(t).$

Example 11
Matrices for
Figure 1.5

For the electric circuit of Figure 1.5 described by Equation (1.9a) and Equation (1.9b), the matrices $\mathbf{A}(t), \mathbf{B}(t), \mathbf{C}(t), \mathbf{D}(t)$ are

$$\mathbf{A}(t) = \begin{bmatrix} 0 & 1/C \\ -1/L & -(R_1/L + R_2/L) \end{bmatrix}, \quad \mathbf{B}(t) = \begin{bmatrix} 0 \\ 1/L \end{bmatrix},$$

$$\mathbf{C}(t) = [-1, -R_1], \quad \mathbf{D}(t) = [1].$$

Dimensions   It is useful to take note of the dimensions of the quantities in (1.11) and (1.12). Writing (1.10) in the form of (1.11) implies that the ordinary rules of matrix addition and multiplication are being followed; that is, addition is defined only for matrices of identical size, and multiplication is defined only for matrices that are conformable. The following rules can be developed, and are illustrated in Figure 1.8.

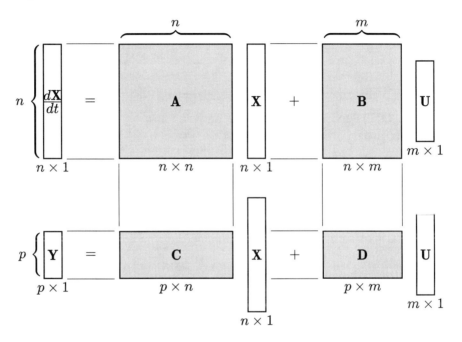

Fig. 1.8   Dimensions of the matrices of a linear system. Dimension m or p or both may be greater than n.

1. The number of system inputs, m, is the dimension of the vector $\mathbf{U}(t)$, and since the terms $\mathbf{B}(t)\,\mathbf{U}(t)$ and $\mathbf{D}(t)\,\mathbf{U}(t)$ appear in the equations, $\mathbf{B}(t)$ and $\mathbf{D}(t)$ must have m columns.

2. The number of system outputs, p, is the dimension of the vector $\mathbf{Y}(t)$; thus the right-hand sides of (1.11b) and (1.12b) must have p rows, implying that $\mathbf{C}(t)$ and $\mathbf{D}(t)$ must have p rows.

3. The number of state variables, n, determines the dimension of vector $\mathbf{X}(t)$ and of $\frac{d}{dt}\mathbf{X}(t)$ in (1.11) or $\mathbf{X}(t+1)$ in (1.12). By the rules of multiplication, $\mathbf{A}(t)$ and $\mathbf{C}(t)$ must therefore have n columns, and $\mathbf{A}(t)$, $\mathbf{B}(t)$ must have n rows.

In summary, $\mathbf{X}$ is of dimension $n \times 1$, $\mathbf{U}$ is $m \times 1$, and $\mathbf{Y}$ is $p \times 1$. The matrix $\mathbf{A}$ has dimension $n \times n$, $\mathbf{B}$ is $n \times m$, $\mathbf{C}$ is $p \times n$, and $\mathbf{D}$ is $p \times m$.

Example 12
Sampled liquid
flow

Discrete-time systems are often obtained by sampling continuous-time systems at regular intervals. Suppose it takes three seconds for water to flow through a pipe. At the k-th second the inlet water temperature $u(k)$ is measured. The output temperature, assuming no heat transfers, is $y(k)$ at time k. A functional diagram of this process is shown in Figure 1.9.

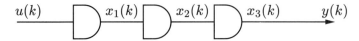

Fig. 1.9   Functional diagram of a sampled flow.

Three state variables are required to write the state-space equations corresponding to the diagram:

$$(1.13a) \qquad \begin{bmatrix} x_1(k+1) \\ x_2(k+1) \\ x_3(k+1) \end{bmatrix} = \begin{bmatrix} 0 & 0 & 0 \\ 1 & 0 & 0 \\ 0 & 1 & 0 \end{bmatrix} \begin{bmatrix} x_1(k) \\ x_2(k) \\ x_3(k) \end{bmatrix} + \begin{bmatrix} 1 \\ 0 \\ 0 \end{bmatrix} u(k)$$

$$(1.13b) \qquad y(k) = \begin{bmatrix} 0 & 0 & 1 \end{bmatrix} \begin{bmatrix} x_1(k) \\ x_2(k) \\ x_3(k) \end{bmatrix} + 0\,u(k).$$

---

3   Time-invariant models

In addition to the property of linearity, which was introduced in previous sections, models can be classified according to whether their properties vary over time.

The right-hand sides of (1.7) and (1.8) are expressed in terms of the entries of $\mathbf{X}$ and $\mathbf{U}$, which may vary with time, but also possibly in terms of t, independently of the values of $\mathbf{X}$, $\mathbf{U}$. In Figure 1.1, if at all instants t the values of the entries of the output and state-update vectors can be computed from the instantaneous values of $\mathbf{X}$ and $\mathbf{U}$ alone, then the system is *time-invariant*. Thus if each of the right-hand sides of (1.7) or (1.8) is not an explicit function of t, then the equations are time-invariant.

To test for time-invariance, first express the model in the form of (1.7) or (1.8) as appropriate, where the right-hand sides are functions of the entries of $\mathbf{X}$ and $\mathbf{U}$, possibly of t, and of no other quantities. If

$$(1.14) \qquad \frac{\partial}{\partial t} f_i(\mathbf{X}, \mathbf{U}, t) = 0, \quad i = 1, \cdots n$$

and

(1.15) $\dfrac{\partial}{\partial t} g_i(\mathbf{X}, \mathbf{U}, t) = 0, \quad i = 1, \cdots p$

then the equations are time-invariant. By definition, the above partial derivatives are obtained by replacing the entries of $\mathbf{X}$ and $\mathbf{U}$ by *constant* parameters, and then taking the ordinary derivative with respect to t.

Example 13
Time-invariant and time-varying resistors

A constant resistor R with input $i(t)$ and output $v(t)$ has model $v(t) = Ri(t)$. Setting the input $i(t)$ to a constant, say α, and differentiating,

$$\frac{d(R\alpha)}{dt} = 0,$$

which confirms that the model is time-invariant. However, if the resistance is some function $R(t)$ of time, then in general,

$$\frac{d(R(t)\alpha)}{dt} \neq 0,$$

and the model is not time-invariant. Circuits with constant components are time-invariant.

Example 14
Time-invariance by choice of state variables

The equation $\frac{d}{dt}x = 2tx$ is time-varying since $\frac{\partial}{\partial t}(2tx) = 2x$, which is not zero for nonzero x. On the other hand, the new choice of state $x_1 = x$, $x_2 = t$ results in the equation

$$\frac{d}{dt}\begin{bmatrix} x_1 \\ x_2 \end{bmatrix} = \begin{bmatrix} 2x_1 x_2 \\ 1 \end{bmatrix},$$

for which the right-hand side is time-invariant. In this example the original equation is linear in the state x, whereas the second set of equations is time-invariant but not linear in x. Thus *the choice of state vector* may determine whether the equations are time-varying or not.

---

4   **Linear, time-invariant (LTI) models**

Time-invariance corresponds to constant right-hand coefficients in (1.10) and to matrices $\mathbf{A}$, $\mathbf{B}$, $\mathbf{C}$, $\mathbf{D}$ with constant entries in (1.11) and (1.12).

If the continuous-time equations can be written in the form

(1.16a) $\dfrac{d}{dt}\mathbf{X}(t) = \mathbf{A}\mathbf{X}(t) + \mathbf{B}\mathbf{U}(t)$

(1.16b) $\mathbf{Y}(t) = \mathbf{C}\mathbf{X}(t) + \mathbf{D}\mathbf{U}(t),$

containing constant matrices $\mathbf{A}$, $\mathbf{B}$, $\mathbf{C}$, $\mathbf{D}$, then the equations are said to be LTI. Similarly, LTI discrete-time equations have the form

(1.17a) $\quad \mathbf{X}(t+1) = \mathbf{A}\mathbf{X}(t) + \mathbf{B}\mathbf{U}(t)$

(1.17b) $\quad\quad \mathbf{Y}(t) = \mathbf{C}\mathbf{X}(t) + \mathbf{D}\mathbf{U}(t).$

Example 15
Shift register

A binary computer circuit containing "D" flip-flops and exclusive-or adders is shown in Figure 1.10. The flip-flops act as delay elements, and when two or

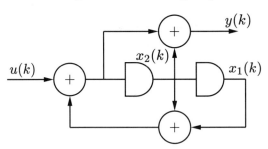

Fig. 1.10 A linear binary circuit.

more are connected sequentially, as shown, they are sometimes called a shift-register. The descriptive equations are

$$\begin{bmatrix} x_1(k+1) \\ x_2(k+1) \end{bmatrix} = \begin{bmatrix} 0 & 1 \\ 1 & 1 \end{bmatrix} \begin{bmatrix} x_1(k) \\ x_2(k) \end{bmatrix} + \begin{bmatrix} 0 \\ 1 \end{bmatrix} u(k)$$

$$y(k) = (u(k) + (x_2(k) + x_1(k))) + x_2(k)$$

$$= u(k) + x_2(k) + x_1(k) + x_2(k) = x_1(k) + u(k).$$

In the last equation, the additions are modulo 2, that is, exclusive-or, so that $x_2(k) + x_2(k) = (1+1)\,x_2(k) = (0)\,x_2(k) = 0.$

5   System properties and model properties

In previous sections, sufficient tests for linearity and time-invariance of equations were given. That is, if the equations can be written in the form of (1.11) or (1.12), they are linear; if the right-hand sides satisfy (1.14) and (1.15), the equations are time-invariant. However, sometimes the input-output properties of a system are not the same as the properties of the equations that model it.

For example, the voltage-current relations for the circuit of Figure 1.11 are nonlinear because of the ideal diodes, which are nonlinear elements; but by symmetry the nonlinearities cancel, so an equivalent model for the input-output relation can be found that is linear.

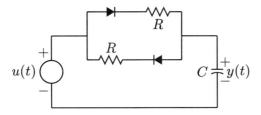

Fig. 1.11   A nonlinear circuit with linear input-output properties.

Similarly, in the circuit of Figure 1.12, let $R_1 = R_2 = 1$, and let $C(t) = L(t) > 0$ for all t. Using the inductor flux-linkage λ and the capacitor charge q as state variables, the equations describing the circuit are

(1.18a) $\dot{\lambda} = -\lambda/L(t) + u(t)$

(1.18b) $\dot{q} = -q/C(t) + u(t)$

(1.18c) $y(t) = \lambda/L(t) - q/C(t) + u(t),$

and these equations are time-varying. Because the equations are linear, the output as a function of the input can be found by setting initial conditions to zero and solving; but since the differential equations above have the same form and initial conditions, and since $L(t) = C(t)$, it follows that the solutions $\lambda(t)$ and $q(t)$ are identical. Thus the output equation reduces to $y = u$, which is a time-invariant description of the input-output relation of the system.

In the above examples the linearity or time-invariance of the input-output relation, sometimes called the *system* properties, differs from the properties of the equations, or *model*. This difference is typically of little importance for models of physical systems, although axiomatic definitions of system linearity and time-invariance based on the input-output relation are sometimes used, rather than considering only the properties of the model equations.

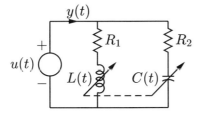

Fig. 1.12   A time-varying circuit with time-invariant input-output properties.

6   Linearized small-signal models

Linear models are often valid representations of nonlinear physical phenomena when small changes in variables are considered. The analysis of electronic filter and amplifier circuits, for example, typically uses linear small-signal representations of the circuit components. This linearization is possible for a large class of systems. A basic circuit example follows, and then the derivation required for more general state-space models.

Example 16
Small-signal
diode model

The small-signal behavior of electronic circuits is often of interest for communications applications. Figure 1.13 shows one of the simplest examples, a diode with input voltage $v = V^o + \delta v$, where V^o is a constant bias voltage and $\delta v(t)$ is a small input signal. The output current is $i(t) = I^o + \delta i$. The state-space equa-

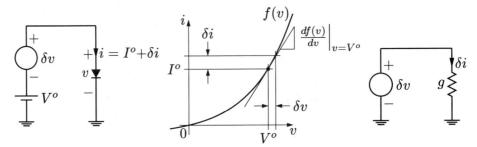

Fig. 1.13   Diode circuit showing output characteristic and small-signal equivalent circuit.

tions require no differentiated quantities, and consist only of the output equation

$$(1.19) \quad i = f(v),$$

where $f(\cdot)$ is the voltage-current characteristic of the diode, as shown in the figure. The operating point is $v = V^o$, $i = I^o = f(V^o)$, as determined by the bias voltage. Writing the output equation (1.19) as a Taylor series gives

$$(1.20) \quad I^o + \delta i = f(V^o + \delta v) = f(V^o) + \left.\frac{df(v)}{dv}\right|_{v=V^o} \delta v + \text{higher-order terms},$$

from which the small-signal equation is

$$(1.21) \quad \delta i = g\,\delta v,$$

where $g = \left.\frac{df(v)}{dv}\right|_{v=V^o}$, as illustrated in the diagram. In this and many other circuit examples, the linear small-signal equivalent circuit is obtained by replacing all voltages and currents by their perturbations, replacing all nonlinear elements by their slopes at the operating point, as shown in the figure, and by keeping linear components unchanged.

General case

Extending the linearization technique in the above example to more general state-space models requires careful notation, but no fundamentally different methods. Given a nonlinear set of state-space equations (1.7), repeated here,

(1.7a) $\quad \dfrac{d}{dt}\mathbf{X}(t) = \mathbf{F}(\mathbf{X}(t), \mathbf{U}(t), t),$

(1.7b) $\quad\quad \mathbf{Y}(t) = \mathbf{G}(\mathbf{X}(t), \mathbf{U}(t), t),$

suppose that a symbolic or tabulated solution $\mathbf{X}^o(t)$ has been found for a known input $\mathbf{U}(t) = \mathbf{U}^o(t)$, producing output $\mathbf{Y}(t) = \mathbf{Y}^o(t)$, over some time interval $t_0 \leq t < t_1$. Such a solution is typically obtained by computer simulation, because nonlinear equations have closed-form solutions only in special cases. Suppose also that solutions "near" the known solution are required over the given interval, for input $\mathbf{U}(t) = \mathbf{U}^o(t) + \delta\mathbf{U}(t)$, writing the state as $\mathbf{X}(t) = \mathbf{X}^o(t) + \delta\mathbf{X}(t)$, and the output as $\mathbf{Y}(t) = \mathbf{Y}^o(t) + \delta\mathbf{Y}(t)$. Then the required solution can be obtained by finding and solving equations for the differences $\delta\mathbf{U}(t)$, $\delta\mathbf{X}(t)$, and $\delta\mathbf{Y}(t)$, where advantage is to be taken of the assumed small size of the entries of these vectors. For a slightly cleaner notation in the following, these quantities will be written $\delta\mathbf{U}$, $\delta\mathbf{X}$, $\delta\mathbf{Y}$, with similar omission of the time-dependence of other vectors.

Recall that a function of several variables, $f(v_1, v_2, \cdots v_\eta)$, which is sufficiently differentiable in the neighborhood of a point $(v_1^o, v_2^o, \cdots v_\eta^o)$, has a Taylor series, which is written as

$$f(v_1^o + \delta v_1, v_2^o + \delta v_2, \cdots v_\eta^o + \delta v_\eta) = f(v_1^o, v_2^o, \cdots v_\eta^o) + \left.\frac{\partial f}{\partial v_1}\right|_o \delta v_1$$

$$+ \left.\frac{\partial f}{\partial v_2}\right|_o \delta v_2 + \cdots \left.\frac{\partial f}{\partial v_\eta}\right|_o \delta v_\eta + \text{higher-order terms},$$

where the notation $\left.\frac{\partial f}{\partial v_i}\right|_o$ means the value calculated by taking the indicated partial derivative and evaluating the resulting formula at $v_1 = v_1^o, \cdots v_\eta = v_\eta^o$.

If we apply the above result to the i-th equation of Equation (1.7a), which is

(1.22) $\quad \dfrac{dx_i}{dt} - f_i(x_1, \cdots x_n, u_1, \cdots u_m, t),$

this equation can be expanded as

(1.23) $\quad \dfrac{d}{dt}(x_i^o + \delta x_i) = f_i(x_1^o, \cdots x_n^o, u_1^o, \cdots u_m^o, t) + \left.\frac{\partial f_i}{\partial x_1}\right|_o \delta x_1 + \cdots \left.\frac{\partial f_i}{\partial x_n}\right|_o \delta x_n$

$$+ \left.\frac{\partial f_i}{\partial u_1}\right|_o \delta u_1 + \cdots \left.\frac{\partial f_i}{\partial u_m}\right|_o \delta u_m + \text{higher-order terms,}$$

where the term $\left.\frac{\partial f_i}{\partial t}\right|_o \delta t$ has been omitted on the assumption that either $\left.\frac{\partial f_i}{\partial t}\right|_o = 0$ or that variations with respect to t are not of interest, as can be justified in a context which is beyond the scope of this discussion since time-invariant systems are of principal importance. The above equation is the i-th row of the vector equation (1.7a) written in series form as

$$(1.24) \quad \frac{d}{dt}\mathbf{X}^o + \frac{d}{dt}\delta\mathbf{X} = \mathbf{F}(\mathbf{X}^o, \mathbf{U}^o, t)$$

$$+ \begin{bmatrix} \left.\frac{\partial f_1}{\partial x_1}\right|_o & \cdots & \left.\frac{\partial f_1}{\partial x_n}\right|_o \\ & \cdots & \\ \left.\frac{\partial f_n}{\partial x_1}\right|_o & \cdots & \left.\frac{\partial f_n}{\partial x_n}\right|_o \end{bmatrix} \begin{bmatrix} \delta x_1 \\ \vdots \\ \delta x_n \end{bmatrix} + \begin{bmatrix} \left.\frac{\partial f_1}{\partial u_1}\right|_o & \cdots & \left.\frac{\partial f_1}{\partial u_m}\right|_o \\ & \cdots & \\ \left.\frac{\partial f_n}{\partial u_1}\right|_o & \cdots & \left.\frac{\partial f_n}{\partial u_m}\right|_o \end{bmatrix} \begin{bmatrix} \delta u_1 \\ \vdots \\ \delta u_m \end{bmatrix}$$

$$+ \text{higher-order terms,}$$

which is often written more compactly as

$$(1.25) \quad \frac{d}{dt}\mathbf{X}^o + \frac{d}{dt}\delta\mathbf{X} = \mathbf{F}(\mathbf{X}^o, \mathbf{U}^o, t) + \left[\frac{\partial \mathbf{F}}{\partial \mathbf{X}}\right]_o \delta\mathbf{X} + \left[\frac{\partial \mathbf{F}}{\partial \mathbf{U}}\right]_o \delta\mathbf{U} + \text{higher-order terms.}$$

The vectors $\mathbf{X}^o$, $\mathbf{U}^o$ correspond to a solution of Equation (1.7a); therefore the leftmost term on the left side of (1.25) equals the leftmost term on the right. Furthermore, provided that the desired solution differs from $\mathbf{X}^o$, $\mathbf{U}^o$ by an amount small enough to allow the higher-order terms to be ignored, then the following equation describes the deviations of the solutions from $\mathbf{X}^o$, $\mathbf{U}^o$:

$$(1.26) \quad \frac{d}{dt}\delta\mathbf{X} = \left[\frac{\partial \mathbf{F}}{\partial \mathbf{X}}\right]_o \delta\mathbf{X} + \left[\frac{\partial \mathbf{F}}{\partial \mathbf{U}}\right]_o \delta\mathbf{U}.$$

The matrix $\left[\frac{\partial \mathbf{F}}{\partial \mathbf{X}}\right]$ is called the *Jacobian matrix* of $\mathbf{F}(\mathbf{X}, \mathbf{U}, t)$ with respect to $\mathbf{X}$, and $\left[\frac{\partial \mathbf{F}}{\partial \mathbf{U}}\right]$ is the Jacobian matrix of $\mathbf{F}(\mathbf{X}, \mathbf{U}, t)$ with respect to $\mathbf{U}$. As before, the subscript means that the entries are evaluated at the known solution $x_1 = x_1^o, \cdots x_n = x_n^o, u_1 = u_1^o, \cdots u_m = u_m^o$.

Similarly, Equation (1.7b) can be expanded as

$$(1.27) \quad \mathbf{Y}^o + \delta\mathbf{Y} = \mathbf{G}(\mathbf{Y}^o, \mathbf{U}^o, t) + \left[\frac{\partial \mathbf{G}}{\partial \mathbf{X}}\right]_o \delta\mathbf{X} + \left[\frac{\partial \mathbf{G}}{\partial \mathbf{U}}\right]_o \delta\mathbf{U} + \text{higher-order terms,}$$

of which the part describing the deviations from the solution is, on dropping the high-order terms,

$$(1.28) \quad \delta\mathbf{Y} = \left[\frac{\partial \mathbf{G}}{\partial \mathbf{X}}\right]_o \delta\mathbf{X} + \left[\frac{\partial \mathbf{G}}{\partial \mathbf{U}}\right]_o \delta\mathbf{U}.$$

Equations (1.26) and (1.28) are called the *linearization* of (1.7a) and (1.7b) respectively, at $\mathbf{X}^o(t)$, $\mathbf{U}^o(t)$.

In practice, the partial derivatives required in the Jacobian matrices may have to be computed numerically when closed-form expressions are unavailable.

Operating point   If Equation (1.7) is time-invariant and if $\mathbf{X}^o(t) = \mathbf{X}_0$, $\mathbf{U}^o(t) = \mathbf{U}_0$ are constant vectors, then all of the partial derivatives required in (1.26) and (1.28) evaluate to constants, resulting in an LTI linearization. The pair of vectors $\mathbf{X}_0$, $\mathbf{U}_0$ is then called the *operating point* of the system. If, however, either of $\mathbf{X}^o(t)$ or $\mathbf{U}^o(t)$ varies with time, then the coefficient matrices in (1.26), (1.28) are time-varying in general.

Summary   In summary, to find a linear approximation to a nonlinear state-space model (1.7), first find a solution $\mathbf{X}^o(t)$, $\mathbf{U}^o(t)$, and then compute

$$\mathbf{A}(t) = \left[\frac{\partial \mathbf{F}}{\partial \mathbf{X}}\right]_o = \begin{bmatrix} \frac{\partial f_1}{\partial x_1}\Big|_o & \cdots & \frac{\partial f_1}{\partial x_n}\Big|_o \\ & \cdots & \\ \frac{\partial f_n}{\partial x_1}\Big|_o & \cdots & \frac{\partial f_n}{\partial x_n}\Big|_o \end{bmatrix},$$

$$\mathbf{B}(t) = \left[\frac{\partial \mathbf{F}}{\partial \mathbf{U}}\right]_o = \begin{bmatrix} \frac{\partial f_1}{\partial u_1}\Big|_o & \cdots & \frac{\partial f_1}{\partial u_m}\Big|_o \\ & \cdots & \\ \frac{\partial f_n}{\partial u_1}\Big|_o & \cdots & \frac{\partial f_n}{\partial u_m}\Big|_o \end{bmatrix},$$

$$\mathbf{C}(t) = \left[\frac{\partial \mathbf{G}}{\partial \mathbf{X}}\right]_o = \begin{bmatrix} \frac{\partial g_1}{\partial x_1}\Big|_o & \cdots & \frac{\partial g_1}{\partial x_n}\Big|_o \\ & \cdots & \\ \frac{\partial g_p}{\partial x_1}\Big|_o & \cdots & \frac{\partial g_p}{\partial x_n}\Big|_o \end{bmatrix},$$

$$\mathbf{D}(t) = \left[\frac{\partial \mathbf{G}}{\partial \mathbf{U}}\right]_o = \begin{bmatrix} \frac{\partial g_1}{\partial u_1}\Big|_o & \cdots & \frac{\partial g_1}{\partial u_m}\Big|_o \\ & \cdots & \\ \frac{\partial g_p}{\partial u_1}\Big|_o & \cdots & \frac{\partial g_p}{\partial u_m}\Big|_o \end{bmatrix},$$

all evaluated at $\mathbf{X}^o(t)$, $\mathbf{U}^o(t)$. Then the deviations from $\mathbf{X}^o(t)$, $\mathbf{U}^o(t)$ are given by the linearization of Equation (1.7):

(1.29a) $$\frac{d}{dt}\delta\mathbf{X} = \mathbf{A}(t)\,\delta\mathbf{X} + \mathbf{B}(t)\,\delta\mathbf{U},$$

(1.29b) $$\delta\mathbf{Y} = \mathbf{C}(t)\,\delta\mathbf{X} + \mathbf{D}(t)\,\delta\mathbf{U}.$$

This approximate model exists provided the required derivatives exist at $\mathbf{X}^o(t)$, $\mathbf{U}^o(t)$, and gives a valid description of the deviations provided the high-order terms in the series can be neglected.

If the nonlinear equations are time-invariant, and the solution $\mathbf{X}^o(t)$, $\mathbf{U}^o(t)$ is constant, then the matrices $\mathbf{A}(t), \mathbf{B}(t), \mathbf{C}(t), \mathbf{D}(t)$ are constant and the linearized system is LTI.

All of the above applies to discrete-time equations of the form of Equation (1.8), with differentiation by time replaced by shift in time.

Example 17
Linearized
chemical
process

In simple cases it may be possible to solve the nonlinear state equations explicitly. Suppose x is the difference between the temperature of a certain chemical process and room temperature. Let u be the rate at which the process is stirred, and suppose that stirring tends to slow the temperature change, which is given by the nonlinear time-invariant equation

$$(1.30) \quad \frac{dx}{dt} = f(x, u) = -ax + bxu.$$

It has been found that the best product is produced by stirring constantly at the rate

$$(1.31) \quad u^o = \frac{a}{2b},$$

which results in temperature trajectory

$$(1.32) \quad x^o(t) = x(0)\, e^{-at/2}$$

for initial condition $x(0)$ at time $t = 0$. Because of inevitable variations it is desirable to know how the system behaves for small changes $\delta x(t)$ and $\delta u(t)$ from the nominal values $x^o(t)$ and $u^o(t)$. There is no separate output equation, but rather the state $x(t)$ is of interest.

The linearized system corresponding to Equation (1.26) is given by

$$(1.33) \quad \frac{d(\delta x)}{dt} = \mathbf{A}(t)\, \delta x + \mathbf{B}(t)\, \delta u,$$

where

$$\mathbf{A}(t) = \left.\frac{\partial f}{\partial x}\right|_{\substack{x=x^o(t) \\ u=u^o(t)}} = (-a + bu)\big|_{\substack{x=x(0)e^{-at/2} \\ u=a/2b}} = -a/2,$$

$$\mathbf{B}(t) = \left.\frac{\partial f}{\partial u}\right|_{\substack{x=x^o(t) \\ u=u^o(t)}} = (bx)\big|_{\substack{x=x(0)e^{-at/2} \\ u=a/2b}} = bx(0)\, e^{-at/2}.$$

In this example, although the original nonlinear system is time-invariant, its linearization about the time-varying solution $x^o = x(0)\, e^{-at/2}$, $u^o = a/2b$ is not time-invariant, since $\mathbf{B}(t)$ is not constant.

Example 18
Constant
operating point

Suppose that in the previous example, the stirring rate is increased to maintain the temperature constant, at $x^o = T$, which is obtained by stirring at rate $u^o = a/b$ after the temperature reaches the value T. The solution, or operating point, is now constant. The linearization about this solution is as calculated previously, except that

$$\mathbf{A}(t) = (-a + bu)|_{\substack{x=T \\ u=a/b}} = 0,$$

$$\mathbf{B}(t) = (bx)|_{\substack{x=T \\ u=a/b}} = bT,$$

which are both constant because the nonlinear system is time-invariant and is linearized at a constant operating point. The linearization is therefore an LTI system.

Example 19
Inverse
pendulum

An apparatus commonly found in control-system laboratories is an inverse pendulum, for which it is desired to exert a force $u(t)$ on the wheeled cart of mass M to balance the mass m vertically, with $M \gg m$, as shown in Figure 1.14.

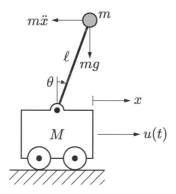

Fig. 1.14   The idealized inverse pendulum.

Ignoring friction and summing torques at the pendulum pivot,

$$m\ell^2\ddot{\theta} = mg\ell \sin\theta - m\ell\ddot{x}\cos\theta,$$

and summing horizontal forces, again ignoring friction, we get

$$u(t) = M\ddot{x} + m\frac{d^2}{dt^2}(x + \ell\sin\theta) = (M + m)\ddot{x} + \ell m(\ddot{\theta}\cos\theta - \dot{\theta}^2\sin\theta).$$

Choosing the state variables to be $x_1 = x$, $x_2 = \dot{x}$, $x_3 = \theta$, $x_4 = \dot{\theta}$ and solving the above equations for $\ddot{x} = \dot{x}_2$, $\ddot{\theta} = \dot{x}_4$ we find the four time-invariant state equations shown, two of which are nonlinear in state variables:

(1.34a)
$$\begin{bmatrix} \dot{x}_1 \\ \dot{x}_3 \end{bmatrix} = \begin{bmatrix} f_1 \\ f_3 \end{bmatrix} = \begin{bmatrix} x_2 \\ x_4 \end{bmatrix}$$

(1.34b)
$$\begin{bmatrix} \dot{x}_2 \\ \dot{x}_4 \end{bmatrix} = \begin{bmatrix} f_2 \\ f_4 \end{bmatrix}$$

$$= \frac{1}{m\ell \cos^2 x_3 - \ell(m+M)} \begin{bmatrix} m\ell \cos x_3 & -\ell \\ -(m+M) & \cos x_3 \end{bmatrix} \begin{bmatrix} g \sin x_3 \\ u + m\ell x_4^2 \sin x_3 \end{bmatrix}.$$

If the operating point is the point of balance, for which all state variables and u are zero, then all partial derivatives evaluate to zero at the operating point, except the following:

$$\left.\frac{\partial f_1}{\partial x_2}\right|_o = 1, \qquad \left.\frac{\partial f_2}{\partial x_3}\right|_o = -mg/M, \qquad \left.\frac{\partial f_2}{\partial u}\right|_o = 1/M,$$

$$\left.\frac{\partial f_3}{\partial x_4}\right|_o = 1, \qquad \left.\frac{\partial f_4}{\partial x_1}\right|_o = g/\ell, \qquad \left.\frac{\partial f_4}{\partial u}\right|_o = -1/\ell M.$$

Then the linearization is

(1.35)
$$\begin{bmatrix} \dot{x}_1 \\ \dot{x}_2 \\ \dot{x}_3 \\ \dot{x}_4 \end{bmatrix} = \begin{bmatrix} 0 & 1 & 0 & 0 \\ 0 & 0 & -mg/M & 0 \\ 0 & 0 & 0 & 1 \\ 0 & 0 & g/\ell & 0 \end{bmatrix} \begin{bmatrix} x_1 \\ x_2 \\ x_3 \\ x_4 \end{bmatrix} + \begin{bmatrix} 0 \\ 1/M \\ 0 \\ -1/M\ell \end{bmatrix} u,$$

which is LTI since the nonlinear system is time-invariant and the operating point is constant.

7 Further study

Although its roots can be traced far back in time, modern analysis of state-space equations dates from approximately 1960, and two classic references of the period are Zadeh and Desoer [58] for linear systems and Minorsky [37] for nonlinear systems. The reader may wish to consult more recent volumes, such as Vidyasagar [54], which heavily uses nonlinear state-space equations, and among the many references for linear systems, those of Antsaklis and Michel [2], Chen [9], DeCarlo [12], Kailath [25], Rugh [46], and Szidarovszky and Bahill [52].

State-space equations are basic vocabulary in the interdisciplinary field of advanced feedback control system analysis and design, for which there have been many excellent textbooks, of which only a few will be mentioned: Åström and Wittenmark [3], Brogan [6], and Friedland [16]. In more specialized fields, there have been books explicitly focused on state-space models, but these models now tend to appear implicitly, or as a means to an end, as in robotics (see Spong and Vidyasagar [48]) or coding (see Dholakia [14]). In other disciplines the state-space model provides a unifying understanding of process dynamics, although the particular state-space form may not be required when a closed-form solution is not being sought. For the computer modeling of large-scale electric circuits, for example, other models are used (see Vlach and Singhal [55]), but much of the background analysis presented here still applies. The same conclusions are true of the general field of simulation; see, for example, Cellier [8].

For a modern motivation and presentation of generalized state-space models, see the publications of Willems [56, 57].

| **8** | **Problems** |

1 Fill in the entries in Table P1.1. Where necessary, choose suitable state variables.

Table P1.1 Systems, variables, equations, matrices.

Device	Input	Output	State	Equations	**A, B, C, D**
$\xrightarrow{i}$ $+\,v\,-$ capacitor C	i	v	q	$\dot{q} = i$ $v = q/C$	0, 1, 1/C, 0
$\xrightarrow{i}$ $+\,v\,-$ capacitor C	i	v	v		
$\xrightarrow{i_1}$ $+\,v_1\,-$ C_1 $\xrightarrow{i_2}$ $+\,v_2\,-$ C_2	$\begin{bmatrix} i_1 \\ i_2 \end{bmatrix}$	v_1			
i — C $\parallel$ R v $+/-$	i	v			
u —[AND]— y	u	y			
u —[+]— y (mod 2)	u	y			

2 How many state variables are required to model the following systems?

(a) A linear resistor with input $v(t)$ and output $i(t)$.

(b) A nonlinear resistor with input $v(t)$ and output $i(t)$, characterized by the equation $i(t) = g(v(t))$.

(c) A frictionless mass in space subject to Newton's second law, with momentum $q(t)$ as output, and force $f(t)$ as input.

(d) The linear model of an automobile suspension shown in Figure P1.2(d).

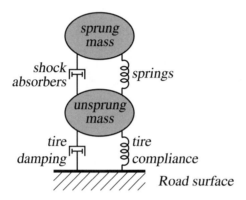

Fig. P1.2(d)   Automobile suspension.

(e) The motion, with respect to the earth, of a rigid satellite revolving around the earth.

(f) A room lighting fixture that can be turned on or off from two separate switches. The switch positions are inputs, and the light intensity is the output.

3 A nonlinear system is described by the following equations with input variables u_1, u_2, and output y:

$$\frac{d^2 v}{dt^2} + (\sin v)\frac{dv}{dt} + u_2 v = u_1 + u_2, \quad y = (\cos v)\, u_2.$$

(a) Is the system described above by these equations time-varying or time-invariant? Explain.

(b) Find state-space matrices $\mathbf{A}$, $\mathbf{B}$, $\mathbf{C}$, $\mathbf{D}$ for the system linearized at the constant operating point defined by $u_1 = 0$, $u_2 = 1$.

4   A simple nonlinear system that exhibits particularly complex behavior, the subject of chaos theory, is given by the equations

$$\dot{x}_1 = 10(-x_1 + x_2)$$
$$\dot{x}_2 = 28x_1 - x_2 - x_1 x_3$$
$$\dot{x}_3 = -8x_3/3 + x_1 x_2.$$

   (a) Determine whether this system is time-varying or time-invariant.

   (b) Find the constant operating points for the system.

   (c) At one of the constant operating points which is not the origin, find the
   matrices of the linearized state-space model.

5   In Figure P1.5, the output $\mathbf{Y}_2$ of system $\mathcal{S}_2$ is the input $\mathbf{U}_1$ of system $\mathcal{S}_1$. Let the

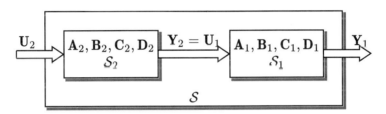

Fig. P1.5   Series connection of two systems.

state vector of system $\mathcal{S}$, composed of $\mathcal{S}_1$ and $\mathcal{S}_2$, be $\mathbf{X} = \begin{bmatrix} \mathbf{X}_1 \\ \mathbf{X}_2 \end{bmatrix}$, where $\mathbf{X}_1$ is
the state of $\mathcal{S}_1$ and $\mathbf{X}_2$ the state of system $\mathcal{S}_2$. Find the $\mathbf{A}$, $\mathbf{B}$, $\mathbf{C}$, $\mathbf{D}$ matrices of
system $\mathcal{S}$ in terms of $\mathbf{A}_1$, $\mathbf{B}_1$, $\mathbf{C}_1$, $\mathbf{D}_1$ of $\mathcal{S}_1$ and $\mathbf{A}_2$, $\mathbf{B}_2$, $\mathbf{C}_2$, $\mathbf{D}_2$ of system $\mathcal{S}_2$.

6   A set of state-space equations is given by

$$\frac{dx_1}{dt} = x_1(u - \beta x_2), \qquad \frac{dx_2}{dt} = x_2(-\alpha + \beta x_1),$$

where the input is u and α and β are positive constants.

   (a) Is this system linear or nonlinear, time-varying or time-invariant?

   (b) Determine the equilibrium points (the constant operating points), for constant input $u = 1$.

   (c) Near the positive equilibrium point, find a linearized state-space model of
   the system.

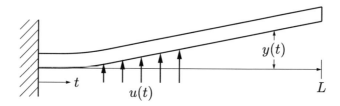

Fig. P1.7   Beam fixed at one end.

7   An elastic beam of length L is rigidly supported as illustrated in Figure P1.7. Let t be the distance from the left end, $y(t)$ the upward deflection of the beam, $v(t)$ the shear force, and $m(t)$ the moment. The upward applied force per unit distance at point t is given by a known function $u(t)$. The bending of the beam is described by the equations

$$\frac{dy}{dt} = y', \quad EI\frac{dy'}{dt} = m(t), \quad \frac{dm}{dt} = v(t), \quad \frac{dv}{dt} = u(t),$$

where E, I are constants.

(a) For input $u(t)$ and output $y(t)$, and choosing the differentiated quantities in the above equations as variables x_1, x_2, x_3, x_4, find matrices **A**, **B**, **C**, **D** of the equations written in the form of (1.16).

(b) In order to solve the equations for $y(t)$, the boundary conditions $y(0) = 0$, $y'(0) = 0$, $v(L) = 0$, $m(L) = 0$ must be used, in addition to the equations. Discuss whether these boundary conditions satisfy the definition of state used in Section 1.1. This system illustrates the subtle fact that the form of the equations alone does not determine whether a state-space model has been obtained.

2

Solution of state-space equations

In order to verify a model of a physical system, or to use it to predict system behavior, it is necessary to solve the model equations.

The word "solution" has more than one working meaning. It can mean a formula defining the output $\mathbf{Y}(t)$ as a function of arguments including time t, initial time t_0, initial state $\mathbf{X}(t_0)$, and the input function $\mathbf{U}(.)$, but formulas for such solutions are not available in general for sets of nonlinear differential or difference equations. For LTI systems, however, suitable formulas can be developed, and involve integrals in the case of continuous-time systems, and sums for discrete-time systems. The second working meaning of the word "solution" is a graphical curve or set of curves, for which the formula may be unknown, which is the response $\mathbf{X}(t)$ or $\mathbf{Y}(t)$ over an interval $t_0 \leq t < t_f$, for specified initial time, initial state, and input over $t_0 \leq t < t_f$. Such specific solution curves are also called solution *trajectories*.

Solution curves for the n simultaneous differential or difference equations of a state-space model are readily obtained by computer, but for LTI systems, significant conclusions about solutions can be found without relying on computation. Closed form expressions can be obtained for solutions and components of solutions, and qualitative properties of the trajectories can be obtained without the approximations inherent in computer simulation. It also turns out that essentially all of the important qualitative properties of solutions are common to both discrete-time and continuous-time models.

Formulas for solutions of the continuous-time state-space equations (1.7)

$$(1.7a) \quad \frac{d}{dt}\mathbf{X}(t) = \mathbf{F}(\mathbf{X}(t), \mathbf{U}(t), t),$$

$$(1.7b) \quad \mathbf{Y}(t) = \mathbf{G}(\mathbf{X}(t), \mathbf{U}(t), t), \quad t \in \mathbb{R},$$

or the discrete-time equations (1.8)

$$(1.8a) \quad \mathbf{X}(t+1) = \mathbf{F}(\mathbf{X}(t), \mathbf{U}(t), t),$$

$$(1.8b) \quad \mathbf{Y}(t) = \mathbf{G}(\mathbf{X}(t), \mathbf{U}(t), t), \quad t \in \mathbb{Z},$$

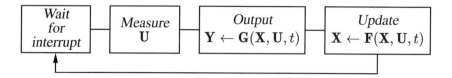

Fig. 2.1   Embedded software loop.

are obtained in their respective time domains by first solving the n simultaneous differential or difference equations for the state $\mathbf{X}$ and then substituting the solution into (1.7b) or (1.8b) to calculate $\mathbf{Y}$.

1   Solution of discrete-time equations

Let the time interval over which solutions are required be $[t_0, t_f)$, where t_f may approach ∞. The past of the system determines the initial state $\mathbf{X}(t_0)$, which is assumed to be known. Then the input is the vector sequence

$$(2.1) \quad \{\mathbf{U}(t)\}_{t_0}^{t_f} = \{\mathbf{U}(t_0),\ \mathbf{U}(t_0+1),\ \cdots \mathbf{U}(t_f)\},$$

where the subscript t_0 and superscript t_f are used to indicate the interval, but may be dropped when the context makes these limits clear. Using the same notation, the solution to be calculated is the sequence $\{\mathbf{X}(t)\}_{t_0}^{t_f}$, from the terms of which the terms of the sequence $\{\mathbf{Y}(t)\}_{t_0}^{t_f}$ can be calculated.

Equation (1.8a) gives a recursive formula for solving the state sequence $\{\mathbf{X}(t)\}_{t_0}^{t_f}$, that is,

$$\mathbf{Y}(t_0) = \mathbf{G}(\mathbf{X}(t_0), \mathbf{U}(t_0), t_0),$$
$$\mathbf{X}(t_0+1) = \mathbf{F}(\mathbf{X}(t_0), \mathbf{U}(t_0), t_0),$$
$$\mathbf{Y}(t_0+1) = \mathbf{G}(\mathbf{X}(t_0+1), \mathbf{U}(t_0+1), t_0+1),$$
$$\mathbf{X}(t_0+2) = \mathbf{F}(\mathbf{X}(t_0+1), \mathbf{U}(t_0+1), t_0+1),$$
$$\mathbf{Y}(t_0+2) = \mathbf{G}(\mathbf{X}(t_0+2), \mathbf{U}(t_0+2), t_0+2),$$
$$\mathbf{X}(t_0+3) = \mathbf{F}(\mathbf{X}(t_0+2), \mathbf{U}(t_0+2), t_0+2),$$

$$\vdots$$

which is typically an efficient method for computation. In the context of embedded software, Figure 2.1 shows the steps in an interrupt-driven software loop, for applications such as real-time control or filtering, which measures the current input, generates the current output, and updates the stored state.

Notation

1. It is common to denote the ℓ-th term of a sequence by using subscript notation rather than parentheses, so that the notations $\{\mathbf{Y}(\ell)\}$ and $\{\mathbf{Y}_\ell\}$ have identical meaning.

2. *Single-sided* sequences of the form $\{\mathbf{Y}(\ell)\}_{t_0}^\infty$ as used above are natural objects for solving dynamical systems as a function of initial conditions and inputs. With the convention that terms $\mathbf{Y}(\ell)$ are identically zero for $\ell < t_0$, these sequences can be considered to be two-sided sequences of the form $\{\mathbf{Y}(\ell)\}_{-\infty}^\infty$.

Example 1
LTI
discrete-time
solution

For the single-input, single-output LTI system

$$\mathbf{A} = \begin{bmatrix} 0 & 1 \\ -1/2 & -3/2 \end{bmatrix}, \quad \mathbf{B} = \begin{bmatrix} 0 \\ 1 \end{bmatrix}, \quad \mathbf{C} = [1, 0], \quad \mathbf{D} = 1,$$

with initial state $\mathbf{X}(0) = \begin{bmatrix} 5 \\ 0 \end{bmatrix}$ and input sequence $\{\mathbf{U}(t)\}_0^\infty = \{1, 1, \cdots\}$, the output sequence can be calculated as

$$\mathbf{Y}(0) = [1, 0]\begin{bmatrix} 5 \\ 0 \end{bmatrix} + 1 = 6,$$

$$\mathbf{X}(1) = \begin{bmatrix} 0 & 1 \\ -1/2 & -3/2 \end{bmatrix}\begin{bmatrix} 5 \\ 0 \end{bmatrix} + \begin{bmatrix} 0 \\ 1 \end{bmatrix} = \begin{bmatrix} 0 \\ -3/2 \end{bmatrix},$$

$$\mathbf{Y}(1) = [1, 0]\begin{bmatrix} 0 \\ -3/2 \end{bmatrix} + 1 = 1,$$

$$\mathbf{X}(2) = \begin{bmatrix} 0 & 1 \\ -1/2 & -3/2 \end{bmatrix}\begin{bmatrix} 0 \\ -3/2 \end{bmatrix} + \begin{bmatrix} 0 \\ 1 \end{bmatrix} = \begin{bmatrix} -3/2 \\ 13/4 \end{bmatrix},$$

$$\mathbf{Y}(2) = [1, 0]\begin{bmatrix} -3/2 \\ 13/4 \end{bmatrix} + 1 = -\frac{1}{2},$$

$$\mathbf{X}(3) = \begin{bmatrix} 0 & 1 \\ -1/2 & -3/2 \end{bmatrix}\begin{bmatrix} -3/2 \\ 13/4 \end{bmatrix} + \begin{bmatrix} 0 \\ 1 \end{bmatrix} = \begin{bmatrix} 13/4 \\ -25/8 \end{bmatrix},$$

$$\vdots$$

Although an arbitrary number of terms of the sequence can be computed numerically, this cannot be regarded as a closed-form solution since a general formula for $\mathbf{Y}(\ell)$ has not been obtained.

1.1   LTI equations

Consider LTI equations of the form of Equation (1.17), repeated here:

(1.17a) $\mathbf{X}(t+1) = \mathbf{AX}(t) + \mathbf{BU}(t),$

(1.17b) $\mathbf{Y}(t) = \mathbf{CX}(t) + \mathbf{DU}(t).$

Starting from known $\mathbf{X}(t_0)$, the recursion for $\mathbf{X}(t)$ gives

$$\mathbf{X}(t_0{+}1) = \mathbf{AX}(t_0) + \mathbf{BU}(t_0),$$
$$\mathbf{X}(t_0{+}2) = \mathbf{A}^2\mathbf{X}(t_0) + \mathbf{ABU}(t_0) + \mathbf{BU}(t_0{+}1),$$
$$\mathbf{X}(t_0{+}3) = \mathbf{A}^3\mathbf{X}(t_0) + \mathbf{A}^2\mathbf{BU}(t_0) + \mathbf{ABU}(t_0{+}1) + \mathbf{BU}(t_0{+}2),$$
$$\vdots$$

so that the solution for $\mathbf{X}$ at time ℓ, $\ell \geq t_0$ can be written by using matrices, as shown:

$$(2.2) \quad \mathbf{X}(\ell) = \mathbf{A}^{\ell-t_0}\mathbf{X}(t_0) + [\,\mathbf{A}^{\ell-t_0-1}\mathbf{B},\ \mathbf{A}^{\ell-t_0-2}\mathbf{B}, \cdots \mathbf{AB},\ \mathbf{B}\,]\begin{bmatrix} \mathbf{U}(t_0) \\ \mathbf{U}(t_0{+}1) \\ \vdots \\ \mathbf{U}(\ell{-}2) \\ \mathbf{U}(\ell{-}1) \end{bmatrix}.$$

Substituting the recursive solution into (1.17b) gives the output as

$$\mathbf{Y}(t_0) = \mathbf{CX}(t_0) + \mathbf{DU}(t_0),$$
$$\mathbf{Y}(t_0{+}1) = \mathbf{CAX}(t_0) + \mathbf{CBU}(t_0) + \mathbf{DU}(t_0{+}1),$$
$$\mathbf{Y}(t_0{+}2) = \mathbf{CA}^2\mathbf{X}(t_0) + \mathbf{CABU}(t_0) + \mathbf{CBU}(t_0{+}1) + \mathbf{DU}(t_0{+}2),$$
$$\vdots$$

or in matrix form, for each time ℓ, $\ell \geq t_0$,

$$(2.3) \quad \mathbf{Y}(\ell) = \mathbf{CA}^{\ell-t_0}\mathbf{X}(t_0)$$

$$+ [\,\mathbf{CA}^{\ell-t_0-1}\mathbf{B},\ \mathbf{CA}^{\ell-t_0-2}\mathbf{B}, \cdots \mathbf{CAB},\ \mathbf{CB},\ \mathbf{D}\,]\begin{bmatrix} \mathbf{U}(t_0) \\ \mathbf{U}(t_0{+}1) \\ \vdots \\ \mathbf{U}(\ell{-}2) \\ \mathbf{U}(\ell{-}1) \\ \mathbf{U}(\ell) \end{bmatrix}.$$

From Equation (2.4), the ℓ-th term of the solution is a function of the initial state, the input sequence, and of $\ell{-}t_0$, but not otherwise of t_0. The initial time t_0 is often taken to be 0 to simplify the expressions.

Inspection of Equation (2.4) leads to several further important observations, summarized in the following sections.

Example 2
State solution Suppose the state at time $t = 3$ for Example 1 is to be calculated. Then Equation (2.2) is

$$\mathbf{X}(3) = \begin{bmatrix} 0 & 1 \\ -1/2 & -3/2 \end{bmatrix}^3 \begin{bmatrix} 5 \\ 0 \end{bmatrix}$$

$$+ \left[\begin{bmatrix} 0 & 1 \\ -1/2 & -3/2 \end{bmatrix}^2 \begin{bmatrix} 0 \\ 1 \end{bmatrix}, \begin{bmatrix} 0 & 1 \\ -1/2 & -3/2 \end{bmatrix} \begin{bmatrix} 0 \\ 1 \end{bmatrix}, \begin{bmatrix} 0 \\ 1 \end{bmatrix} \right] \begin{bmatrix} 1 \\ 1 \\ 1 \end{bmatrix}.$$

Example 3
Output solution

Rather than the state, suppose the output at time $t = 3$ for Example 1 is to be calculated. Then Equation (2.4) is

$$\mathbf{Y}(3) = [1, \, 0] \begin{bmatrix} 0 & 1 \\ -1/2 & -3/2 \end{bmatrix}^3 \begin{bmatrix} 5 \\ 0 \end{bmatrix}$$

$$+ \left[[1, \, 0] \begin{bmatrix} 0 & 1 \\ -1/2 & -3/2 \end{bmatrix}^2 \begin{bmatrix} 0 \\ 1 \end{bmatrix}, \, [1, \, 0] \begin{bmatrix} 0 & 1 \\ -1/2 & -3/2 \end{bmatrix} \begin{bmatrix} 0 \\ 1 \end{bmatrix}, \, [1, \, 0] \begin{bmatrix} 0 \\ 1 \end{bmatrix}, 1 \right] \begin{bmatrix} 1 \\ 1 \\ 1 \\ 1 \end{bmatrix}$$

$$= [3/4, \, 7/4] \begin{bmatrix} 5 \\ 0 \end{bmatrix} + [-3/2, \, 1, \, 0, \, 1] \begin{bmatrix} 1 \\ 1 \\ 1 \\ 1 \end{bmatrix} = 15/4 + 1/2 = 17/4.$$

1.2   Free response

For any time ℓ, the output $\mathbf{Y}(\ell)$ is the sum of two components, the free response discussed in this section, and the forced response discussed in the next section.

In Equation (2.4), the component

(2.4) $\mathbf{Y}_{\text{free}}(\ell) = \mathbf{CA}^{\ell - t_0} \mathbf{X}(t_0)$

is the response of the system to initial state $\mathbf{X}(t_0)$ with $\{\mathbf{U}(t)\}_{t_0}^{\ell} = \{0, 0, \cdots 0\}$. This component is called the *free* or *zero-input* response at time ℓ.

For fixed time ℓ and initial time t_0, the response $\mathbf{Y}_{\text{free}}(\ell)$ is the product of a constant matrix times the initial state. Therefore $\mathbf{Y}_{\text{free}}(\ell)$ is a linear function of the initial state. That is, given responses

$$\mathbf{Y}_{\text{free}}^{(1)}(\ell) = (\mathbf{CA}^{\ell - t_0}) \mathbf{X}^{(1)}(t_0), \quad \mathbf{Y}_{\text{free}}^{(2)}(\ell) = (\mathbf{CA}^{\ell - t_0}) \mathbf{X}^{(2)}(t_0)$$

to initial states $\mathbf{X}^{(1)}(t_0)$, $\mathbf{X}^{(2)}(t_0)$ respectively, then the free response to any initial state $\alpha \mathbf{X}^{(1)}(t_0) + \beta \mathbf{X}^{(2)}(t_0)$ is

$$(\mathbf{CA}^{\ell - t_0})(\alpha \mathbf{X}^{(1)}(t_0) + \beta \mathbf{X}^{(2)}(t_0)) = \alpha \mathbf{Y}_{\text{free}}^{(1)}(\ell) + \beta \mathbf{Y}_{\text{free}}^{(2)}(\ell).$$

In other words, provided the input sequence is zero, the technique of superposition can be used to calculate free responses.

Considering only the state vector rather than the output vector, as above, from (2.2) the free component of the response is

(2.5) $\mathbf{X}_{\text{free}}(\ell) = \mathbf{A}^{\ell - t_0} \mathbf{X}(t_0).$

State transition matrix

From Equation (2.5), in the absence of inputs the state at time $t = \ell$ is obtained by multiplying the state at time $t = t_0$ by the matrix $\mathbf{A}^{\ell - t_0}$. This matrix is called the *state-transition matrix* of this time-invariant discrete-time system. Thus for LTI discrete-time systems, the state-transition matrix exists and is unique for $\ell \geq t_0$; but if $\ell < t_0$, then $\mathbf{A}^{\ell - t_0} = (\mathbf{A}^{-1})^{|\ell - t_0|}$ exists only if $\mathbf{A}$ has an inverse. If a system is linear but time-varying, a state-transition matrix also can be defined, and to allow for such a generalized context, the state-transition matrix is often given the symbol $\mathbf{\Phi}(\ell, t_0)$.

Example 4
Linearity of free response

In the previous example, the free response at time $t = 3$ for $\mathbf{X}^{(1)}(0) = \begin{bmatrix} 5 \\ 0 \end{bmatrix}$ is

$$\mathbf{Y}_{\text{free}}^{(1)}(3) = [1,\ 0] \begin{bmatrix} 0 & 1 \\ -1/2 & -3/2 \end{bmatrix}^3 \begin{bmatrix} 5 \\ 0 \end{bmatrix} = 15/4.$$

Suppose $\mathbf{X}^{(2)}(0) = \begin{bmatrix} 0 \\ -1 \end{bmatrix}$, for which, at $t = 3$,

$$\mathbf{Y}_{\text{free}}^{(2)}(3) = [1,\ 0] \begin{bmatrix} 0 & 1 \\ -1/2 & -3/2 \end{bmatrix}^3 \begin{bmatrix} 0 \\ -1 \end{bmatrix} = -7/4.$$

Then any other initial state $\mathbf{X}(0)$ can be expressed as a linear combination of the previous two initial states:

$$\mathbf{X}(0) = \alpha \begin{bmatrix} 5 \\ 0 \end{bmatrix} + \beta \begin{bmatrix} 0 \\ -1 \end{bmatrix},$$

so that at $t = 3$, the free response for initial state $\mathbf{X}(0)$ will be

$$\mathbf{Y}(3) = \alpha(15/4) + \beta(-7/4).$$

1.3 Forced response

In Equation (2.4), the component

$$(2.6) \quad \mathbf{Y}_{\text{forced}}(\ell) = [\,\mathbf{CA}^{\ell - t_0 - 1}\mathbf{B},\ \mathbf{CA}^{\ell - t_0 - 2}\mathbf{B}, \cdots \mathbf{CAB},\ \mathbf{CB},\ \mathbf{D}\,] \begin{bmatrix} \mathbf{U}(t_0) \\ \mathbf{U}(t_0 + 1) \\ \vdots \\ \mathbf{U}(\ell - 2) \\ \mathbf{U}(\ell - 1) \\ \mathbf{U}(\ell) \end{bmatrix}$$

is the response of the system to input sequence $\{\mathbf{U}(t)\}$ for zero initial state $\mathbf{X}(t_0)$. This component is called the *forced* or *zero-state* response at time ℓ.

For fixed time ℓ and initial time t_0, because the forced response is the product of a constant matrix times the vectors of the input sequence, the forced response is a linear function of the input. For example, doubling the input values doubles the forced response. Superposition can be used to calculate forced responses.

Example 5
Forced response

From Example 1 and the previous calculations, the forced response at $t = 3$ for input $\{U(t)\}_0^\infty = \{1, 1, \cdots\}$, as before, is

$$\mathbf{Y}_{\text{forced}}(3) = [-3/2, 1, 0, 1] \begin{bmatrix} 1 \\ 1 \\ 1 \\ 1 \end{bmatrix} = 1/2,$$

and multiplying the input sequence by a constant μ, so that $\{U(t)\}_0^\infty = \{\mu, \mu, \cdots\}$, will result in

$$\mathbf{Y}_{\text{forced}}(3) = [-3/2, 1, 0, 1] \begin{bmatrix} \mu \\ \mu \\ \mu \\ \mu \end{bmatrix} = \mu/2.$$

Example 6
Complete response

The matrices of a discrete-time LTI system are given as

$$\mathbf{A} = \begin{bmatrix} 0.963064 & 0.074082 \\ -0.666736 & 0.518573 \end{bmatrix}, \quad \mathbf{B} = \begin{bmatrix} 0.004104 \\ 0.074082 \end{bmatrix}, \quad \mathbf{C} = [9, 0], \quad \mathbf{D} = 0.$$

Figure 2.2 illustrates that the complete response $\{\mathbf{Y}(k)\}$ is the sum of the free response $\{\mathbf{Y}_{\text{free}}(k)\}$ and forced response $\{\mathbf{Y}_{\text{forced}}(k)\}$, here computed for initial state $\mathbf{X}(0) = \begin{bmatrix} 0 \\ 1 \end{bmatrix}$ and input $\{U(k)\} = \{0.5\}$.

1.4   Weighting sequence

Let the sequence $\{\mathbf{H}(k)\}_0^\infty$ be defined as

(2.7) $\{\mathbf{H}(0), \mathbf{H}(1), \mathbf{H}(2), \cdots \mathbf{H}(k), \cdots\} = \{\mathbf{D}, \mathbf{CB}, \mathbf{CAB}, \cdots \mathbf{CA}^{k-1}\mathbf{B}, \cdots\}.$

This sequence is called the *weighting-matrix* sequence or *impulse-response* sequence, or sometimes the *Markov* sequence of the system. If we set $t_0 = 0$ for notational simplicity, then for $\mathbf{X}(0) = 0$, at any time $\ell \geq 0$, from (2.4) the response $\mathbf{Y}_{\text{forced}}(\ell)$ is the weighted sum of the inputs:

(2.8) $\mathbf{Y}_{\text{forced}}(\ell) = \mathbf{H}(\ell)\mathbf{U}(0) + \cdots \mathbf{H}(1)\mathbf{U}(\ell-1) + \mathbf{H}(0)\mathbf{U}(\ell)$

$$= \sum_{k=0}^{\ell} \mathbf{H}(\ell - k)\, \mathbf{U}(k).$$

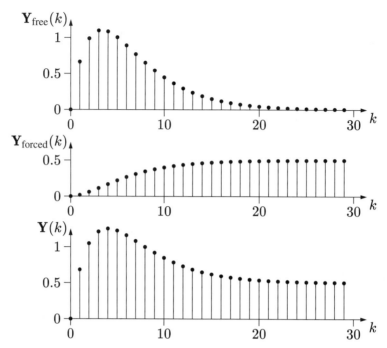

Fig. 2.2   Free, forced, and complete response.

Equivalence   An extremely useful conclusion can now be reached: the forced response of system (1.17) is determined uniquely by the weighting-matrix sequence $\{\mathbf{H}(k)\}_0^\infty$
and the input, so that any two such systems with identical weighting matrix sequences have identical forced responses for identical inputs. The systems are
said to be *externally equivalent,* since their external, or input-output, behaviors
are identical. There is an apparent difficulty here in that, given two weighting-
matrix sequences, an infinite number of terms would have to be checked in order
to determine that they are identical. In fact, as will be shown in Chapter 5, for
a system of state dimension n, the sequence $\{\mathbf{H}(k)\}_0^{2n}$ specifies the forced behavior exactly, so that only a finite number of terms must be checked.

Example 7   Consider the single-input, single-output system with two state variables de-
Equivalent   scribed by the matrices
systems

$$\mathbf{A} = \begin{bmatrix} 1 & 1 \\ 0 & 1 \end{bmatrix}, \quad \mathbf{B} = \begin{bmatrix} 1 \\ 0 \end{bmatrix}, \quad \mathbf{C} = [1, 0], \quad \mathbf{D} = 1.$$

The weighting sequence $\{\mathbf{H}(k)\}$ is determined conveniently by recursively calculating the sequence of products $\mathbf{A}^{k-1}\mathbf{B}$ and at each recursion multiplying by

C to obtain the **H**(k), as shown:

$$\mathbf{H}(0) = \mathbf{D} = 1, \qquad \mathbf{B} = \begin{bmatrix} 1 \\ 0 \end{bmatrix},$$

$$\mathbf{H}(1) = \mathbf{CB} = 1, \qquad \mathbf{AB} = \begin{bmatrix} 1 \\ 0 \end{bmatrix},$$

$$\vdots \qquad\qquad\qquad \vdots$$

$$\mathbf{H}(k) = \mathbf{CA}^{k-1}\mathbf{B} = 1, \quad \mathbf{A}^k\mathbf{B} = \mathbf{A}(\mathbf{A}^{k-1}\mathbf{B}) = \begin{bmatrix} 1 \\ 0 \end{bmatrix}.$$

Now consider the single-input, single-output system with one state variable described by the matrices

$$\mathbf{A}' = 1, \quad \mathbf{B}' = 1, \quad \mathbf{C}' = 1, \quad \mathbf{D}' = 1.$$

For this system the weighting sequence $\{\mathbf{H}'(k)\}$ is obtained from

$$\mathbf{H}'(0) = \mathbf{D}' = 1, \qquad\qquad \mathbf{B}' = 1,$$
$$\mathbf{H}'(1) = \mathbf{C}'\mathbf{B}' = 1, \qquad\qquad \mathbf{A}'\mathbf{B}' - 1,$$

$$\vdots \qquad\qquad\qquad\qquad \vdots$$

$$\mathbf{H}'(k) = \mathbf{C}'(\mathbf{A}')^{k-1}\mathbf{B}' = 1, \quad (\mathbf{A}')^k\mathbf{B}' = 1.$$

Thus the forced responses of the two systems are identical since they have identical weighting-matrix sequences.

1.5   Impulse response

The impulse-response of a single-input system is the response to an impulse, with zero initial conditions. However, when the number m of inputs is greater than 1, the impulse response is no longer defined to be the response to a single impulse. Consider the discrete-time system (1.17) with m inputs, and suppose the following m experiments are performed:

Experiment 1   The *unit impulse sequence* $\{1, 0, 0, \cdots\}$ is applied to the first input, all other inputs remaining zero. The input vector sequence is

$$\{\mathbf{U}^{(1)}(k)\} = \{\mathbf{e}_1, 0, 0, \cdots\},$$

where the notation $\mathbf{e}_i$ means a vector of appropriate dimension, in this case, m, containing zeros as entries except for a 1 in the i-th row; that is, $\mathbf{e}_i$ is the i-th column of the identity matrix $\mathbf{I}_m$. The entries of the forced-response sequence are given by the formula

$$\mathbf{Y}^{(1)}(\ell) = \sum_{k=0}^{\ell} \mathbf{H}(\ell-k)\,\mathbf{U}^{(1)}(k)$$
$$= \mathbf{H}(\ell)\,\mathbf{e}_1 + \mathbf{H}(\ell-1)\,0 + \cdots \mathbf{H}(0)\,0$$
$$= \mathbf{H}(\ell)\,\mathbf{e}_1$$
$$= \text{1st column of } \mathbf{H}(\ell).$$

Experiment i Similarly, for $i = 2, \cdots m$, the unit impulse sequence $\{1, 0, 0, \cdots\}$ is applied to the i-th input, all other inputs remaining zero. The input vector sequence is

$$\{\mathbf{U}^{(i)}(k)\} = \{\mathbf{e}_i,\, 0,\, 0 \cdots\},$$

and the formula for the terms of the forced response is

$$\mathbf{Y}^{(i)}(\ell) = \mathbf{H}(\ell)\,\mathbf{e}_i = i\text{-th column of } \mathbf{H}(\ell).$$

The results of the previous experiments are combined as follows. The *impulse response sequence* is defined to be the sequence of which the k-th term is a matrix containing columns $\mathbf{Y}^{(1)}(k)$, $\mathbf{Y}^{(2)}(k)$, $\cdots \mathbf{Y}^{(m)}(k)$. From the above experiments, the k-th term of this sequence will be the matrix

$$[\,\mathbf{H}(k)\,\mathbf{e}_1,\ \mathbf{H}(k)\,\mathbf{e}_2, \cdots \mathbf{H}(k)\,\mathbf{e}_m\,],$$

which is precisely $\mathbf{H}(k)$. For this reason, the matrix sequence $\{\mathbf{H}(k)\}_0^{\infty}$ is called the impulse-response sequence.

Example 8
Impulse-response sequence   For the system of Example 1, the impulse-response sequence is
$$\mathbf{H}(k) = \{\mathbf{D},\ \mathbf{CB},\ \mathbf{CAB}, \cdots\}$$
$$= \{1,\ 0,\ 1,\ -3/2,\ 7/4,\ -15/8,\ 31/16,\ -63/32 \cdots\}.$$

Example 9
Weighting sequence   The system of Example 6 has weighting sequence $\{\mathbf{H}(k)\}$, which can be plotted as in Figure 2.3 since the $\mathbf{H}(k)$ are scalars. This sequence is therefore also the impulse response.

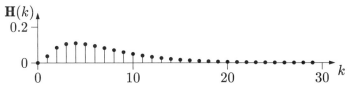

Fig. 2.3   Weighting sequence for the system of Example 6.

1.6   Convolution

As shown in Section 1.4, the formula

$$(2.9) \quad \mathbf{Y}_{\text{forced}}(\ell) = \sum_{k=0}^{\ell} \mathbf{H}(\ell - k)\, \mathbf{U}(k)$$

gives the forced response at any time ℓ as a weighted sum of the input-sequence terms. The summation in (2.9) is of a form known as a *convolution sum*. A sequence $\{\mathbf{Y}(\ell)\}_0^\infty$ of which the terms satisfy (2.9) with respect to two sequences $\{\mathbf{H}(\ell)\}_0^\infty$, $\{\mathbf{U}(\ell)\}_0^\infty$ is called the *convolution* of these two sequences, written

$$(2.10) \quad \{\mathbf{Y}(\ell)\} = \{\mathbf{H}(\ell)\} * \{\mathbf{U}(\ell)\},$$

where the limits may be, in general, from $-\infty$ to ∞.

As illustrated below for $\ell = 3$,

$$\cdots \quad \mathbf{H}(4) \quad \mathbf{H}(3) \quad \mathbf{H}(2) \quad \mathbf{H}(1) \quad \mathbf{H}(0) \quad 0 \quad \cdots$$
$$\cdots \quad 0 \quad \mathbf{U}(0) \quad \mathbf{U}(1) \quad \mathbf{U}(2) \quad \mathbf{U}(3) \quad \mathbf{U}(4) \quad \cdots$$

the convolution sum corresponds to a computation that first reverses the order of the sequence $\{\mathbf{H}(k)\}$, then shifts it to the right by ℓ places, and then sums the nonzero products of the overlapping terms to compute the ℓ-th term of the convolution. In computing the products, it is important to preserve the correct order of matrix multiplication, since in general, $\mathbf{H}(i)\mathbf{U}(j) \neq \mathbf{U}(j)\mathbf{H}(i)$.

Example 10
Convolution

For the system and input of Example 1, the forced response at $t = 3$ can be obtained by reversing, shifting and multiplying as shown,

$$
\begin{array}{ccccccccc}
\cdots & -15/8 & 7/4 & -3/2 & 1 & 0 & 1 & 0 & 0 & \cdots \\
\cdots & 0 & 0 & 1 & 1 & 1 & 1 & 1 & 1 & \cdots, \\
\hline
\cdots & 0 & 0 & -3/2 & 1 & 0 & 1 & 0 & 0 & \cdots
\end{array}
$$

from which, summing the last line, $\mathbf{Y}_{\text{forced}}(3) = 1/2$.

2   **Solution of continuous-time equations**

In the previous sections, the solutions of the state equation (1.8a) for vectors $\mathbf{X}(\ell)$ depended only on the existence of the function $\mathbf{F}(\cdots)$ and its arguments. However, the n simultaneous differential equations in the continuous-time vector equation (1.7a) may not have solutions, or the solutions may not be unique.

Assume that an initial vector $\mathbf{X}(t_0) = \mathbf{X}_0$ is given, as well as the input $\mathbf{U}(t)$ on the interval $t_0 \leq t < t_f$, where t_f may be ∞. Since $\mathbf{U}(t)$ is known, let it be substituted into (1.7a) to give

(2.11) $$\frac{d}{dt}\mathbf{X} = \mathbf{F}(\mathbf{X}, t),$$

where, temporarily, the symbol $\mathbf{F}(\cdot, \cdot)$ is taken to indicate the result of the above substitution.

Definition   By a *solution* of (2.11) is meant a function $\mathbf{X}(t) : \mathbb{R} \to \mathbb{R}^n$ that is, a function producing $\mathbf{X}$ in $\mathbb{R}^n$ given t in $\mathbb{R}$ that satisfies (2.11) over $[t_0, t_f)$ and that has the value $\mathbf{X}_0$ at t_0.

If (2.11) is a model of a physical system, then the existence of solutions is justified on physical grounds; otherwise a modeling blunder has been committed. On the other hand, during the modeling process it is important to be able to test whether equations that have been postulated as a representation of a physical system have solutions, and if so, whether the solutions are unique.

Example 11
Euler's forward
method

Numerical computer solutions of (2.11) approximate $\mathbf{X}(\cdot)$ over the time interval of interest by a sequence of computed vectors $\{\mathbf{X}(k)\}$, for some finite set of values of k. One of the simplest numerical techniques is to compute the terms $\mathbf{X}(k)$ sequentially by Euler's forward method, which is derived by noting that

(2.12) $$\frac{d}{dt}\mathbf{X} = \lim_{h \to 0} \frac{1}{h}(\mathbf{X}(t+h) - \mathbf{X}(t)).$$

Then this formula, with nonzero step-size h, is substituted on the left side of (2.11) to generate the sequence

(2.13) $$\mathbf{X}(0) = \mathbf{X}_0, \quad \mathbf{X}(k+1) = \mathbf{X}(k) + h\,\mathbf{F}(\mathbf{X}(k), kh),$$

so that each term $\mathbf{X}(k)$ is an approximation of the solution at time $t = kh$. The simplicity of this method shows that approximate solution curves are readily available, but the results may be meaningless if the differential equation does not have a solution, if the solution is not unique, or if the approximation is inadequate, as is too often the case for the above method. Provided a solution exists, the numerical technique best used for a given simulation will be determined by questions of accuracy and efficiency.

2.1	**Existence and uniqueness**

Since signals containing abrupt changes are common, the entries of $\mathbf{U}(t)$ or of $\mathbf{F}(\mathbf{X}, t)$ may be discontinuous functions of time, but it will be assumed that the number of points of discontinuity is finite per unit interval, so that these functions are piecewise-continuous.

Existence   Let the entries of $\mathbf{F}(\mathbf{X}, t)$ be continuous functions of t and of the entries of $\mathbf{X}$ in a nonzero neighborhood including $\mathbf{X}_0, t_0$. Then for a nonzero range of values $t-t_0$ there exists a continuous and differentiable function $\mathbf{X}(t)$ that is a solution; that is, $\mathbf{X}(t)$ satisfies (2.11), and $\mathbf{X}(t_0) = \mathbf{X}_0$.

Uniqueness   Let a solution to (2.11) exist, and furthermore, let the rate of change of the entries of $\mathbf{F}(\mathbf{X}, t)$ with respect to changes in the entries of $\mathbf{X}$ be bounded by a finite constant L, for all $\mathbf{X}, t$ in a nonzero neighborhood containing $\mathbf{X}_0, t_0$. This condition is called the Lipschitz condition, with Lipschitz constant L. Then the solution is unique in a nonzero neighborhood of $\mathbf{X}_0, t_0$.

A detailed proof and the extension of the above results to arbitrary time intervals are beyond the scope of the present discussion, since we shall be concerned with LTI systems, which have unique solutions under easily verifiable conditions. However, notice that the condition for uniqueness requires $\mathbf{F}(\mathbf{X}, t)$ to be continuous and to have finite rate of change with respect to the entries of $\mathbf{X}$, but not with respect to t. In fact, functions containing step discontinuities with respect to time are allowed.

Example 12
Nonexistent
solution

The real differential equation

$$\frac{dx}{dt} = 1/(1-t), \quad x(0) = 0$$

has solution

$$x(t) = \log_e \frac{1}{1-t},$$

which does not exist on an interval that includes values of $t \geq 1$.

Example 13
Multiple
solutions

The differential equation

$$\frac{dx}{dt} = x^{1/5}, \quad x(0) = 0$$

has infinitely many solutions, including $x(t) = 0$ and $x(t) = (4t/5)^{5/4}$, and more generally, for any $c \geq 0$,

$$x(t) = \begin{cases} 0, & 0 \leq t \leq c, \\ (4(t-c)/5)^{5/4}, & t > c \end{cases}.$$

Example 14
Lipschitz
condition

The function in Example 13 does not satisfy a Lipschitz condition, since the slope (with respect to x) of the right-hand side of the differential equation is infinite at $x = 0$.

2.2   LTI continuous-time equations

From the previous section, Let $\mathbf{U}(t)$ be bounded and piecewise-continuous. Then because the entries of $\mathbf{A}$, $\mathbf{B}$ in (1.16), repeated here,

$$(1.16a) \quad \frac{d}{dt}\mathbf{X}(t) = \mathbf{A}\mathbf{X}(t) + \mathbf{B}\mathbf{U}(t),$$

$$(1.16b) \quad \mathbf{Y}(t) = \mathbf{C}\mathbf{X}(t) + \mathbf{D}\mathbf{U}(t),$$

are assumed finite, the Lipschitz condition is satisfied. Therefore, unique, piecewise-differentiable solutions $\mathbf{X}(t)$ of (1.16a) exist for arbitrary $\mathbf{X}(t_0)$, t_0, and a piecewise-continuous output solution $\mathbf{Y}(t)$ is obtained when a solution $\mathbf{X}(t)$ is substituted into (1.16b).

Rather than finding the complete solution and examining its parts as for the discrete-time case, the components of the complete solution will be found and examined individually.

2.3   Free response of continuous-time LTI systems

With $\mathbf{U}(t) = 0$ in (1.16a), the initial conditions determine the response. It turns out that a closed-form expression can be obtained in this LTI case. The equation to be solved is

$$(2.14) \quad \frac{d}{dt}\mathbf{X}(t) = \mathbf{A}\mathbf{X}(t), \quad \mathbf{X}(0) = \mathbf{X}_0.$$

First note that if the entries of $\mathbf{A}$ are finite then the Lipschitz condition of Section 2.1 is satisfied for all t, and a unique solution exists in a neighborhood of $\mathbf{X}_0, t_0$ including all finite positive and negative values of $t - t_0$. Therefore the differential equation can be solved both forward and backward in time.

Assume that the solution $\mathbf{X}(t)$ is an infinitely differentiable vector function of t. This assumption will allow a formula for a candidate solution to be found. Then, checking that the formula satisfies Equation (2.14) will verify that the formula is *a* solution. The Lipschitz condition will guarantee that it is *the* unique solution for given $\mathbf{X}(0)$.

By assumption, the series for the solution is

$$(2.15) \quad \mathbf{X}(t) = \mathbf{X}(0) + \frac{t}{1!}\frac{d}{dt}\mathbf{X}(t)\bigg|_0 + \frac{t^2}{2!}\frac{d^2}{dt^2}\mathbf{X}(t)\bigg|_0 + \cdots,$$

where the subscript is used to indicate that the derivatives are evaluated at $\mathbf{X}(0)$. The required derivatives are obtainable from (2.14) as

$$\frac{d}{dt}\mathbf{X} = \mathbf{A}\mathbf{X}$$

$$\frac{d^2}{dt^2}\mathbf{X} = \frac{d}{dt}\mathbf{A}\mathbf{X} = \mathbf{A}\frac{d}{dt}\mathbf{X} = \mathbf{A}^2\mathbf{X}$$

$$\vdots$$

$$\frac{d^i}{dt^i}\mathbf{X} = \mathbf{A}^i\mathbf{X}.$$

Thus $\mathbf{X}(t)$ can be written as

$$(2.16) \quad \mathbf{X}(t) = \mathbf{X}(0) + \frac{t}{1!}\mathbf{A}\mathbf{X}(0) + \frac{t^2}{2!}\mathbf{A}^2\mathbf{X}(0) + \cdots$$

$$= \left(\mathbf{I} + \frac{t}{1!}\mathbf{A} + \frac{t^2}{2!}\mathbf{A}^2 + \cdots\right)\mathbf{X}(0).$$

Matrix exponential Because the solution exists and is unique for finite $\mathbf{X}(0)$ and $t\mathbf{A}$, the series within parentheses in the above equation must converge. By simple comparison with the series $e^{ta} = 1 + \frac{t}{1!}a + \frac{t^2}{2!}a^2 + \cdots$, we can write the above series by using the notation

$$(2.17) \quad e^{t\mathbf{A}} = \mathbf{I} + \frac{t}{1!}\mathbf{A} + \frac{t^2}{2!}\mathbf{A}^2 + \cdots,$$

which allows the solution of (2.14) to be written compactly as

$$(2.18) \quad \mathbf{X}(t) = e^{t\mathbf{A}}\mathbf{X}(0).$$

The exponential in the above formula is an example of a function of a matrix, such as will be studied in more detail in Chapter 7.

For nonzero initial time t_0, the free solution of (1.16a) becomes, instead of (2.18),

$$(2.19) \quad \mathbf{X}(t) = e^{(t-t_0)\mathbf{A}}\mathbf{X}(t_0).$$

It is possible to investigate (2.18) as the solution of (2.14) by checking that it satisfies the initial condition and the differential equation, using (2.17) as the definition of the exponential. First, substituting $t = 0$ into (2.16) gives $\mathbf{X}(0) = \mathbf{X}_0$. Then, differentiating term by term,

$$(2.20) \quad \frac{d}{dt}e^{t\mathbf{A}} = \frac{d}{dt}\left(\mathbf{I} + \frac{t}{1!}\mathbf{A} + \frac{t^2}{2!}\mathbf{A}^2 + \cdots\right) = 0 + \mathbf{A} + \frac{t}{1!}\mathbf{A}^2 + \frac{t^2}{2!}\mathbf{A}^3 + \cdots = e^{t\mathbf{A}}\mathbf{A} = \mathbf{A}e^{t\mathbf{A}},$$

we see that, postmultiplying by $\mathbf{X}(0)$, the differential equation (2.14) is satisfied.

Term-by-term comparison of the expansions shows that if $\mathbf{AB} = \mathbf{BA}$, then

(2.21) $\quad e^{\mathbf{A}+\mathbf{B}} = e^{\mathbf{A}} e^{\mathbf{B}},$

so the matrix exponential function is analogous in every important way to the scalar exponential function.

State transition Comparing (2.19) with the discrete-time solution (2.5), the state transition matrix for an LTI continuous-time system is $\mathbf{\Phi}(t, t_0) = e^{(t-t_0)\mathbf{A}}$, which exists, in contrast with the discrete-time case, for $t < t_0$ since the solution of the differential equation exists for finite $(t-t_0)\mathbf{A}$, as concluded previously.

Linearity Finally, from (2.19), for fixed t and t_0, the free response of the state $\mathbf{X}(t)$ is a linear function of the initial state $\mathbf{X}(t_0)$ since the response is a product of the initial state multiplied by the constant matrix $e^{(t-t_0)\mathbf{A}}$. Substituting this in (1.16b) with $\mathbf{U}(t) = 0$ gives

(2.22) $\quad \mathbf{Y}_{\text{free}}(t) = \mathbf{C} e^{(t-t_0)\mathbf{A}} \mathbf{X}(t_0),$

implying that the free response of the output $\mathbf{Y}(t)$ is a linear function of the initial state, and therefore that superposition of initial states can be used to calculate free responses.

Example 15
Finite series for $e^{t\mathbf{A}}$

If $\mathbf{A}^k = 0$ for some k, then the series for $e^{t\mathbf{A}}$ is finite. For example, if $\mathbf{A} = \begin{bmatrix} 0 & \alpha \\ 0 & 0 \end{bmatrix}$ then $\mathbf{A}^2 = 0$, and the series is

$$e^{t\mathbf{A}} = \begin{bmatrix} 1 & 0 \\ 0 & 1 \end{bmatrix} + \frac{t}{1!} \begin{bmatrix} 0 & \alpha \\ 0 & 0 \end{bmatrix} = \begin{bmatrix} 1 & \alpha t \\ 0 & 1 \end{bmatrix}.$$

Example 16
$e^{t\mathbf{A}}$ is nonsingular

Since $e^{t\mathbf{A}}$ and $e^{-t\mathbf{A}}$ exist for finite $t\mathbf{A}$, then from (2.21) we have

$$e^{t\mathbf{A}} e^{-t\mathbf{A}} = e^{0\mathbf{A}} = \mathbf{I},$$

which implies that $e^{t\mathbf{A}}$ is always nonsingular for finite $t\mathbf{A}$, and reduces in the scalar case to noting that $e^{ta} \neq 0$ for finite ta.

Example 17
Diagonal matrix

The matrix exponential $e^{t\mathbf{A}}$ for the diagonal matrix $t\mathbf{A} = \begin{bmatrix} t\alpha & 0 \\ 0 & t\beta \end{bmatrix}$ is given by the series

$$e^{t\mathbf{A}} = \begin{bmatrix} 1 & 0 \\ 0 & 1 \end{bmatrix} + \frac{t}{1!} \begin{bmatrix} \alpha & 0 \\ 0 & \beta \end{bmatrix} + \frac{t^2}{2!} \begin{bmatrix} \alpha^2 & 0 \\ 0 & \beta^2 \end{bmatrix} + \frac{t^3}{3!} \begin{bmatrix} \alpha^3 & 0 \\ 0 & \beta^3 \end{bmatrix} \cdots$$

$$= \begin{bmatrix} e^{t\alpha} & 0 \\ 0 & e^{t\beta} \end{bmatrix},$$

which is a diagonal matrix containing the scalar functions of the diagonal entries of $t\mathbf{A}$. More generally, the exponential of a diagonal matrix is a diagonal matrix.

Example 18
sin and cos
functions

Consider the system for which the matrix $\mathbf{A}$ and initial state $\mathbf{X}(0)$ are

$$\mathbf{A} = \begin{bmatrix} 0 & \omega \\ -\omega & 0 \end{bmatrix}, \quad \mathbf{X}(0) = \begin{bmatrix} 1 \\ 0 \end{bmatrix}.$$

The state-transition matrix will be calculated:

$$
\begin{aligned}
e^{t\mathbf{A}} &= \begin{bmatrix} 1 & 0 \\ 0 & 1 \end{bmatrix} + \frac{t}{1!}\begin{bmatrix} 0 & \omega \\ -\omega & 0 \end{bmatrix} + \frac{t^2}{2!}\begin{bmatrix} -\omega^2 & 0 \\ 0 & -\omega^2 \end{bmatrix} + \frac{t^3}{3!}\begin{bmatrix} 0 & -\omega^3 \\ \omega^3 & 0 \end{bmatrix} \\
&\quad + \frac{t^4}{4!}\begin{bmatrix} \omega^4 & 0 \\ 0 & \omega^4 \end{bmatrix} + \frac{t^5}{5!}\begin{bmatrix} 0 & \omega^5 \\ -\omega^5 & 0 \end{bmatrix} + \cdots \\
&= \begin{bmatrix} 1 - \frac{t^2\omega^2}{2!} + \frac{t^4\omega^4}{4!} \cdots & t\omega - \frac{t^3\omega^3}{3!} + \frac{t^5\omega^5}{5!} \cdots \\ -t\omega + \frac{t^3\omega^3}{3!} - \frac{t^5\omega^5}{5!} \cdots & 1 - \frac{t^2\omega^2}{2!} + \frac{t^4\omega^4}{4!} \cdots \end{bmatrix} \\
&= \begin{bmatrix} \cos\omega t & \sin\omega t \\ -\sin\omega t & \cos\omega t \end{bmatrix}.
\end{aligned}
$$

To find $\mathbf{X}(t)$ the initial state will be multiplied by the state-transition matrix $\Phi(t,0) = e^{t\mathbf{A}}$:

$$\mathbf{X}(t) = e^{t\mathbf{A}}\mathbf{X}(0) = \begin{bmatrix} \cos\omega t & \sin\omega t \\ -\sin\omega t & \cos\omega t \end{bmatrix}\begin{bmatrix} 1 \\ 0 \end{bmatrix} = \begin{bmatrix} \cos\omega t \\ -\sin\omega t \end{bmatrix}.$$

Example 19
Exponentially
weighted sin
and cos

An expression for $e^{t\mathbf{A}}$ for the matrix $\mathbf{A} = \begin{bmatrix} \alpha & \omega \\ -\omega & \alpha \end{bmatrix}$ is difficult to write by direct expansion, but can be obtained by noting that $t\mathbf{A}$ is the sum of two terms,

$$t\mathbf{A} = \begin{bmatrix} t\alpha & 0 \\ 0 & t\alpha \end{bmatrix} + \begin{bmatrix} 0 & t\omega \\ -t\omega & 0 \end{bmatrix},$$

and that these terms commute; that is,

$$\begin{bmatrix} t\alpha & 0 \\ 0 & t\alpha \end{bmatrix}\begin{bmatrix} 0 & t\omega \\ -t\omega & 0 \end{bmatrix} = \begin{bmatrix} 0 & t\omega \\ -t\omega & 0 \end{bmatrix}\begin{bmatrix} t\alpha & 0 \\ 0 & t\alpha \end{bmatrix},$$

so that from (2.21) and Examples 17 and 18,

$$
\begin{aligned}
e^{t\mathbf{A}} &= e^{\left(\begin{bmatrix} t\alpha & 0 \\ 0 & t\alpha \end{bmatrix} + \begin{bmatrix} 0 & t\omega \\ -t\omega & 0 \end{bmatrix}\right)} = e^{t\begin{bmatrix} \alpha & 0 \\ 0 & \alpha \end{bmatrix}}e^{t\begin{bmatrix} 0 & \omega \\ -\omega & 0 \end{bmatrix}} \\
&= \begin{bmatrix} e^{t\alpha} & 0 \\ 0 & e^{t\alpha} \end{bmatrix}\begin{bmatrix} \cos\omega t & \sin\omega t \\ -\sin\omega t & \cos\omega t \end{bmatrix} = (e^{t\alpha})\begin{bmatrix} \cos\omega t & \sin\omega t \\ -\sin\omega t & \cos\omega t \end{bmatrix}.
\end{aligned}
$$

Unless the matrix exponential can be calculated in closed form as in the above examples, the solution $\mathbf{X}(t)$ of (2.14) must be obtained by a method that is either numerical or approximate. Some possibilities include the following:

- solve (2.14) directly by a numerical method,
- compute $e^{t\mathbf{A}}$ by a numerical method for different values of t, and postmultiply by $\mathbf{X}(0)$,
- compute $e^{\tau\mathbf{A}}$ for some small τ, enabling the computation of $\mathbf{X}(\tau)$, $\mathbf{X}(2\tau)$, $\cdots \mathbf{X}(q\tau)$, using the formula $\mathbf{X}((k+1)\tau) = e^{\tau\mathbf{A}}\mathbf{X}(k\tau)$.

2.4   Complete response of continuous-time LTI systems

Several methods may be used to derive a solution for (1.7a), and hence (1.7b) for nonzero $\mathbf{U}(t)$. One is to postulate the form of the solution, and verify that it satisfies the differential equation and the initial condition, as follows.

The scalar differential equation $\dot{x} = a\,x$, analogous to (2.14) with initial time t_0, has solution $x(t) = e^{(t-t_0)a}x(t_0)$, analogous to (2.19). By any of a number of elementary methods, the scalar equation

$$(2.23) \quad \frac{dx}{dt} = ax + bu$$

with scalar input $u(t)$ can be shown to have solution

$$(2.24) \quad x(t) = e^{(t-t_0)a}x(t_0) + \int_{t_0}^{t} e^{(t-\tau)a}b\,u(\tau)\,d\tau.$$

Solution   By analogy, the solution of (1.7a) will be postulated as

$$(2.25) \quad \mathbf{X}(t) = e^{(t-t_0)\mathbf{A}}\mathbf{X}(t_0) + \int_{t_0}^{t} e^{(t-\tau)\mathbf{A}}\mathbf{B}\,\mathbf{U}(\tau)\,d\tau,$$

with the assumption that $\mathbf{U}(t)$ is sufficiently smooth for the integral to exist. First, the initial condition will be checked. For $t = t_0$, (2.25) is

$$(2.26) \quad \mathbf{X}(t_0) = e^{(t_0-t_0)\mathbf{A}}\mathbf{X}(t_0) + \int_{t_0}^{t_0} e^{(t-\tau)\mathbf{A}}\mathbf{B}\,\mathbf{U}(\tau)\,d\tau = \mathbf{I}\mathbf{X}(t_0) + 0 = \mathbf{X}(t_0),$$

so the right-hand side of (2.25) is a formula that satisfies the required initial condition.

To check that the formula on the right side of Equation (2.25) satisfies the differential equation, note that $e^{(t-\tau)\mathbf{A}} = e^{t\mathbf{A}}e^{-\tau\mathbf{A}}$, and because the variable of integration is τ, the factor $e^{t\mathbf{A}}$ can be removed from the integral, as shown:

$$(2.27) \quad \int_{t_0}^{t} e^{(t-\tau)\mathbf{A}}\mathbf{B}\,\mathbf{U}(\tau)\,d\tau = e^{t\mathbf{A}}\int_{t_0}^{t} e^{-\tau\mathbf{A}}\mathbf{B}\,\mathbf{U}(\tau)\,d\tau.$$

Now (2.25) will be differentiated,

(2.28) $\dfrac{d}{dt}\mathbf{X}(t) = \dfrac{d}{dt}\left(e^{(t-t_0)\mathbf{A}}\mathbf{X}(t_0) + e^{t\mathbf{A}}\displaystyle\int_{t_0}^{t} e^{-\tau\mathbf{A}}\mathbf{B}\,\mathbf{U}(\tau)\,d\tau \right)$

$= \left(\dfrac{d}{dt}e^{t\mathbf{A}} \right) e^{-t_0\mathbf{A}}\mathbf{X}(t_0) + \left(\dfrac{d}{dt}e^{t\mathbf{A}} \right)\displaystyle\int_{t_0}^{t} e^{-\tau\mathbf{A}}\mathbf{B}\,\mathbf{U}(\tau)\,d\tau$

$+ e^{t\mathbf{A}}\dfrac{d}{dt}\displaystyle\int_{t_0}^{t} e^{-\tau\mathbf{A}}\mathbf{B}\,\mathbf{U}(\tau)\,d\tau$

$= \mathbf{A}\,e^{t\mathbf{A}}e^{-t_0\mathbf{A}}\mathbf{X}(t_0) + \mathbf{A}e^{t\mathbf{A}}\displaystyle\int_{t_0}^{t} e^{-\tau\mathbf{A}}\mathbf{B}\,\mathbf{U}(\tau)\,d\tau + e^{t\mathbf{A}}e^{-t\mathbf{A}}\mathbf{B}\,\mathbf{U}(t)$

$= \mathbf{A}\left(e^{(t-t_0)\mathbf{A}}\mathbf{X}(t_0) + e^{t\mathbf{A}}\displaystyle\int_{t_0}^{t} e^{-\tau\mathbf{A}}\mathbf{B}\,\mathbf{U}(\tau)\,d\tau \right) + \mathbf{B}\mathbf{U}(t)$

$= \mathbf{A}\mathbf{X}(t) + \mathbf{B}\mathbf{U}(t),$

showing that (2.25) satisfies (1.7a), and hence is a solution.

In the above development it was assumed that the integral in (2.25) exists. The exponential is continuous everywhere, implying that the solution will be defined for any $\mathbf{U}(t)$ with entries that are piecewise-continuous, with steps or impulses allowed at the discontinuities, if any. If, for example, $\mathbf{U}(t)$ contains an impulse at $t = 0$, then the formula is correct provided $t_0 < 0 < t$. In such cases, the initial time is often taken to be $t_0 = 0-$, that is, a vanishingly small time prior to $t = 0$.

Output solution   With a solution (2.25) for (1.7a) in hand, the solution for $\mathbf{Y}(t)$ is obtained by substitution into (1.7b) as

(2.29) $\mathbf{Y}(t) = \mathbf{C}\left(e^{(t-t_0)\mathbf{A}}\mathbf{X}(t_0) + \displaystyle\int_{t_0}^{t} e^{(t-\tau)\mathbf{A}}\mathbf{B}\,\mathbf{U}(\tau)\,d\tau \right) + \mathbf{D}\mathbf{U}(t).$

Example 20
Sifting property of impulse
For the scalar system (2.23), let $u(t) = \delta(t)$, the unit impulse, and suppose it is known that $x(0-) = 0$. Recall the *sifting property* of the impulse, which is that if $a < \tau < b$ and $f(t)$ is continuous at $t = \tau$, then $\int_a^b f(t)\delta(t-\tau)\,dt = f(\tau)$. Then for any $t > 0$, from (2.24), the solution (2.25) is

$x(t) = e^{(t-0-)a}x(0-) + \displaystyle\int_{0-}^{t} e^{(t-\tau)a}b\,\delta(\tau)\,d\tau = e^{ta}0 + e^{ta}b = e^{ta}b.$

Example 21
Complete response
For the system with matrices and initial state given as

$\mathbf{A} = \begin{bmatrix} -1 & 0 \\ 0 & -2 \end{bmatrix}, \quad \mathbf{B} = \begin{bmatrix} 1 \\ 1 \end{bmatrix}, \quad \mathbf{C} = [1, 0], \quad \mathbf{D} = 0, \quad \mathbf{X}(0) = \begin{bmatrix} 1 \\ 0 \end{bmatrix},$

let the input be $\cos t$. One way to compute the output $y(t)$ is first to find the state-transition matrix,

$$e^{t\mathbf{A}} = \begin{bmatrix} e^{-t} & 0 \\ 0 & e^{-2t} \end{bmatrix},$$

so that, by substituting in (2.29), the output is

$$y(t) = [1,\, 0] \begin{bmatrix} e^{-t} & 0 \\ 0 & e^{-2t} \end{bmatrix} \begin{bmatrix} 1 \\ 0 \end{bmatrix}$$

$$+ \int_0^t [1,\, 0] \begin{bmatrix} e^{-(t-\tau)} & 0 \\ 0 & e^{-2(t-\tau)} \end{bmatrix} \begin{bmatrix} 1 \\ 1 \end{bmatrix} \cos \tau \, d\tau$$

$$= e^{-t} + \int_0^t e^{-(t-\tau)} \cos \tau \, d\tau.$$

Performing the integral, we get

$$y(t) = e^{-t} + \frac{1}{2}(\cos t + \sin t - e^{-t}) = \frac{1}{2}(\cos t + \sin t + e^{-t}).$$

2.5   Forced response

From (2.29), the complete response $\mathbf{Y}(t)$ consists of the free response as given in (2.22), plus the forced response, which is

(2.30) $$\mathbf{Y}_{\text{forced}}(t) = \int_{t_0}^t \mathbf{C}e^{(t-\tau)\mathbf{A}}\mathbf{B}\,\mathbf{U}(\tau)\,d\tau + \mathbf{D}\mathbf{U}(t),$$

from which several conclusions can be deduced.

Linearity   The first conclusion is that the forced response is a linear function of the input $\mathbf{U}(t)$, since for all real α, β, if piecewise-continuous inputs $\mathbf{U}_1(t)$ and $\mathbf{U}_2(t)$ produce forced outputs $\mathbf{Y}_1(t)$ and $\mathbf{Y}_2(t)$ respectively, then the input $\alpha\mathbf{U}_1(t) + \beta\mathbf{U}_2(t)$ produces forced output $\alpha\mathbf{Y}_1(t) + \beta\mathbf{Y}_2(t)$.

Equivalence   If we expand the exponential in the integral in (2.30),

$$\mathbf{Y}_{\text{forced}}(t) = \int_{t_0}^t \left(\mathbf{CB} + \frac{t-\tau}{1!}\mathbf{CAB} + \frac{(t-\tau)^2}{2!}\mathbf{CA}^2\mathbf{B} + \cdots \right) \mathbf{U}(\tau)\,d\tau$$

(2.31) $$+ \mathbf{D}\mathbf{U}(t)$$

$$= \int_{t_0}^t \left(\mathbf{H}(1) + \frac{t-\tau}{1!}\mathbf{H}(2) + \frac{(t-\tau)^2}{2!}\mathbf{H}(3) + \cdots \right) \mathbf{U}(\tau)\,d\tau$$

$$+ \mathbf{H}(0)\mathbf{U}(t),$$

where the matrices $\mathbf{H}(k)$ are exactly as defined previously in (2.7) for the analysis of the forced response of discrete-time systems, showing that the forced response is determined by the Markov sequence $\mathbf{H}(0) = \mathbf{D}$, $\mathbf{H}(k) = \mathbf{CA}^{k-1}\mathbf{B}$, $k > 0$.

As a consequence of (2.31), two continuous-time LTI systems (1.16) have identical forced responses for identical inputs if and only if they have identical Markov sequences $\{\mathbf{H}(k)\}_0^\infty$.

Example 22
Forced
response by
superposition

The system of Example 21, with input $\cos t$ as before, has forced response

$$y_{\text{forced}}(t) = \frac{1}{2}(\cos t + \sin t - e^{-t}),$$

so, by using superposition, if the input is a multiple $\alpha \cos t$ of the previous quantity, the forced response to it is

$$y_{\text{forced}}(t) = \frac{\alpha}{2}(\cos t + \sin t - e^{-t}).$$

Example 23
Complete
response

The matrices of a continuous-time LTI system are given as

$$\mathbf{A} = \begin{bmatrix} 0 & 1 \\ -9 & -6 \end{bmatrix}, \quad \mathbf{B} = \begin{bmatrix} 0 \\ 1 \end{bmatrix}, \quad \mathbf{C} = [9, 0], \quad \mathbf{D} = 0.$$

Figure 2.4 illustrates that the complete response $\mathbf{Y}(t)$ is the sum of the free response $\mathbf{Y}_{\text{free}}(t)$ and forced response $\mathbf{Y}_{\text{forced}}(t)$, computed for initial state $\mathbf{X}(0) = \begin{bmatrix} 0 \\ 1 \end{bmatrix}$ and input $\mathbf{U}(t) = 0.5$.

2.6   Continuous-time impulse response

Provided the entries of the product $\mathbf{DU}(t)$ are continuous for $t > t_0$, the term $\mathbf{DU}(t)$ in (2.30) can be moved under the integral sign, using the properties of impulses, to give

(2.32) $$\mathbf{Y}_{\text{forced}}(t) = \int_{t_0}^{t} \left(\mathbf{C}e^{(t-\tau)\mathbf{A}}\mathbf{B} + \mathbf{D}\delta(t - \tau) \right) \mathbf{U}(\tau)\,d\tau.$$

This has the form of a convolution integral, so the *impulse response matrix* $\mathbf{H}(t)$, also called the *weighting matrix,* is defined to be the term in parentheses with $t-\tau$ replaced by t, as shown,

(2.33) $$\mathbf{H}(t) = \mathbf{C}e^{t\mathbf{A}}\mathbf{B} + \mathbf{D}\delta(t),$$

and with this definition, the forced response is the convolution of the impulse-response matrix and the input vector, as is true for other LTI systems.

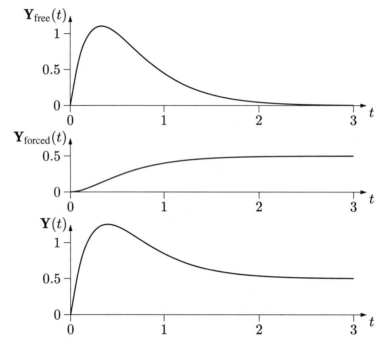

Fig. 2.4 Continuous-time free, forced, and complete response.

As for the discrete-time case, the impulse-response matrix $\mathbf{H}(t)$ is not the response to a single input vector, but each of its columns $\mathbf{H}(t)\mathbf{e}_i$ is the forced response to a unit impulse at the i-th input at time $t = 0$, that is, to input vector $\mathbf{e}_i\delta(t)$.

Example 24
Impulse-response matrix

For the system previously considered in Example 21, the state-transition matrix was calculated as

$$e^{t\mathbf{A}} = \begin{bmatrix} e^{-t} & 0 \\ 0 & e^{-2t} \end{bmatrix},$$

and the impulse-response matrix is therefore

$$\mathbf{H}(t) = \mathbf{C}e^{t\mathbf{A}}\mathbf{B} + \mathbf{D}\delta(t) = [\,1, 0\,] \begin{bmatrix} e^{-t} & 0 \\ 0 & e^{-2t} \end{bmatrix} \begin{bmatrix} 1 \\ 1 \end{bmatrix} + 0 = e^{-t}.$$

Example 25
Equivalent system

The system $\mathbf{A} = -1$, $\mathbf{B} = 1$, $\mathbf{C} = 1$, $\mathbf{D} = 0$ has impulse-response matrix $\mathbf{H}(t) = e^{-t}$ and is therefore externally equivalent to the system in the above example.

2.7   Continuous-time convolution

From the definition of the impulse-response matrix (2.33), and by setting $t_0 = 0$ in (2.30) for simplicity, the forced-response Equation (2.30) can be written in the form of a convolution integral as

$$(2.34) \quad \mathbf{Y}_{\text{forced}}(t) = \int_0^t \mathbf{H}(t - \tau)\mathbf{U}(\tau)\,d\tau,$$

or by change of variable $t - \tau = s$, as

$$(2.35) \quad \mathbf{Y}_{\text{forced}}(t) = \int_t^0 \mathbf{H}(s)\mathbf{U}(t-s)(-ds) = \int_0^t \mathbf{H}(s)\mathbf{U}(t-s)\,ds.$$

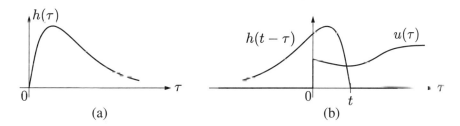

Fig. 2.5   The scalar convolution process: (a) impulse response; (b) integrand.

Recognizing that the response can be computed by either of the above integrals, we can also use the following notation:

$$(2.36) \quad \mathbf{Y}_{\text{forced}}(t) = \mathbf{H}(t) * \mathbf{U}(t).$$

For scalar systems the consequence of (2.34) and (2.35) is that (2.36) is sometimes written $y(t) = h(t) * u(t) = u(t) * h(t)$, but for matrices $\mathbf{H}(t), \mathbf{U}(t)$, the order of multiplication cannot be reversed in general.

An intuitive explanation of (2.34) can be given for a scalar system as illustrated in Figure 2.5. First note that the variable of integration is τ, so the horizontal axis in the figure is in units of τ. Using the argument $t - \tau$ instead of t in $h(\cdot)$ corresponds to replacing t by τ, reversing $h(\cdot)$ in time τ, and then shifting by t. Figure 2.5(a) shows $h(\tau)$, which is identically zero to the left of the origin. The reversed and shifted function is then multiplied by $u(\tau)$, as shown in Figure 2.5(b), and the area under the product is computed from 0 to t.

A second interpretation of the convolution process is shown in Figure 2.6. Suppose the input $u(t)$ is approximated as the sum of a set of pulse functions $\delta u(t)$, each zero everywhere except for the interval $\tau < t < \tau + \delta\tau$, where

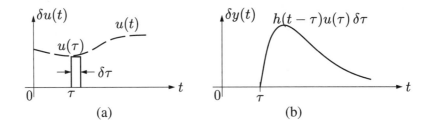

Fig. 2.6 Response to a pulse: (a) input pulse; (b) approximate response.

the height is $u(\tau)$, as shown in Figure 2.6(a). The pulse $\delta u(t)$ produces output $\delta y(t)$. Then for $t > \tau + \delta\tau$, with s as the variable of integration, the convolution integral becomes

$$(2.37) \quad \delta y(t) = \int_0^t h(t-s)u(s)\,ds = \int_\tau^{\tau+\delta\tau} h(t-s)u(\tau)\,ds \simeq h(t-\tau)u(\tau)\,\delta\tau,$$

as shown in Figure 2.6(b). For the input pulse specified, the output $\delta y(t)$ is approximately the impulse response $h(t)$ delayed to begin at time τ, and weighted by the factor $u(\tau)\delta\tau$. The response $y(t)$ to the complete function $u(t)$ is a sum of weighted and delayed responses of the above form, and in the limit as $\delta\tau \to 0$, this sum becomes the convolution integral.

Example 26
Continuous-
time
convolution

Figure 2.7 illustrates the convolution process for a system described by the equations

$$\frac{dx}{dt} = -x/T + 1u, \quad y = x + 0u,$$

with input $u(t)$, which is a pulse of height 1 and duration a, where T and a are constants. Row (a) in the figure shows the input $u(t)$ and the impulse response $h(t)$. Row (b) shows the response to a narrow pulse of width $\delta\tau$ at time $t = \tau$, and row (c) shows the sum of the responses to the narrow pulses, which, when added together, make up the actual input.

3   Discretization

For modeling, control, and signal processing, the accuracy with which discrete-time systems approximate continuous-time systems, or the reverse, may require investigation. Numerous measures of validity of approximation are possible,

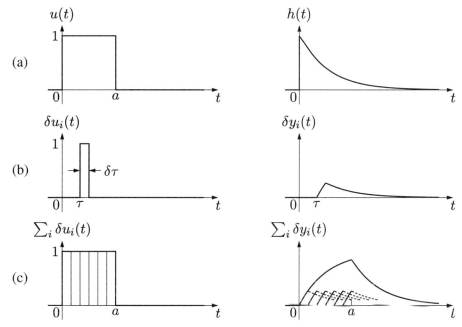

Fig. 2.7   Convolution as a sum of pulse responses: (a) input function and impulse response; (b) pulse component of the input and its response; (c) the sum of pulses and the sum of responses.

depending on the context. Here a common and important approximation will be given, requiring only knowledge of the solution of continuous-time and discrete-time state-space equations.

Suppose the input $\{\hat{\mathbf{U}}(k)\}$ of a discrete-time system is the sequence of values $\mathbf{U}(kT)$ of the continuous-system input, sampled with period T. It is required to design the discrete-time system so that its output sequence approximates the output of the continuous-time system. The result is called a discretization of the continuous system.

As shown in Figure 2.8, let the continuous-time system be described by

(2.38a) $\dot{\mathbf{X}}(t) = \mathbf{A}\mathbf{X}(t) + \mathbf{B}\tilde{\mathbf{U}}(t)$

(2.38b) $\mathbf{Y}(t) = \mathbf{C}\mathbf{X}(t) + \mathbf{D}\tilde{\mathbf{U}}(t),$

and the discrete-time system to be found by

(2.39a) $\hat{\mathbf{X}}(k{+}1) = \mathbf{F}\hat{\mathbf{X}}(k) + \mathbf{G}\hat{\mathbf{U}}(k)$

(2.39b) $\hat{\mathbf{Y}}(k) = \mathbf{H}\hat{\mathbf{X}}(k) + \mathbf{L}\hat{\mathbf{U}}(k).$

It is possible to find a discrete-time system that approximates the contin-

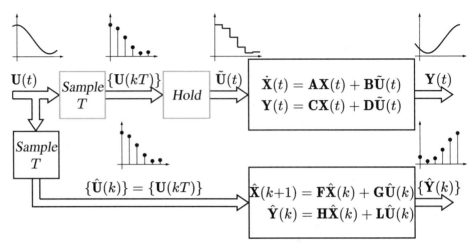

Fig. 2.8   Discretization.

uous-time system in the sense that the discrete output $\hat{\mathbf{Y}}(k)$ equals the continuous output $\mathbf{Y}(kT)$ exactly at each sample time kT for inputs that approximate the actual input. Replace the input $\mathbf{U}(t)$ to the continuous system by $\tilde{\mathbf{U}}(t)$, such that $\tilde{\mathbf{U}}(t) = \mathbf{U}(kT)$ over every interval $kT \leq t < (k+1)T$. That is, $\mathbf{U}(t)$ is passed through a fictitious sampler and zero-order hold as shown in the figure.

From Equation (2.25) the solution for the continuous state at the next sample time $(k+1)T$ in terms of the state at the current sample time kT and the input $\tilde{\mathbf{U}}(t)$ over this interval is

(2.40) $\mathbf{X}((k+1)T) = e^{((k+1)T - kT)\mathbf{A}}\mathbf{X}(kT) + \displaystyle\int_{kT}^{(k+1)T} e^{((k+1)T - \tau)\mathbf{A}}\mathbf{B}\,\tilde{\mathbf{U}}(\tau)\,d\tau,$

or, with change of variable $s = (k+1)T - \tau$ and constant input $\tilde{\mathbf{U}}(t) = \mathbf{U}(kT)$,

(2.41) $\mathbf{X}((k+1)T) = e^{T\mathbf{A}}\mathbf{X}(kT) + \displaystyle\int_{T}^{0} e^{s\mathbf{A}}(-ds)\,\mathbf{B}\mathbf{U}(kT)$

$= e^{T\mathbf{A}}\mathbf{X}(kT) + \displaystyle\int_{0}^{T} e^{\tau\mathbf{A}}d\tau\mathbf{B}\,\mathbf{U}(kT),$

where the variable of integration has been written as τ in the final right-hand side. Comparing this equation to (2.39a), we see that if

(2.42a) $\hat{\mathbf{X}}(0) = \mathbf{X}(0), \quad \mathbf{F} = e^{T\mathbf{A}}, \quad \mathbf{G} = \left(\displaystyle\int_{0}^{T} e^{\tau\mathbf{A}}d\tau\right)\mathbf{B},$

then $\{\hat{\mathbf{X}}(k)\} = \{\mathbf{X}(kT)\}$, and from (2.38b) and (2.39b), if in addition,

(2.42b) $\mathbf{H} = \mathbf{C}, \quad \mathbf{L} = \mathbf{D},$

then $\{\hat{\mathbf{Y}}(k)\} = \{\mathbf{Y}(kT)\}$.

In summary, this method of discretization requires the computation of the matrix exponential $e^{T\mathbf{A}}$ and its integral $\int_0^T e^{\tau\mathbf{A}}d\tau$ as in (2.42a), which is usually accomplished numerically. Then the discrete system output exactly matches the sampled continuous output, provided the continuous input is constant between sampling points, and provided the initial conditions are identical. Small errors of initial condition or in the matrices themselves may destroy the accuracy of the approximation if the continuous system is unstable, that is, if $e^{t\mathbf{A}}$ contains values that increase without limit as t becomes large.

Example 27
Discretization

Suppose, for sampling period $T = 0.1$, the discretization of the system in Example 21 is to be found. From the above,

$$\mathbf{H} = [1, 0], \quad \mathbf{L} = 0,$$

the matrix $\mathbf{F}$ is

$$\mathbf{F} = e^{0.1 \begin{bmatrix} 1 & 0 \\ 0 & -2 \end{bmatrix}} = \begin{bmatrix} e^{-0.1} & 0 \\ 0 & e^{-0.2} \end{bmatrix}.$$

Since $\mathbf{A}$ is diagonal and $\int_0^T e^{\alpha\tau}d\tau = \frac{1}{\alpha}(e^{\alpha T} - 1)$ for nonzero α, the matrix $\mathbf{G}$ is

$$\mathbf{G} = \int_0^{0.1} e^{\tau \begin{bmatrix} -1 & 0 \\ 0 & -2 \end{bmatrix}} d\tau \begin{bmatrix} 1 \\ 1 \end{bmatrix} = \begin{bmatrix} 1-e^{-0.1} & 0 \\ 0 & \frac{1}{2}(1-e^{-0.2}) \end{bmatrix}\begin{bmatrix} 1 \\ 1 \end{bmatrix}$$

$$= \begin{bmatrix} 1-e^{-0.1} \\ \frac{1}{2}(1-e^{-0.2}) \end{bmatrix}.$$

Example 28
Computing the integral

The definition of the matrix exponential leads to at least two ways of computing $\mathcal{I} = \int_0^T e^{\tau\mathbf{A}}d\tau$. First, using the definition of $e^{T\mathbf{A}}$, we find

(2.43) $$\mathcal{I} = \int_0^T e^{\tau\mathbf{A}}d\tau = \int_0^T \left(\mathbf{I} + \frac{\tau}{1!}\mathbf{A} + \frac{\tau^2}{2!}\mathbf{A}^2 + \cdots\right)d\tau$$

$$= T\mathbf{I} + \frac{T^2}{2!}\mathbf{A} + \frac{T^3}{3!}\mathbf{A}^2 + \cdots,$$

which is a series solution for the integral. Furthermore, if $\mathbf{A}$ is nonsingular then

(2.44) $$\mathcal{I} = \mathbf{A}^{-1}\left(e^{T\mathbf{A}} - \mathbf{I}\right),$$

which can be verified by expanding $e^{T\mathbf{A}}$ in (2.44) and comparing the result to (2.43).

4 | **Further study**

The straightforward computations required for solving discrete-time state-space equations can be refined for efficiency or precision, when used for real-time control (see Houpis and Lamont [22]), or for filtering (see, for example, King et al. [27]). Such references should also be consulted for discrete-transform methods, quantization errors, and finite computer word-length effects.

A standard reference for the classical topic of solving (2.11) with known initial values is Coddington and Levinson [11], but texts such as Kreyszig [29] provide an introduction. Numerical methods for these equations, known as *initial value problems,* can be found in books of numerical recipes, such as Press et al. [43], in specialized volumes such as Roberts [44] or Mattheij and Molenaar [36], in scientific computing libraries, and implicitly in simulation software.

The matrix $e^{T\mathbf{A}}$ with fixed real T and $\mathbf{A}$ appears in the discretization calculation and elsewhere. The reader should consult Moler and Van Loan [39] before reinventing a method for the computation, and then a reference such as Golub and Van Loan [19].

5 | **Problems**

1 Compute the state-transition matrix $\mathbf{\Phi}(k,0) = \mathbf{A}^k$ for systems with the following matrices: (a) $\mathbf{A} = 1/2$, (b) $\mathbf{A} = 2$, (c) $\mathbf{A} = \begin{bmatrix} 1/2 & 0 \\ 0 & 1/3 \end{bmatrix}$.

2 An LTI system has free-response sequence $\{1, 2, 3, 4, \cdots\}$ for initial state $\mathbf{X}_0 = \begin{bmatrix} 2 \\ 1 \end{bmatrix}$ and free-response sequence $\{1, 1, 1, \cdots\}$ for initial state $\mathbf{X}_0 = \begin{bmatrix} 1 \\ 0 \end{bmatrix}$. What will the free response be for initial state $\mathbf{X}_0 = \begin{bmatrix} 0 \\ 1 \end{bmatrix}$?

3 A discrete-time LTI system with matrix $\mathbf{A} = \begin{bmatrix} 1 & 2 \\ 0 & \alpha \end{bmatrix}$, with $\alpha \neq 0$, has free state response $\mathbf{X} = \begin{bmatrix} 1 \\ 0 \end{bmatrix}$ at time $k = 4$.

(a) What will the state be at time $k = 7$?

(b) What was the state at time $k = 2$?

(c) How would the answers to these two questions change if $\alpha = 0$?

4   Applying an input which is a unit step given by $\text{step}(t) = \begin{cases} 0, & t < 0 \\ 1, & t \geq 0 \end{cases}$ to an

LTI system produces the forced response $\begin{bmatrix} \dfrac{1 - e^{-3t}}{te^{-3t}} \end{bmatrix} \text{step}(t)$. What will be the

forced response of the system for a unit impulse input $u(t) = \delta(t)$?

5   For very large systems or for hand calculation, it is sometimes desirable to minimize the number of algebraic operations in a calculation. There are mnq scalar multiplications (or multiplication-addition pairs, sometimes called "flops") required to calculate the product $\mathbf{MN}$ of $m \times n$ matrix $\mathbf{M}$ and $n \times q$ matrix $\mathbf{N}$. Determine the number of scalar multiplications required to calculate $\mathbf{X}(\ell)$ in (2.2) by calculating the powers of $\mathbf{A}$ separately, as written, for $t_0 = 0$. Compare this number with recursive application of (1.17a) to calculate successively $\mathbf{X}(1) = \mathbf{AX}(0) + \mathbf{BU}(0)$, $\mathbf{X}(2) = \mathbf{AX}(1) + \mathbf{BU}(1)$, and so on, up to $\mathbf{X}(\ell)$. Compare these two numbers when, in (2.2), $\mathbf{X}$ is 1000×1, $\mathbf{U}$ is 50×1, and $t_0 = 0$, $\ell = 100$.

6   An LTI system has forced response $\{1, 2, 3, 4, \cdots\}$ for input $\{1, 2, 0, 0, \cdots\}$ and forced response $\{0, -1, -2, -3, \cdots\}$ for input $\{4, 1, 0, 0, \cdots\}$. What will the forced response be for input $\{4, 3, 5, 2, 0, 0, \cdots\} = \{4, 1, 0, 0, \cdots\} + 2\{0, 1, 2, 0, \cdots\} + \{0, 0, 1, 2, 0, 0, \cdots\}$?

7   A discrete-time system with one input, that is, $m = 1$, has the property $\mathbf{D} = 0$. For initial state $\mathbf{X}(0) = \mathbf{B}$, how does the free response differ from the weighting sequence?

8   Find the convolution sequence $\{\mathbf{Y}_k\} = \{\mathbf{H}_k\} * \{\mathbf{U}_k\}$ for the following sequences

(a) $\{\mathbf{H}_k\} = \{1, 3, 3, 1, 0, 0, \cdots\}$, $\{\mathbf{U}_k\} = \{1, 2, 1, 0, 0, \cdots\}$.

(b) $\{\mathbf{H}_k\} = \{4, 1, 7, 8, 0, 0, \cdots\}$, $\{\mathbf{U}_k\} = \{0, 0, 1, 0, 0, \cdots\}$.

(c) $\{\mathbf{H}_k\} = \left\{ \begin{bmatrix} 0 & 1 \\ 1 & 0 \end{bmatrix}, \begin{bmatrix} 2 & 0 \\ 0 & 2 \end{bmatrix}, \begin{bmatrix} 0 & 1 \\ 1 & 0 \end{bmatrix}, \begin{bmatrix} 0 & 0 \\ 0 & 0 \end{bmatrix}, \begin{bmatrix} 0 & 0 \\ 0 & 0 \end{bmatrix}, \cdots \right\}$,

$\{\mathbf{U}_k\} = \left\{ \begin{bmatrix} 1 \\ 0 \end{bmatrix}, \begin{bmatrix} 0 \\ 2 \end{bmatrix}, \begin{bmatrix} 1 \\ 1 \end{bmatrix}, \begin{bmatrix} 0 \\ 0 \end{bmatrix}, \begin{bmatrix} 0 \\ 0 \end{bmatrix}, \cdots \right\}$.

9 Compute the impulse-response sequence of

(a) the discrete-time real system $\mathbf{A} = \begin{bmatrix} 1/2 & 2 \\ 0 & 1 \end{bmatrix}$, $\mathbf{B} = \begin{bmatrix} 1 \\ 0 \end{bmatrix}$, $\mathbf{C} = [2, 0]$, $\mathbf{D} = 1$.

(b) the binary system $\mathbf{A} = \begin{bmatrix} 1 & 0 \\ 1 & 1 \end{bmatrix}$, $\mathbf{B} = \begin{bmatrix} 1 & 1 \\ 1 & 0 \end{bmatrix}$, $\mathbf{C} = [0, 1]$, $\mathbf{D} = 0$.

10 Computational efficiency is sometimes an issue when solutions of state-space equations are required, especially for large systems. One measure of computational cost is to count the number of scalar multiplications required. It requires n scalar multiplications to compute a product $\mathbf{MN}$ when $\mathbf{M}$ is $1 \times n$ and $\mathbf{N}$ is $n \times 1$, and the cost of more general matrix multiplications can be calculated from this basic result. For a discrete-time system, two methods of computing $\mathbf{X}(\ell)$ have been given, in recursive formula (1.17a), and in formula (2.2). Find the approximate ratio of the cost of these two formulas in computing $\mathbf{X}(\ell)$, given $\mathbf{X}(0)$ and $\{\mathbf{U}(k)\}_0^{\ell-1}$, if $\ell = 1000$, $n = 100$, $m = 1$.

11 Show whether the system $\mathbf{A} = \begin{bmatrix} 1 & 0 \\ 2 & 1/2 \end{bmatrix}$, $\mathbf{B} = \begin{bmatrix} 0 \\ 1 \end{bmatrix}$, $\mathbf{C} = [1, 2]$, $\mathbf{D} = 1$ has identical forced behavior to the system of question 9(a).

12 Find $e^{t\mathbf{A}}$ for the following matrices $\mathbf{A}$, and verify in each case that the result is a nonsingular matrix:

(a) $\begin{bmatrix} 0 & 2 \\ 0 & 0 \end{bmatrix}$, (b) $\begin{bmatrix} 2 & 0 \\ 0 & 2 \end{bmatrix}$, (c) $\begin{bmatrix} 0 & 3 \\ -3 & 0 \end{bmatrix}$, (d) $\begin{bmatrix} 2 & 3 \\ -3 & 2 \end{bmatrix}$.

13 For the system with matrix $\mathbf{A} = \begin{bmatrix} -2 & 1 \\ 0 & -2 \end{bmatrix}$,

(a) compute the state-transition matrix $\mathbf{\Phi}(t, 0) = e^{t\mathbf{A}}$,

(b) compute the free response $\mathbf{X}(t)$ for initial state $\mathbf{X}(0) = \begin{bmatrix} 1 \\ 1 \end{bmatrix}$,

(c) if the free response is $\mathbf{X}(10) = \begin{bmatrix} 1 \\ 2 \end{bmatrix}$ at $t = 10$, what was it at $t = 9$?

14 Consider the continuous-time system for which $\mathbf{A} = \begin{bmatrix} -1 & 0 \\ 0 & -2 \end{bmatrix}$, $\mathbf{B} = \begin{bmatrix} 1 \\ 1 \end{bmatrix}$, $\mathbf{C} = [1, 0]$, $\mathbf{D} = 0$.

(a) Find the free response to initial state $X(0) = \begin{bmatrix} 0 \\ 1 \end{bmatrix}$.

(b) Find the forced response to the input $U(t) = \delta(t)$.

(c) Find the complete response to input $U(t) = \delta(t)$ for initial state $X(0) = \begin{bmatrix} 0 \\ 1 \end{bmatrix}$.

(d) Find the forced response to input $U(t) = e^{-3t}$.

15   Show that $e^{A+B} = e^A e^B$ if $AB = BA$.

16   Compute the weighting matrix for the continuous-time system
$$A = \begin{bmatrix} -1 & 0 \\ 0 & -2 \end{bmatrix}, B = \begin{bmatrix} 1 & 0 \\ 1 & 1 \end{bmatrix}, C = [1, 1], D = [0,0].$$

17   For the differential equations
$$\dot{X} = \begin{bmatrix} 1 & 1 \\ 0 & 2 \end{bmatrix} X + \begin{bmatrix} 1 \\ 0 \end{bmatrix},$$

find the value of $X(t_0)$ that should be selected at $t_0 = 0$ such that at $t = 1$ the solution is $X(1) = \begin{bmatrix} 1 \\ 1 \end{bmatrix}$.

18   A system has weighting matrix $H(t) = \begin{bmatrix} e^{-t} \\ e^{-2t} \end{bmatrix}$. Compute the forced response to input $U(t) = t$.

19   Show that the system
$$\dot{X} = \begin{bmatrix} A_1 & A_2 \\ 0 & A_3 \end{bmatrix} X + \begin{bmatrix} B_1 \\ 0 \end{bmatrix} U, \quad Y = [C_1, C_2] X + DU,$$

where A_1 is square and the partitions of B and C are conformable, and the system
$$\dot{X}' = A_1 X' + B_1 U, \quad Y = C_1 X' + DU,$$

have the same forced behavior.

20   Find the discretization of the system $A = -10, B = 1, C = 3, D = 0$, sampled with period $T = 0.001$.

21 Use a computer to verify that the discretization of the system of Example 23 with sampling interval $T = 0.1$ is the system of Example 6. Show that since the input is constant, the graphs of Figure 2.2 contain the sampled values of the graphs of Figure 2.4.

22 Computing the matrix exponential $e^{T\mathbf{A}}$ by summing the series $e^{T\mathbf{A}} = \mathbf{I} + \frac{1}{1!}(T\mathbf{A}) + \frac{1}{2!}(T\mathbf{A})^2 + \cdots$ at first glance may appear to be hopeless when $T\mathbf{A}$ contains large entries, since many terms must be summed before additional terms become sufficiently small. However, although there is no single reliable method for all $T\mathbf{A}$, the series method becomes surprisingly robust when combined with certain tricks, one of which is to note that

$$e^{T\mathbf{A}} = \left(e^{(T/2^k)\mathbf{A}}\right)^{2^k}.$$

The size of the entries in $T\mathbf{A}$ are reduced to less than 0.5, say, by dividing exactly by 2^k. Then $e^{(T/2^k)\mathbf{A}}$ is computed efficiently, using the series, and the result is squared k times. In the simplest case, $T\mathbf{A}$ is a scalar. For $T\mathbf{A} = 0.5$, determine how many terms of the series have to be computed before the term $\frac{1}{r!}(T\mathbf{A})^r$ has magnitude less than 10^{-16}, the approximate floating-point precision of many computers.

23 Find the discretization of the system $\mathbf{A} = \begin{bmatrix} \alpha & \omega \\ -\omega & \alpha \end{bmatrix}$, $\mathbf{B} = \begin{bmatrix} 0 \\ 1 \end{bmatrix}$, $\mathbf{C} = [3, -1]$, $\mathbf{D} = 0$, sampled with period T.

24 Because, as in Example 16, the matrix $\mathbf{F} = e^{T\mathbf{A}}$ is nonsingular, this fact has occasionally led to the conclusion that discrete-time systems always have non-singular matrices $\mathbf{F}$ in the equation $\mathbf{X}_{k+1} = \mathbf{F}\mathbf{X}_k + \mathbf{G}\mathbf{U}_k$. Consider Example 12 and Example 10 of Chapter 1, and discuss why such a conclusion is naïve.

3

Transform methods

The equations of the LTI state-space models (1.16) and (1.17) are linear and contain constant coefficients. Therefore these equations can be solved by using Laplace transforms in the case of (1.16), and $\mathcal{Z}$-transforms in the case of (1.17).

The notation to be used for the Laplace transform of $f(t)$ is

(3.1) $\hat{f}(s) = \mathcal{L}\{f(t)\} = \int_{0-}^{\infty} f(t)e^{-st}dt,$

where $\hat{f}(s)$ denotes the transform of $f(t)$ and the lower limit of integration is shown as $0-$ to emphasize that steps or impulses at $t = 0$ are allowed. Sometimes, when the context makes the meaning clear, the caret is omitted, so that $\hat{f}(s)$ becomes simply $f(s)$, and functions of s are taken to be transform functions. The transform of a vector is the vector of transforms; thus:

(3.2) $\mathcal{L}\{\mathbf{X}(t)\} = \begin{bmatrix} \mathcal{L}\{x_1(t)\} \\ \vdots \\ \mathcal{L}\{x_n(t)\} \end{bmatrix} = \begin{bmatrix} \hat{x}_1(s) \\ \vdots \\ \hat{x}_n(s) \end{bmatrix} = \hat{\mathbf{X}}(s),$

and the result is sometimes written as $\mathbf{X}(s)$ when the meaning is clear.

In contrast to the above, in which functions of continuous time are converted to functions of a variable s, in the discrete-time case, the transform of a sequence is

(3.3) $\hat{f}(z) = \mathcal{Z}\{f(t)\} = \mathcal{Z}\{f(0),\, f(1),\, f(2),\, \cdots\} = \sum_{t=0}^{\infty} f(t)\, z^{-t},$

where sequences are taken to be identically zero for terms with negative indices, so that the sum is over nonnegative t. Again, the result is sometimes simply written as $f(z)$ when the meaning is clear. Using subscript notation instead of parentheses, Equation (3.3) can be rewritten as

(3.4) $\hat{f}(z) = \mathcal{Z}\{f_k\} = \mathcal{Z}\{f_0,\, f_1,\, f_2,\, \cdots\} = \sum_{t=0}^{\infty} f_k\, z^{-t}.$

The subscript notation is used in Table 3.1, which summarizes some important properties of Laplace and $\mathcal{Z}$ transforms. Table 3.2 shows some useful elementary Laplace transform pairs, and Table 3.3 on page 68 shows some useful $\mathcal{Z}$-transform pairs.

Table 3.1 Transform properties. The unit step function is $\text{step}(t) = \{0$ for $t < 0$, 1 for $t \geq 0\}$. The equalities involving limits are true provided the limits exist.

Property	Laplace transform	$\mathcal{Z}$ transform
Function	$f(t)$, $t \in \mathbb{R}$	$\{f_k\}_0^\infty$, $k \in \mathbb{Z}$
Notation	$\mathcal{L}\{f(t)\} = \hat{f}(s)$	$\mathcal{Z}\{f_k\} = \hat{f}(z)$
Definition	$\hat{f}(s) = \int_{0-}^\infty f(t)e^{-st}dt$	$\hat{f}(z) = \sum\limits_{k=0}^\infty f_k z^{-k}$
Linearity	$\mathcal{L}\{\alpha f(t) + \beta g(t)\}$ $= \alpha \hat{f}(s) + \beta \hat{g}(s)$	$\mathcal{Z}\{\alpha f_k + \beta g_k\}$ $= \alpha \hat{f}(z) + \beta \hat{g}(z)$
$\mathcal{L}$: derivative $\mathcal{Z}$: left shift	$\mathcal{L}\{\frac{d}{dt}f(t)\} = s\hat{f}(s) - f(0-)$	$\mathcal{Z}\{f_1, f_2, \cdots\} = \mathcal{Z}\{f_{k+1}\}$ $= z\hat{f}(z) - z f_0$
$\mathcal{L}$: integral $\mathcal{Z}$: delay	$\mathcal{L}\{\int_{0-}^t f(\tau)\,d\tau\} = s^{-1}\hat{f}(s)$	$\mathcal{Z}\{0, f_0, f_1, \cdots\}$ $= \mathcal{Z}\{f_{k-1}\} = z^{-1}\hat{f}(z)$
Multiplication by time	$\mathcal{L}\{t\,f(t)\} = -\frac{d}{ds}\hat{f}(s)$	$\mathcal{Z}\{k\,f_k\} = -z\frac{d}{dz}\hat{f}(z)$
Multiplication by exponent or power	$\mathcal{L}\{e^{-at}f(t)\} = \hat{f}(s+a)$	$\mathcal{Z}\{\beta^k f_k\} = \hat{f}(z/\beta)$
Time shift	$\mathcal{L}\{f(t-\tau)\,\text{step}(t-\tau)\}$ $= e^{-\tau s}\hat{f}(s)$	$\mathcal{Z}\{f_{k-q}\} = z^{-q}\hat{f}(z)$
Convolution	$\mathcal{L}\{\int_{0-}^t f(t-\tau)\,g(\tau)\,d\tau$ $= \hat{f}(s)\,\hat{g}(s)$	$\mathcal{Z}\{\sum\limits_{r=0}^k f_{k-r}g_r\} = \hat{f}(z)\,\hat{g}(z)$
Initial value	$\lim\limits_{t\to 0+} f(t) = \lim\limits_{s\to\infty} s\,\hat{f}(s)$	$f_0 = \lim\limits_{z\to\infty} f(z)$
Final value	$\lim\limits_{t\to\infty} f(t) = \lim\limits_{s\to 0} s\,\hat{f}(s)$	$\lim\limits_{k\to\infty} f_k = \lim\limits_{z\to 1}(z-1)\hat{f}(z)$

This chapter requires basic knowledge of matrix determinants and inverses, covered in more detail in Sections 2 and 4 of Chapter 5.

Table 3.2   Elementary Laplace transform pairs.

Transform	Time function
1	$\delta(t)$
$\dfrac{1}{s}$	$\text{step}(t)$
$\dfrac{1}{(s-a)}$	e^{at}
$\dfrac{1}{(s-a)^q}$	$\dfrac{t^{q-1}}{(q-1)!}e^{at}$
$\dfrac{s+\alpha}{(s+\alpha)^2 + \omega^2}$	$e^{-\alpha t}\cos(\omega t)$
$\dfrac{\omega}{(s+\alpha)^2 + \omega^2}$	$e^{-\alpha t}\sin(\omega t)$

1   Continuous-time models

In order to solve Equation (1.16), repeated here,

(1.16a) $\dfrac{d}{dt}\mathbf{X}(t) = \mathbf{A}\mathbf{X}(t) + \mathbf{B}\mathbf{U}(t)$

(1.16b) $\mathbf{Y}(t) = \mathbf{C}\mathbf{X}(t) + \mathbf{D}\mathbf{U}(t),$

with initial value $\mathbf{X}(0) = \mathbf{X}_0$, both equations are transformed; first (1.16a):

(3.5) $\mathcal{L}\{\dfrac{d}{dt}\mathbf{X}(t)\} = s\hat{\mathbf{X}}(s) - \mathbf{X}_0 = \mathcal{L}\{\mathbf{A}\mathbf{X}(t) + \mathbf{B}\mathbf{U}(t)\}$

$= \mathbf{A}\mathcal{L}\{\mathbf{X}(t)\} + \mathbf{B}\mathcal{L}\{\mathbf{U}(t)\} = \mathbf{A}\hat{\mathbf{X}}(s) + \mathbf{B}\hat{\mathbf{U}}(s),$

and then the output equation (1.16b):

(3.6) $\mathcal{L}\{\mathbf{Y}(t)\} = \mathcal{L}\{\mathbf{C}\mathbf{X}(t) + \mathbf{D}\mathbf{U}(t)\} = \mathbf{C}\mathcal{L}\{\mathbf{X}(t)\} + \mathbf{D}\mathcal{L}\{\mathbf{U}(t)\} = \mathbf{C}\hat{\mathbf{X}}(s) + \mathbf{D}\hat{\mathbf{U}}(s).$

Collecting terms to isolate the unknown $\hat{\mathbf{X}}(s)$ in (3.5) gives

(3.7) $(s\mathbf{I} - \mathbf{A})\,\hat{\mathbf{X}}(s) = \mathbf{X}_0 + \mathbf{B}\hat{\mathbf{U}}(s).$

The determinant of $(s\mathbf{I} - \mathbf{A})$ is a polynomial of degree n in s (as discussed in detail in Section 1.3) and is thus not identically zero, so this matrix has an inverse, and the solution can be written symbolically as

(3.8) $\hat{\mathbf{X}}(s) = (s\mathbf{I} - \mathbf{A})^{-1}\mathbf{X}_0 + (s\mathbf{I} - \mathbf{A})^{-1}\mathbf{B}\hat{\mathbf{U}}(s),$

showing that the Laplace transform of the solution for $\mathbf{X}(t)$ is a function of the initial state $\mathbf{X}_0$, as well as of the input transform $\hat{\mathbf{U}}(s)$.

To solve for $\hat{\mathbf{Y}}(s)$, the above solution is substituted in (3.6), giving

(3.9) $\hat{\mathbf{Y}}(s) = \mathbf{C}(s\mathbf{I}-\mathbf{A})^{-1}\mathbf{X}_0 + \left(\mathbf{C}(s\mathbf{I}-\mathbf{A})^{-1}\mathbf{B} + \mathbf{D}\right)\hat{\mathbf{U}}(s).$

Inspection of (3.8) and (3.9), and comparison with time-domain solutions (2.25) and (2.29) respectively, yield several useful conclusions, which are discussed in the following sections.

Example 1
Complete
solution

For the system of Example 21 of Chapter 2, repeated here:

$$\mathbf{A} = \begin{bmatrix} -1 & 0 \\ 0 & -2 \end{bmatrix}, \quad \mathbf{B} = \begin{bmatrix} 1 \\ 1 \end{bmatrix}, \quad \mathbf{C} = [1, 0], \quad \mathbf{D} = 0,$$

$$\mathbf{X}(0) = \begin{bmatrix} 1 \\ 0 \end{bmatrix}, \quad \mathbf{U}(t) = \cos t,$$

the transform of the state is, from Equation (3.8),

$$\hat{\mathbf{X}}(s) = \begin{bmatrix} s+1 & 0 \\ 0 & s+2 \end{bmatrix}^{-1} \begin{bmatrix} 1 \\ 0 \end{bmatrix} + \begin{bmatrix} s+1 & 0 \\ 0 & s+2 \end{bmatrix}^{-1} \begin{bmatrix} 1 \\ 1 \end{bmatrix} \mathcal{L}\{\cos t\}$$

$$= \begin{bmatrix} (s+1)^{-1} \\ 0 \end{bmatrix} + \begin{bmatrix} (s+1)^{-1} \\ (s+2)^{-1} \end{bmatrix} \frac{s}{s^2+1},$$

and the transform of the output is, from Equation (3.9),

$$\hat{\mathbf{Y}}(s) = [1, 0]\begin{bmatrix} s+1 & 0 \\ 0 & s+2 \end{bmatrix}^{-1} \begin{bmatrix} 1 \\ 0 \end{bmatrix} + \left([1, 0]\begin{bmatrix} s+1 & 0 \\ 0 & s+2 \end{bmatrix}^{-1} \begin{bmatrix} 1 \\ 1 \end{bmatrix} + 0\right) \frac{s}{s^2+1}$$

$$= \frac{1}{s+1} + \frac{s}{(s+1)(s^2+1)}.$$

1.1 **Free response**

Comparing (3.9) with its time-domain equivalent (2.29), the transformed free-response component of $\mathbf{Y}$ is

(3.10) $\mathcal{L}\{\mathbf{Y}_{\text{free}}(t)\} = \mathcal{L}\{\mathbf{C}e^{-t\mathbf{A}}\mathbf{X}_0\} = \mathbf{C}(s\mathbf{I}-\mathbf{A})^{-1}\mathbf{X}_0.$

Because this equation and (3.8) are true for arbitrary $\mathbf{X}_0$, it is possible to reach the theoretical conclusion that the transform of the state-transition matrix $\mathbf{\Phi}(t, 0) = e^{t\mathbf{A}}$ is

(3.11) $\mathcal{L}\{e^{t\mathbf{A}}\} = (s\mathbf{I}-\mathbf{A})^{-1}.$

This equivalence can be derived by an alternative route. First, write the formal expansion of $(s\mathbf{I}-\mathbf{A})^{-1}$ as

(3.12) $(s\mathbf{I}-\mathbf{A})^{-1} = s^{-1}\mathbf{I} + s^{-2}\mathbf{A} + s^{-3}\mathbf{A}^2 + \cdots,$

which can be verified by premultiplying both sides of the above equation by $(s\mathbf{I} - \mathbf{A})$ and comparing terms. Then, take the transform of the expansion of $e^{t\mathbf{A}}$ to get

$$(3.13) \quad \mathcal{L}\{e^{t\mathbf{A}}\} = \mathcal{L}\{\mathbf{I} + \frac{t}{1!}\mathbf{A} + \frac{t^2}{2!}\mathbf{A}^2 + \cdots\} = s^{-1}\mathbf{I} + s^{-2}\mathbf{A} + s^{-3}\mathbf{A}^2 + \cdots,$$

which is the required series in s.

Equation (3.11) has theoretical interest, but it also provides an alternative for symbolic computation of responses.

Example 2
Free response

The transform of the free response for Example 1 is, from (3.10),

$$\hat{\mathbf{Y}}_{\text{free}}(s) = [1, 0]\begin{bmatrix} s+1 & 0 \\ 0 & s+2 \end{bmatrix}^{-1}\begin{bmatrix} 1 \\ 0 \end{bmatrix} = \frac{1}{s+1}.$$

Example 3
$e^{t\mathbf{A}}$ by Laplace transform

The function $e^{t\mathbf{A}}$ for $\mathbf{A} = \begin{bmatrix} 0 & 1 \\ -1/4 & -1 \end{bmatrix}$ is difficult to write in closed from by inspection of its series, but it can be computed as

$$e^{t\mathbf{A}} = \mathcal{L}^{-1}\{(s\mathbf{I} - \mathbf{A})^{-1}\} = \mathcal{L}^{-1}\begin{bmatrix} s & -1 \\ 1/4 & s+1 \end{bmatrix}^{-1} = \mathcal{L}^{-1}\frac{1}{(s+1/2)^2}\begin{bmatrix} s+1 & 1 \\ -1/4 & s \end{bmatrix}$$

$$= \begin{bmatrix} (1+t/2)e^{-t/2} & te^{-t/2} \\ (-t/4)e^{-t/2} & (1-t/2)e^{-t/2} \end{bmatrix}.$$

1.2   Forced response and transfer matrix

From (3.9) the transformed forced response is

$$(3.14) \quad \mathcal{L}\{\mathbf{Y}_{\text{forced}}(t)\} = \mathcal{L}\{\int_{t_0}^{t} \mathbf{C}e^{(t-\tau)\mathbf{A}}\mathbf{B}\,\mathbf{U}(\tau)\,d\tau + \mathbf{D}\mathbf{U}(t)\}$$

$$= \left(\mathbf{C}(s\mathbf{I} - \mathbf{A})^{-1}\mathbf{B} + \mathbf{D}\right)\hat{\mathbf{U}}(s),$$

or

$$(3.15) \quad \hat{\mathbf{Y}}_{\text{forced}}(s) = \hat{\mathbf{H}}(s)\,\hat{\mathbf{U}}(s),$$

where, by definition, the transfer function *matrix* is

$$(3.16) \quad \hat{\mathbf{H}}(s) = \mathbf{C}(s\mathbf{I} - \mathbf{A})^{-1}\mathbf{B} + \mathbf{D}.$$

From (3.11), the inverse transform of $\hat{\mathbf{H}}(s)$ is

$$(3.17) \quad \mathcal{L}^{-1}\hat{\mathbf{H}}(s) = \mathbf{C}e^{t\mathbf{A}}\mathbf{B} + \mathbf{D}\delta(t) = \mathbf{H}(t),$$

which is the impulse-response matrix given in Equation (2.33). Thus, from a comparision of (2.34) and (3.14), the convolution theorem is seen to hold even when the transfer function is a matrix: the transform of the convolution $\mathbf{H}(t) * \mathbf{U}(t)$ is the product of the transform $\hat{\mathbf{H}}(s)$ of $\mathbf{H}(t)$ and the transform $\hat{\mathbf{U}}(s)$ of $\mathbf{U}(t)$.

Example 4
Forced
response

The transform of the forced response for Example 1 is, from (3.14),

$$\hat{\mathbf{Y}}_{\text{forced}}(s) = \left([1, 0] \begin{bmatrix} s+1 & 0 \\ 0 & s+2 \end{bmatrix}^{-1} \begin{bmatrix} 1 \\ 0 \end{bmatrix} + 0 \right) \frac{s}{s^2+1} = \frac{s}{(s+1)(s^2+1)}.$$

1.3 Properties of the transfer matrix

The formula (3.16) for the transfer matrix will be examined in more detail, in order to demonstrate some important properties of state-space models. As described in Chapter 5, one formula for the inverse of an n-square matrix (that is, an $n \times n$ matrix) $\mathbf{Q}$ is

$$(3.18) \quad \mathbf{Q}^{-1} = \frac{1}{\det(\mathbf{Q})} \operatorname{adj}(\mathbf{Q}),$$

where $\det(\mathbf{Q})$ is the determinant of $\mathbf{Q}$, and $\operatorname{adj}(\mathbf{Q})$ is the n-square adjoint matrix, of which each entry ij is $(-1)^{i+j} \det(\mathbf{Q}_{ji})$, where $\mathbf{Q}_{ji}$ is $\mathbf{Q}$ with row j and column i removed. Writing the entries of $\mathbf{Q}$ as q_{ij}, one formula for the determinant is given as

$$(3.19) \quad \det(\mathbf{Q}) = \sum \pm q_{1i} q_{2j} \cdots q_{nk},$$

where each product contains one entry from every row of $\mathbf{Q}$, from columns $i, j, \cdots k$, where the latter list of n integers takes on all permutations of the integers $1, 2, \cdots n$. The sign is negative for odd permutations, and positive for even permutations.

In (3.16) the inverse is

$$(3.20) \quad (s\mathbf{I}-\mathbf{A})^{-1} = \frac{1}{\det(s\mathbf{I}-\mathbf{A})} \operatorname{adj}(s\mathbf{I}-\mathbf{A}),$$

and the required determinant is

$$(3.21) \quad \det(s\mathbf{I}-\mathbf{A}) = \det \begin{bmatrix} s-a_{11} & -a_{12} & \cdots & -a_{1n} \\ -a_{21} & s-a_{22} & \ddots & -a_{2n} \\ \vdots & \ddots & \ddots & \vdots \\ -a_{n1} & \cdots & -a_{n,n-1} & s-a_{nn} \end{bmatrix}.$$

For this matrix, one of the products in (3.19) is $(s-a_{11})(s-a_{22})\cdots(s-a_{nn})$ and this is the only product of degree n in s. Therefore the determinant is a polynomial,

(3.22) $\det(s\mathbf{I}-\mathbf{A}) = s^n + a_1 s^{n-1} + \cdots a_{n-1}s + a_n,$

of degree n, where the a_i are constants. The entries of $\mathrm{adj}(s\mathbf{I}-\mathbf{A})$, by contrast, contain polynomials of degree at most $n-1$, since these entries are signed determinants of $n-1$-square submatrices of $(s\mathbf{I}-\mathbf{A})$.

Dimensions   From (3.16), $\hat{\mathbf{H}}(s)$ has p rows and m columns where by convention $\mathbf{Y}$ in (3.15) is a p-vector, and $\mathbf{U}$ is an m-vector. If we denote polynomial (3.22) by $\phi(s)$, the ij-th entry of $\hat{\mathbf{H}}(s)$ has the form

(3.23) $\hat{h}_{ij}(s) = \dfrac{b_1^{ij} s^{n-1} + b_2^{ij} s^{n-2} + \cdots b_n^{ij}}{\phi(s)} + d_{ij}$

$= \dfrac{d_{ij}\phi(s) + b_1^{ij} s^{n-1} + b_2^{ij} s^{n-2} + \cdots b_n^{ij}}{\phi(s)},$

where the high-degree coefficient b_1^{ij} may be zero, in which case the polynomial $b_1^{ij} s^{n-1} + b_2^{ij} s^{n-2} + \cdots b_n^{ij}$ may have degree less than $n-1$.

Properness   For any LTI continuous-time state-space system with defining matrices $\mathbf{A}$, $\mathbf{B}$, $\mathbf{C}$, $\mathbf{D}$, the entries of the transfer matrix $\hat{\mathbf{H}}(s) = \mathbf{D} + \mathbf{C}(s\mathbf{I}-\mathbf{A})^{-1}\mathbf{B}$ are rational functions of s, in which the degree of each numerator does not exceed the degree of the corresponding denominator. Such a matrix is said to be *proper*. From (3.23), each entry ij has numerator degree equal to the denominator degree if and only if d_{ij} is nonzero. A rational matrix for which the entries have degrees strictly less than the corresponding denominator degrees is *strictly proper*, and the transfer matrix of a state-space model is strictly proper if and only if $\mathbf{D} = 0$.

Example 5
Improper
transfer function
The transfer function $\hat{h}(s) = s/1$ cannot be the transfer function of any state-space model, because its numerator has degree 1 and its denominator has degree 0. A rational transfer function with numerator of higher degree than the denominator, or a rational matrix containing such an entry, is called *improper*.

Markov
sequence
From (3.23), each entry $\hat{h}_{ij}(s)$ of $\hat{\mathbf{H}}(s)$ can be formally expanded, using long division, into the form

(3.24) $\hat{h}_{ij}(s) = d_{ij} + h_{ij}(1)s^{-1} + h_{ij}(2)s^{-2} + \cdots,$

so by long division (or use of (3.12)), $\hat{\mathbf{H}}(s)$ can be written as the formal series

(3.25) $\hat{\mathbf{H}}(s) = \mathbf{H}(0) + s^{-1}\mathbf{H}(1) + s^{-2}\mathbf{H}(2) + \cdots,$

where $\mathbf{H}(0) = \mathbf{D}$ and the ij entry of $\mathbf{H}(k)$ is $h_{ij}(k)$. Comparison of this expansion with the transform of the impulse-response

$$(3.26) \quad \mathcal{L}\{\mathbf{H}(t)\} = \mathcal{L}\{\mathbf{D}\delta(t) + \mathbf{CB} + \frac{t}{1!}\mathbf{CAB} + \frac{t^2}{2!}\mathbf{CA}^2\mathbf{B} + \cdots\}$$
$$= \mathbf{H}(0) + s^{-1}\mathbf{H}(1) + s^{-2}\mathbf{H}(2) + \cdots,$$

shows that the $\mathbf{H}(k)$ of (3.25) are the Markov matrices previously defined in (2.7).

Example 6
Properness

In Example 1 the polynomial $\phi(s)$ is

$$\phi(s) = \det \begin{bmatrix} s+1 & 0 \\ 0 & s+2 \end{bmatrix} = (s+1)(s+2) = s^2 + 3s + 2,$$

and the transfer matrix is

$$\hat{\mathbf{H}}(s) = [\,1, 0\,]\begin{bmatrix} s+1 & 0 \\ 0 & s+2 \end{bmatrix}^{-1}\begin{bmatrix} 1 \\ 1 \end{bmatrix} + 0 = \frac{1}{s+1},$$

which is strictly proper. If the example is modified so that $\mathbf{D} = \beta \neq 0$, then the transfer matrix becomes

$$\hat{\mathbf{H}}(s) = [\,1, 0\,]\begin{bmatrix} s+1 & 0 \\ 0 & s+2 \end{bmatrix}^{-1}\begin{bmatrix} 1 \\ 1 \end{bmatrix} + \beta = \frac{\beta(s+1) + 1}{s+1},$$

which is proper.

Example 7
Markov
sequence

By long division, the transfer function of the above example (with $\mathbf{D} = 0$) can be expanded formally as

$$\hat{\mathbf{H}}(s) = 0 + s^{-1} - s^{-2} + s^{-3} - s^{-4}\cdots,$$

so the Markov sequence is

$$\{\mathbf{H}(k)\} = \{0, 1, -1, 1, -1, \cdots\}.$$

---

2 · Discrete-time models

The transform-solution properties of continuous-time models carry over almost directly to discrete-time models. To solve Equation (1.17), repeated here,

$$(1.17a) \quad \mathbf{X}(t+1) = \mathbf{AX}(t) + \mathbf{BU}(t)$$
$$(1.17b) \quad \mathbf{Y}(t) = \mathbf{CX}(t) + \mathbf{DU}(t),$$

note first that both equations apply at every time $t \geq 0$, and hence these equations apply term-wise to the sequence equations

(3.27a) $\{\mathbf{X}(1), \mathbf{X}(2), \cdots\} = \{\mathbf{A}\mathbf{X}(0)+\mathbf{B}\mathbf{U}(0), \mathbf{A}\mathbf{X}(1)+\mathbf{B}\mathbf{U}(1), \cdots\}$
$$= \mathbf{A}\{\mathbf{X}(0), \mathbf{X}(1), \cdots\} + \mathbf{B}\{\mathbf{U}(0), \mathbf{U}(1), \cdots\},$$

(3.27b) $\{\mathbf{Y}(0), \mathbf{Y}(1), \cdots\} = \mathbf{C}\{\mathbf{X}(0), \mathbf{X}(1), \cdots\} + \mathbf{D}\{\mathbf{U}(0), \mathbf{U}(1), \cdots\},$

or, using abbreviated notation,

(3.28a) $\{\mathbf{X}(t+1)\}_0^\infty = \mathbf{A}\{\mathbf{X}(t)\}_0^\infty + \mathbf{B}\{\mathbf{U}(t)\}_0^\infty,$

(3.28b) $\{\mathbf{Y}(t)\}_0^\infty = \mathbf{C}\{\mathbf{X}(t)\}_0^\infty + \mathbf{D}\{\mathbf{U}(t)\}_0^\infty.$

Taking the $\mathcal{Z}$-transform of (3.28a) yields

(3.29) $\mathcal{Z}\{\mathbf{X}(t+1)\} = z\hat{\mathbf{X}}(z) - z\mathbf{X}(0) = \mathbf{A}\hat{\mathbf{X}}(z) + \mathbf{B}\hat{\mathbf{U}}(z),$

and the transform of (3.28b) is

(3.30) $\hat{\mathbf{Y}}(z) = \mathbf{C}\hat{\mathbf{X}}(z) + \mathbf{D}\hat{\mathbf{U}}(z).$

From Equation (3.29), $\hat{\mathbf{X}}(z)$ is

(3.31) $\hat{\mathbf{X}}(z) = (z\mathbf{I}-\mathbf{A})^{-1}z\mathbf{X}_0 + (z\mathbf{I}-\mathbf{A})^{-1}\mathbf{B}\hat{\mathbf{U}}(z),$

which can be substituted into (3.30) to give the transformed output as a function of initial state $\mathbf{X}_0$ and input transform $\hat{\mathbf{U}}(z)$,

(3.32) $\hat{\mathbf{Y}}(z) = \mathbf{C}(z\mathbf{I}-\mathbf{A})^{-1}z\mathbf{X}_0 + \left(\mathbf{C}(z\mathbf{I}-\mathbf{A})^{-1}\mathbf{B} + \mathbf{D}\right)\hat{\mathbf{U}}(z).$

Comparing this formula with (3.9), we see that the first right-hand side term of (3.32) contains an additional factor z, but otherwise the formula is identical, with z replacing s, to the continuous-time transform solution (3.9).

Tables 3.2 and 3.3 show some useful elementary transform pairs and their properties.

Example 8
Complete
output response

For the discrete-time Example 1 of Chapter 2, repeated here,

$$\mathbf{A} = \begin{bmatrix} 0 & 1 \\ -1/2 & -3/2 \end{bmatrix}, \quad \mathbf{B} = \begin{bmatrix} 0 \\ 1 \end{bmatrix}, \quad \mathbf{C} = [1, 0], \quad \mathbf{D} = 1,$$

with $\mathbf{X}(0) = \begin{bmatrix} 5 \\ 0 \end{bmatrix}, \quad \{\mathbf{U}(t)\}_0^\infty = \{1, 1, \cdots\},$

the transform of the output is, from Equation (3.32),

$$\hat{\mathbf{Y}}(z) = [1, 0]\begin{bmatrix} z & -1 \\ 1/2 & z+3/2 \end{bmatrix}^{-1} z \begin{bmatrix} 5 \\ 0 \end{bmatrix}$$

$$+ \left([1, 0]\begin{bmatrix} z & -1 \\ 1/2 & z+3/2 \end{bmatrix}^{-1}\begin{bmatrix} 0 \\ 1 \end{bmatrix}+1\right)\mathcal{Z}\{1, 1\cdots\}$$

$$= \frac{5z(z+3/2)}{(z+1/2)(z+1)} + \left(\frac{1}{(z+1/2)(z+1)}+1\right)\frac{z}{z-1}.$$

Table 3.3 Elementary $\mathcal{Z}$-transform pairs. The sequence $\{\delta_k\}$ is called the unit pulse (or impulse) sequence.

Transform	*Sequence*
1	$\{\delta_k\} = \{1,\, 0,\, 0,\, \cdots\}$
z^{-i}	$\{\delta_{k-i}\} = \{0,\, \cdots 0,\, 1,\, 0,\, \cdots\}$ (i-th term is 1)
$\dfrac{z}{z-1}$	$\{1\}$
$\dfrac{z}{z-a}$	$\{a^k\}$
$\dfrac{z}{(z-a)^q}$	$\left\{ \begin{array}{ll} 0 & k < q-1 \\ \binom{k}{q-1} a^{k-q+1} & k \geq q-1 \end{array} \right\}$

2.1 Free response

The first right-hand term of (3.32) is the transform of the free response to initial state $\mathbf{X}_0$.

Just as for continuous-time systems, the transform of the state-transition matrix can be derived from this term. Expanding the term in negative powers of z, as in (3.12), gives

$$(3.33) \quad \hat{\mathbf{Y}}_{\text{free}}(z) = \mathbf{C}(z^{-1}\mathbf{I} + z^{-2}\mathbf{A} + z^{-3}\mathbf{A}^2 + \cdots)\, z\mathbf{X}_0$$
$$= \mathbf{C}(\mathbf{I} + z^{-1}\mathbf{A} + z^{-2}\mathbf{A}^2 + \cdots)\, \mathbf{X}_0,$$

in which, from the definition of the $\mathcal{Z}$-transform, the series in parentheses is the transform of the state-transition matrix sequence $\{\mathbf{A}^t\}_0^\infty$; that is,

$$(3.34) \quad \mathcal{Z}\{\mathbf{A}^t\} = (z\mathbf{I} - \mathbf{A})^{-1} z,$$

which is the discrete-time equivalent of (3.11).

Example 9
Free response
by long division

By long division, the free response in the previous example can be written

$$\frac{5(z + 3/2)}{(z + 1/2)(z + 1)} = 0 + 5z^{-1} + 0 - \frac{5}{2}z^{-3} + \frac{15}{4}z^{-4} - \frac{35}{8}z^{-5} + \cdots,$$

so the free response to the given initial state is the sequence

$$\{\mathbf{Y}_{\text{free}}(t)\} = \left\{0,\, 5,\, 0,\, -\frac{5}{2},\, \frac{15}{4},\, -\frac{35}{8},\, \cdots\right\}.$$

2.2   Forced response

The second right-hand term of (3.32) gives

(3.35) $\hat{\mathbf{Y}}_{\text{forced}}(z) = \hat{\mathbf{H}}(z)\,\hat{\mathbf{U}}(z),$

where $\hat{\mathbf{H}}(z) = \mathbf{C}(z\mathbf{I}-\mathbf{A})^{-1}\mathbf{B} + \mathbf{D}$, which, with the replacement of s by z, is algebraically identical to the continuous-time transfer matrix $\hat{\mathbf{H}}(s)$.

Properness   Consequently, all discrete-time state-space models have proper transfer matrices.

Example 10
Impulse response sequence by partial fractions

The forced response in Example 8 is the second term on the right-hand side, in which the transfer function is the factor in parentheses. The impulse-response (Markov) sequence for the system is therefore the inverse transform of this factor, which can be expanded by long division as in the previous example, or by partial fractions as follows:

$$\frac{1}{(z+1/2)(z+1)} + 1 = z\left(\frac{z^2 + (3/2)z + 3/2}{z(z+1/2)(z+1)}\right)$$

$$= z\left(\frac{3}{z} + \frac{-4}{z+1/2} + \frac{2}{z+1}\right) = 3 + \frac{-4z}{z+1/2} + \frac{2z}{z+1},$$

so that the closed form for the Markov sequence is

$$\{\mathbf{H}(k)\} = \left\{\begin{array}{ll} 1, & k = 0 \\ -4(-\tfrac{1}{2})^k + 2(-1)^k, & k > 0 \end{array}\right\},$$

which is the sequence of Example 8 of Chapter 2.

3   Further study

The Laplace transform is standard fare, but Guest [20] is a good reference for distributions and other material that many undergraduate texts leave out.

Introductions to the $\mathcal{Z}$-transform appear in many books on digital control, signal analysis, and digital filtering. Two good references are Lathi [30] and Oppenheim and Willsky [40].

4   Problems

The sets of matrices shown below are used in the following problems. These matrices will represent continuous-time or discrete-time systems depending on the context.

$$(S_1)\ \mathbf{A} = [1/4], \quad \mathbf{B} = [1, 2], \quad \mathbf{C} = \begin{bmatrix} 1 \\ 0 \\ 1 \end{bmatrix}, \quad \mathbf{D} = \begin{bmatrix} 0 & 0 \\ 2 & 0 \\ 0 & 1 \end{bmatrix}$$

$$(S_2)\ \mathbf{A} = \begin{bmatrix} 1/4 & 0 \\ 0 & -1 \end{bmatrix}, \quad \mathbf{B} = \begin{bmatrix} 1 & 0 \\ 0 & 1 \end{bmatrix}, \quad \mathbf{C} = [2, 1], \quad \mathbf{D} = 0$$

$$(S_3)\ \mathbf{A} = \begin{bmatrix} 1/4 & 1 \\ 0 & 1/4 \end{bmatrix}, \quad \mathbf{B} = \begin{bmatrix} 1 \\ 0 \end{bmatrix}, \quad \mathbf{C} = [2, 0], \quad \mathbf{D} = 0$$

$$(S_4)\ \mathbf{A} = \begin{bmatrix} 1/4 & 0 & 0 \\ 1 & 1/3 & -1 \\ 0 & 0 & 2 \end{bmatrix}, \quad \mathbf{B} = \begin{bmatrix} 1 \\ 0 \\ 0 \end{bmatrix}, \quad \mathbf{C} = [1, 0, 1], \quad \mathbf{D} = 0$$

1   Using Laplace transforms, find the forced, free, and complete response of the following:

(a) system S_1 with $\mathbf{X}(0) = 3$, $\mathbf{U}(t) = \begin{bmatrix} \delta(t) \\ 0 \end{bmatrix}$,

(b) system S_2 with $\mathbf{X}(0) = \begin{bmatrix} 0 \\ 2 \end{bmatrix}$, $\hat{\mathbf{U}}(s) = \begin{bmatrix} 1/s \\ 0 \end{bmatrix}$,

(c) system S_3 with $\mathbf{X}(0) = \begin{bmatrix} 0 \\ 2 \end{bmatrix}$, $\mathbf{U}(t) = e^{-t}$,

(d) system S_4 with $\mathbf{X}(0) = \begin{bmatrix} 0 \\ 2 \\ -1 \end{bmatrix}$, $\mathbf{U}(t) = e^{-t}$.

2   Compute the transfer matrix $\hat{\mathbf{H}}(s)$ for each of S_1, S_2, S_3, S_4, and in each case, compute the weighting matrix $\mathbf{H}(t) = \mathcal{L}^{-1}\hat{\mathbf{H}}(s)$.

3   Using $\mathcal{Z}$-transforms, find the forced, free, and complete response of the following:

(a) system S_1 with $\mathbf{X}(0) = 3$, $\{\mathbf{U}(t)\}_0^\infty = \left\{ \begin{bmatrix} 1 \\ 0 \end{bmatrix}, \begin{bmatrix} 0 \\ 0 \end{bmatrix}, \begin{bmatrix} 0 \\ 0 \end{bmatrix}, \cdots \right\}$,

(b) system S_2 with $\mathbf{X}(0) = \begin{bmatrix} 0 \\ 2 \end{bmatrix}$, $\hat{\mathbf{U}}(z) = \begin{bmatrix} z/(z-1) \\ 0 \end{bmatrix}$,

(c) system S_3 with $\mathbf{X}(0) = \begin{bmatrix} 0 \\ 2 \end{bmatrix}$, $\hat{\mathbf{U}}(z) = z/(z-1/3)$,

(d) system S_4 with $\mathbf{X}(0) = \begin{bmatrix} 0 \\ 2 \\ -1 \end{bmatrix}$, $\hat{\mathbf{U}}(z) = z/(z-1/2)$.

4   Compute the transfer matrix $\hat{\mathbf{H}}(z)$ for each of $\mathcal{S}_1$, $\mathcal{S}_2$, $\mathcal{S}_3$, $\mathcal{S}_4$, and in each case, compute the weighting matrix sequence $\{\mathbf{H}(t)\} = \mathcal{Z}^{-1}\hat{\mathbf{H}}(z)$.

5   For the binary system shown in Figure P3.5, in which all additions of system variables are mod-2 (exclusive-or),

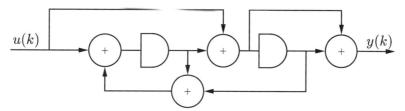

Fig. P3.5   Binary circuit.

  (a) find the transfer-matrix $\hat{\mathbf{H}}(z)$,
  (b) find the transform of the free response as a function of the state variables at initial time $t = 0$,
  (c) find the impulse response sequence by inverting $\hat{\mathbf{H}}(z)$,
  (d) find the forced response to the input that has transform
      $$\hat{U}(z) = (z^3 + z^2 + z + 1)/(z^3),$$
  (e) find the forced response above, using convolution.

6   A system defined over the integers mod 2 has matrices
$$\mathbf{A} = \begin{bmatrix} 1 & 1 \\ 0 & 1 \end{bmatrix}, \quad \mathbf{B} = \begin{bmatrix} 0 \\ 1 \end{bmatrix}, \quad \mathbf{C} = [1, 1], \quad \mathbf{D} = 1.$$

  (a) Find the transfer matrix $\hat{\mathbf{H}}(z)$.
  (b) Find the complete response for initial state $\mathbf{X}(0) = \begin{bmatrix} 1 \\ 0 \end{bmatrix}$ and input sequence $\{0, 1, 1, 0, 0, \cdots\}$.
  (c) Compute the weighting sequence $\{\mathbf{H}_k\}_0^\infty$.

7   Find the transfer matrix $\hat{\mathbf{H}}(s)$ for
$$\mathbf{A} = \begin{bmatrix} 0 & 0 & -4 \\ 1 & 0 & 7 \\ 0 & 1 & 11 \end{bmatrix}, \quad \mathbf{B} = \begin{bmatrix} 1 \\ 3 \\ -2 \end{bmatrix}, \quad \mathbf{C} = [0, 0, 1], \quad \mathbf{D} = 0.$$

Hint: note that not all rows of the inverse $(s\mathbf{I} - \mathbf{A})^{-1}$ are required in the calculation.

8   Solve Problem 7 for the system described by matrices

$$\mathbf{A} = \begin{bmatrix} 0 & 0 & -4 \\ 1 & 0 & 7 \\ 0 & 1 & 11 \end{bmatrix}, \quad \mathbf{B} = \begin{bmatrix} 1 \\ 3 \\ -2 \end{bmatrix}, \quad \mathbf{C} = [0, 0, 1], \quad \mathbf{D} = 0.$$

The system **A** matrices are in different companion forms in the two cases.

4
Writing state-space equations

Mathematical models written as equations in state-space form can be obtained systematically for many classes of dynamical systems. The examples in this chapter illustrate some dynamical systems that will be of interest, and systematic ways to obtain state-space models for them.

State-space equations, or in the case of LTI systems, the coefficient matrices ($\mathbf{A}$, $\mathbf{B}$, $\mathbf{C}$, $\mathbf{D}$), are often collectively called a *realization* of the information used to derive them.

Electric circuits will be treated first. Such systems are described by using graphs, which emphasize topology, or system layout. The Euler-Lagrange equations for mechanical systems will then be summarized, as a class of models in which energy has primary emphasis. Brief examples of aggregation as used in the biological and social sciences will be given. In the domain of signal processing, digital filters and binary computer circuits of a particular type, both described by operational diagrams, will be discussed, with their continuous-time analog, containing operational elements often built by using operational amplifiers. It also will be shown how to rewrite high-order differential equations as several first-order equations, as required by the state-space form.

There is generally more than one choice of state-space model possible, and for LTI systems the direct, observable, and controllable realizations will be considered, with brief mention of factorized models, all for single-input, single-output (SISO) systems.

Finally, state-space models corresponding to systems with multiple inputs and outputs (MIMO systems) will be developed, starting from the transfer matrix.

This chapter uses some elementary matrix terminology that is reviewed in detail in Chapter 5.

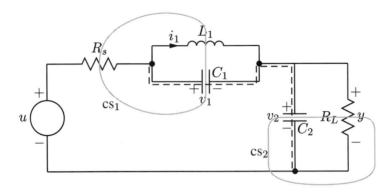

Fig. 4.1   A circuit with one input, one output, and three energy-storage elements.

1   Graph-based methods: Electric circuits

A method for finding state-space models for electric circuits will be given. Electric circuits are only one example of physical systems representable by graphs, and analogous graphs can be used to model mechanical, thermodynamic, hydraulic, and other systems. The method will be illustrated by the example in Figure 4.1 which contains a single independent input $u(t)$, a single output $y(t)$, and three elements that require derivatives in their characterizing equations.

An electric circuit is a graph, that is, a set of nodes connected by a set of branches. It is usually convenient to first perform source transformations as necessary so that every voltage source is in series with a nonsource, and every current source is in parallel with a nonsource. When identifying nodes and loops of the graph in the following procedure, a voltage source together with a nonsource in series are treated as a single branch, and a current source together with a nonsource in parallel are treated as a single branch. Then in Figure 4.1 there are five branches and three nodes.

1. A connected subgraph containing all the graph nodes but no loops is called a tree of the graph. Choose a tree that includes all the capacitors but no inductors. Such a tree is shown by the dashed line in Figure 4.1. The branches not in the tree are called links.

2. Choose the capacitor voltages and the inductor currents as state variables. Alternatively, charge can be substituted for one or more capacitor voltages, and flux linkages for one or more inductor currents.

3. If the removal of a set of branches, leaving all the nodes, reduces the original graph to exactly two subgraphs, the set of removed branches is called a cut-set. A cut-set containing links and exactly one tree branch is a fun-

damental cut-set. Two fundamental cut-sets are shown by the light lines cs_1 and cs_2 in the figure. By Kirchhoff's current law, the algebraic sum of the branch currents for any cut-set is zero. For each capacitor C_j, write the Kirchhoff equation for its fundamental cut-set,

(4.1)   capacitor current $= -\sum$ link currents

expressing all right-hand side quantities in terms of input and state variables. The links of the fundamental cut-set cs_1 for C_1 are the source branch and L_1, and the equation summing the currents to zero becomes, on division by C_1,

(4.2) $$\frac{dv_1}{dt} = -\frac{1}{C_1}\left(i_1 + \frac{v_2 + v_1 - u}{R_s}\right).$$

For C_2 the fundamental cut-set links are the source branch and R_L, giving

(4.3) $$\frac{dv_2}{dt} = -\frac{1}{C_2}\left(\frac{v_2}{R_L} + \frac{v_2 + v_1 - u}{R_s}\right).$$

4. A loop consisting of tree branches and one link is called a fundamental loop. For each inductor, sum the fundamental loop voltages to zero in an equation of the form

(4.4)   inductor voltage $= \sum$ tree-branch voltages,

expressing all right-hand side quantities in terms of input and state variables. In the example the fundamental loop for L_1 contains tree branch C_1, giving, on division by L_1,

(4.5) $$\frac{di_1}{dt} = \frac{1}{L_1}v_1.$$

5. Write equations for the outputs as functions only of input or state variables. For the example the single required equation is

(4.6) $y = v_2.$

Example 1
Circuit
(4.7a)

The equations of Figure 4.1, written in vector-matrix form, become

$$\begin{bmatrix} dv_1/dt \\ dv_2/dt \\ di_1/dt \end{bmatrix} = \begin{bmatrix} -1/(C_1 R_s) & -1/(C_1 R_s) & -1/C_1 \\ -1/(C_2 R_s) & -1/(C_2 R_s)-1/(C_2 R_L) & 0 \\ 1/L_1 & 0 & 0 \end{bmatrix} \begin{bmatrix} v_1 \\ v_2 \\ i_1 \end{bmatrix}$$
$$+ \begin{bmatrix} 1/(C_1 R_s) \\ 1/(C_2 R_s) \\ 0 \end{bmatrix} u$$

(4.7b)

$$y = \begin{bmatrix} 0 & 1 & 0 \end{bmatrix} \begin{bmatrix} v_1 \\ v_2 \\ i_1 \end{bmatrix} + \begin{bmatrix} 0 \end{bmatrix} u.$$

Notes

1. It may not be possible to immediately express the equation right-hand sides in terms of only input and state variables. The example of Figure 4.2 illustrates such a case. Additional variables, normally tree-branch voltages and link currents, may be temporarily introduced; then the fundamental cut-set equations for these tree branches and the fundamental loop equations for these links are used to eliminate the temporary variables from the state-space equations.

2. If the circuit is linear and contains no capacitor-only loops or inductor-only cut-sets, the procedure described above always can be performed. Circuits with essential capacitor-only loops or inductor-only cut-sets do not have directly obtainable state-space models, but are of little practical importance.

3. Nonlinearities may prevent the expression of the equations in the forms required, but if as is often the case the charge-voltage relation for each capacitor can be written $v_j = f_j(q_j)$, and if the flux-current relation for each inductor can be written $i_k = f_k(\phi_k)$, then the choice of fluxes ϕ_k and charges q_j as state variables often produces the correct form.

Example 2
Extra cut-set equation

If Figure 4.1 is modified to insert inductor L_2 in series with C_2, and the tree is then chosen to include R_L, Figure 4.2 results. In this case, one of the tree

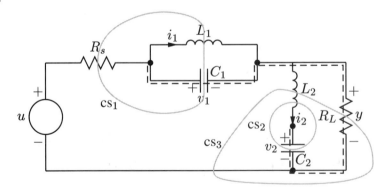

Fig. 4.2 Example requiring a resistive tree-branch cut-set equation.

branches is resistor R_L, and the voltage y across this resistor appears in the equations, as follows:

(cs1) $$C_1 \frac{dv_1}{dt} + i_1 + \frac{v_1 + y - u}{R_s} = 0,$$

(cs2) $$C_2 \frac{dv_2}{dt} = i_2,$$

(loop 1) $$L_1 \frac{di_1}{dt} = v_1,$$

(loop 2) $$L_2 \frac{di_2}{dt} + v_2 = y,$$

from which the voltage y is eliminated by writing the cut-set equation for R_L,

(cs3) $$\frac{y}{R_L} + i_2 + \frac{v_1 + y - u}{R_s} = 0,$$

and solving for y.

2   Energy-based methods: Euler-Lagrange equations

A second very general technique is typically applicable when expressions for potential and kinetic energy can be written for a system subject to Newton's laws. The characteristic example of the technique is the analysis of mechanical systems subject to holonomic constraints, as follows. If the system consists of a set of k particles, possibly infinite in number, at positions given by vectors $\mathbf{r}_1, \mathbf{r}_2, \cdots \mathbf{r}_k$, subject to equality constraints of the form

$$g_i(\mathbf{r}_1, \mathbf{r}_2, \cdots \mathbf{r}_k) = 0, \quad i = 1, \cdots,$$

also possibly infinite in number, then the constraints are called *holonomic*. The constraints defining the relative positions of the particles in a finite mechanical member are an example of such an infinite set, and the constraints equating member end-point positions at the joints of a mechanical robot arm are an example of a finite set. Electrical, hydraulic, and thermodynamic systems can be analyzed analogously, as well as mixed systems containing components from more than one of these classes.

Let a set $q_1, q_2, \cdots q_n$ of independent variables be identified, from which the positions of all elements of the system can be determined. These variables are called *generalized coordinates,* and their time derivatives $\dot{q}_1, \dot{q}_2, \cdots \dot{q}_n$ are *generalized velocities.*

Let the total kinetic energy of the system be written as the quantity

(4.8) $\quad T = T(q_1, \cdots q_n, \dot{q}_1, \cdots \dot{q}_n, t),$

and let the total potential energy, typically due to gravity and elastic potential energy, be

(4.9) $\quad V = V(q_1, \cdots q_n, t).$

Then the scalar quantity

(4.10) $\quad L = T - V$

is called the Lagrangian of the system.

The system is said to have n degrees of freedom since it is characterized by the n generalized coordinates. Then the equations of motion for the system are given by the n second-order Euler-Lagrange differential equations

(4.11) $\quad \dfrac{d}{dt}\left(\dfrac{\partial L}{\partial \dot{q}_i}\right) - \dfrac{\partial L}{\partial q_i} + \dfrac{\partial D}{\partial \dot{q}_i} = f_i, \quad i = 1, \cdots n,$

where the scalar D is half the rate at which energy is dissipated as heat. The quantity f_i is the net external force associated with the q_i coordinate, and is defined to be positive when it acts in the direction of increasing q_i. The coordinates q_i need not have dimensions of length, and the f_i need not have dimensions of force, but the products $q_i f_i$ must always have dimensions of work.

The techniques to be discussed later in Section 6 can be used to obtain first-order equations from the second-order Euler-Lagrange equations, but normally it suffices to let the q_i and $\dot{q}_i$ be state variables, resulting in $2n$ state equations.

For a large class of problems, Equations (4.11) can be written in standard matrix form containing the vector $\mathbf{Q} = [q_i]$ of generalized coordinates and force vector $\mathbf{F} = [f_i]$.

To begin, writing the n equations (4.11) together gives

(4.12) $\quad \dfrac{d}{dt}\begin{bmatrix} \dfrac{\partial L}{\partial \dot{q}_1} \\ \vdots \\ \dfrac{\partial L}{\partial \dot{q}_n} \end{bmatrix} - \begin{bmatrix} \dfrac{\partial L}{\partial q_1} \\ \vdots \\ \dfrac{\partial L}{\partial q_n} \end{bmatrix} + \begin{bmatrix} \dfrac{\partial D}{\partial \dot{q}_1} \\ \vdots \\ \dfrac{\partial D}{\partial \dot{q}_n} \end{bmatrix} = \begin{bmatrix} f_1 \\ \vdots \\ f_n \end{bmatrix},$

or, in abbreviated form,

(4.13) $\quad \dfrac{d}{dt}\dfrac{\partial L}{\partial \dot{\mathbf{Q}}} - \dfrac{\partial L}{\partial \mathbf{Q}} + \dfrac{\partial D}{\partial \dot{\mathbf{Q}}} = \mathbf{F},$

where a vector of derivatives of a scalar is called a *gradient* of the scalar.

| 2.1 | **Quadratic forms** |

In the context of LTI systems, L and D in the previous equations typically have the form $(1/2)\mathbf{X}^T\mathbf{M}\mathbf{X}$, where $\mathbf{X}$ is a vector and $\mathbf{M}$ is a symmetric matrix which is either constant or independent of $\mathbf{X}$. Such a scalar is called a *quadratic form,* and its gradient with respect to $\mathbf{X}$ is

(4.14) $$\frac{\partial}{\partial \mathbf{X}}\left(\frac{1}{2}\mathbf{X}^T\mathbf{M}\mathbf{X}\right) = \mathbf{M}\mathbf{X},$$

which can be verified as follows. Differentiating by the k-th entry x_k of $\mathbf{X}$,

(4.15) $$\frac{\partial}{\partial x_k}\left(\frac{1}{2}\mathbf{X}^T\mathbf{M}\mathbf{X}\right) = \frac{1}{2}\frac{\partial}{\partial x_k}\sum_{i=1}^{n} x_i \times (i\text{-th entry of }\mathbf{M}\mathbf{X})$$

$$= \frac{1}{2}\frac{\partial}{\partial x_k}\sum_{i=1}^{n} x_i \left(\sum_{j=1}^{n} m_{ij}x_j\right) = \frac{1}{2}\sum_{j=1}^{n} m_{kj}x_j + \frac{1}{2}\sum_{i=1}^{n} x_i m_{ik}.$$

Since $\mathbf{M}$ is symmetric, $m_{ik} = m_{ki}$, and the second sum equals the first, so that

(4.16) $$\frac{\partial}{\partial x_k}\left(\frac{1}{2}\mathbf{X}^T\mathbf{M}\mathbf{X}\right) = \sum_{j=1}^{n} m_{kj}x_j = k\text{-th entry of }\mathbf{M}\mathbf{X}$$

as required.

Example 3
Gradient of
quadratic form

For the quadratic form

$$f = \frac{1}{2}\mathbf{X}^T \begin{bmatrix} a_{11} & a_{12} \\ a_{12} & a_{22} \end{bmatrix} \mathbf{X} = \frac{1}{2}[x_1,\, x_2] \begin{bmatrix} a_{11}x_1 + a_{12}x_2 \\ a_{12}x_1 + a_{22}x_2 \end{bmatrix}$$

$$= \frac{1}{2}a_{11}x_1^2 + a_{12}x_1 x_2 + \frac{1}{2}a_{22}x_2^2$$

the gradient is

$$\frac{\partial f}{\partial \mathbf{X}} = \begin{bmatrix} \partial f/\partial x_1 \\ \partial f/\partial x_2 \end{bmatrix} = \begin{bmatrix} a_{11}x_1 + a_{12}x_2 \\ a_{12}x_1 + a_{22}x_2 \end{bmatrix} = \begin{bmatrix} a_{11} & a_{12} \\ a_{12} & a_{22} \end{bmatrix} \begin{bmatrix} x_1 \\ x_2 \end{bmatrix}.$$

| 2.2 | **Standard matrix form** |

In Equation (4.13), regardless of whether the system is LTI, if the kinetic energy can be written

(4.17) $$T = \frac{1}{2}\dot{\mathbf{Q}}^T\mathbf{M}(\mathbf{Q})\,\dot{\mathbf{Q}},$$

and if the potential energy $V = V(\mathbf{Q})$ is independent of $\dot{\mathbf{Q}}$, then with some detail which is omitted there, (4.13) can be shown to have the standard matrix form

(4.18) $\mathbf{M}(\mathbf{Q})\,\ddot{\mathbf{Q}} + \mathbf{C}(\mathbf{Q}, \dot{\mathbf{Q}})\,\dot{\mathbf{Q}} + \mathbf{K}(\mathbf{Q}) = \mathbf{F}'$,

where $\mathbf{F}'$ includes the forces due to energy dissipation and external independent forces.

For LTI systems the above equations are particularly simple. Assume that $\mathbf{M}$ in (4.17) is constant, that the potential energy is

(4.19) $V(\mathbf{Q}) = \dfrac{1}{2}\mathbf{Q}^T\mathbf{K}\mathbf{Q}$,

and that half the dissipation rate is

(4.20) $D = \dfrac{1}{2}\dot{\mathbf{Q}}^T\mathbf{C}\dot{\mathbf{Q}}$.

Then the Euler-Lagrange equations are

(4.21) $\dfrac{d}{dt}\left(\dfrac{\partial}{\partial\dot{\mathbf{Q}}}\left(\dfrac{1}{2}\dot{\mathbf{Q}}^T\mathbf{M}\dot{\mathbf{Q}} - \dfrac{1}{2}\mathbf{Q}^T\mathbf{K}\mathbf{Q}\right)\right) - \dfrac{\partial}{\partial\mathbf{Q}}\left(\dfrac{1}{2}\dot{\mathbf{Q}}^T\mathbf{M}\dot{\mathbf{Q}} - \dfrac{1}{2}\mathbf{Q}^T\mathbf{K}\mathbf{Q}\right)$

$+ \dfrac{\partial}{\partial\dot{\mathbf{Q}}}\left(\dfrac{1}{2}\dot{\mathbf{Q}}^T\mathbf{C}\dot{\mathbf{Q}}\right) = \mathbf{F}$,

which become, using (4.14),

(4.22) $\mathbf{M}\,\ddot{\mathbf{Q}} + \mathbf{C}\,\dot{\mathbf{Q}} + \mathbf{K}\,\mathbf{Q} = \mathbf{F}$,

where $\mathbf{M}$ is called the mass matrix, $\mathbf{C}$ is the damping matrix, and $\mathbf{K}$ is the stiffness matrix. If $\mathbf{M}$ is singular then fewer than $2n$ state equations are required. Otherwise let the state vector contain subvector $\mathbf{X}^{(1)} = \mathbf{Q}$ and subvector $\mathbf{X}^{(2)} = \dot{\mathbf{Q}}$, so that the state equations are

(4.23) $\begin{bmatrix} \dot{\mathbf{X}}^{(1)} \\ \dot{\mathbf{X}}^{(2)} \end{bmatrix} = \begin{bmatrix} 0 & \mathbf{I} \\ -\mathbf{M}^{-1}\mathbf{K} & -\mathbf{M}^{-1}\mathbf{C} \end{bmatrix}\begin{bmatrix} \mathbf{X}^{(1)} \\ \mathbf{X}^{(2)} \end{bmatrix} + \begin{bmatrix} 0 \\ \mathbf{M}^{-1} \end{bmatrix}\mathbf{F}$.

Example 4
Newton's second law

Consider a mass m subject to force f in the absence of gravity. The potential energy V is zero, the kinetic energy is $T = m\dot{q}^2/2$, and the describing equation is

$$\frac{d}{dt}\left(\frac{\partial}{\partial\dot{q}}\left(\frac{m\dot{q}^2}{2}\right)\right) - \frac{\partial}{\partial q}\left(\frac{m\dot{q}^2}{2}\right) + 0 = \frac{d}{dt}(m\dot{q}) + 0 = m\ddot{q} = f.$$

Let $x_1 = q$ and $x_2 = \dot{q}$ to obtain

$$\begin{bmatrix} \dot{x}_1 \\ \dot{x}_2 \end{bmatrix} = \begin{bmatrix} x_2 \\ f/m \end{bmatrix} = \begin{bmatrix} 0 & 1 \\ 0 & 0 \end{bmatrix}\begin{bmatrix} x_1 \\ x_2 \end{bmatrix} + \begin{bmatrix} 0 \\ 1/m \end{bmatrix}f.$$

Example 5
Second-order
mechanical
system

The mass, spring, dashpot system shown in Figure 4.3 has one degree of free-dom, and generalized coordinate x such that the system is at rest at $x = 0$.

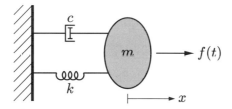

Fig. 4.3   Mass, spring, dashpot system with one degree of freedom.

The potential energy is $V = kx^2/2$, the kinetic energy is $T = m\dot{x}^2/2$, and the rate of energy dissipation is $c\dot{x}^2$. Therefore the Euler-Lagrange model of the system is

$$\frac{d}{dt}\left(\frac{\partial}{\partial \dot{x}}\left(\frac{m\dot{x}^2}{2} - \frac{kx^2}{2}\right)\right) - \frac{\partial}{\partial x}\left(\frac{m\dot{x}^2}{2} - \frac{kx^2}{2}\right) + \frac{\partial}{\partial \dot{x}}\left(\frac{c\dot{x}^2}{2}\right) = f(t).$$

The result is the second-order differential equation

$$m\ddot{x} + c\dot{x} + kx = f(t).$$

Example 6
Two-link robot
arm

Figure 4.4 is a simplified robot arm with two rigid links and two joints. The link masses are assumed to be insignificant compared to masses m_1, m_2 of the middle joint and the tool respectively. Assuming no friction, the dissipation term

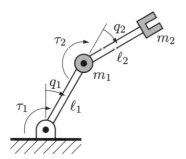

Fig. 4.4   Simplified robot arm with two degrees of freedom

is $D = 0$, and since the generalized coordinates are angles q_1 and q_2, the vector
$\mathbf{F} = \begin{bmatrix} \tau_1 \\ \tau_2 \end{bmatrix}$ contains torques. From the diagram, the potential energy is

$$V = m_1 g \ell_1 \cos q_1 + m_2 g \left(\ell_1 \cos q_1 + \ell_2 \cos(q_1 + q_2)\right),$$

which is independent of $\dot{\mathbf{Q}}$. The kinetic energy of m_1 is $(m_1/2)(\ell_1\dot{q}_1)^2$. The coordinates x, y of m_2 relative to the base are

$$x = \ell_1 \sin q_1 + \ell_2 \sin(q_1 + q_2), \quad y = \ell_1 \cos q_1 + \ell_2 \cos(q_1 + q_2)$$

and the squared velocity of m_2 is $\dot{x}^2 + \dot{y}^2$, from which the kinetic energy of m_2 can be obtained as

$$\frac{m_2}{2}(\dot{x}^2 + \dot{y}^2) = \frac{m_2}{2}\ell_1^2\dot{q}_1^2 + \frac{m_2}{2}\ell_2^2(\dot{q}_1 + \dot{q}_2)^2 + m_2\ell_1\ell_2\dot{q}_1(\dot{q}_1 + \dot{q}_2)\cos q_2.$$

Combining the above expression with the kinetic energy of m_1, we can write the total kinetic energy in the form of (4.17) as

$$T = \frac{1}{2}\dot{\mathbf{Q}}^T \begin{bmatrix} m_1\ell_1^2 + m_2(\ell_1^2 + \ell_2^2 + 2\ell_1\ell_2\cos q_2) & m_2(\ell_2^2 + \ell_1\ell_2\cos q_2) \\ m_2(\ell_2^2 + \ell_1\ell_2\cos q_2) & m_2\ell_2^2 \end{bmatrix} \dot{\mathbf{Q}}$$

$$= \frac{1}{2}\dot{\mathbf{Q}}^T\mathbf{M}(\mathbf{Q})\,\dot{\mathbf{Q}}.$$

Written in detail, the Lagrangian is

$$T - V = (\frac{m_1}{2}\ell_1^2 + \frac{m_2}{2}(\ell_1^2 + \ell_2^2 + 2\ell_1\ell_2\cos q_2))\,\dot{q}_1^2 + m_2(\ell_2^2 + \ell_1\ell_2\cos q_2)\,\dot{q}_1\dot{q}_2$$

$$+ \frac{m_2}{2}\ell_2^2\dot{q}_2^2 - m_1 g\ell_1\cos q_1 - m_2 g\,(\ell_1\cos q_1 + \ell_2\cos(q_1 + q_2)).$$

Differentiating as required in (4.11) gives, first with respect to q_1,

$$(m_1\ell_1^2 + m_2(\ell_1^2 + \ell_2^2 + 2\ell_1\ell_2\cos q_2))\,\ddot{q}_1 + m_2(\ell_2^2 + \ell_1\ell_2\cos q_2)\,\ddot{q}_2$$

$$- (m_2\ell_1\ell_2\sin q_2)(2\dot{q}_1\dot{q}_2 + \dot{q}_2^2) - (m_1 + m_2)\,g\ell_1\sin q_1$$

$$- m_2 g\ell_2\sin(q_1 + q_2) = \tau_1,$$

and then with respect to q_2,

$$m_2(\ell_2^2 + \ell_1\ell_2\cos q_2)\,\ddot{q}_1 + m_2\ell_2^2\ddot{q}_2 + m_2\ell_1\ell_2\dot{q}_1^2\sin q_2 - m_2 g\ell_2\sin(q_1 + q_2)$$

$$= \tau_2.$$

Equations in the form of (4.13) can be obtained by using $\mathbf{M}(\mathbf{Q})$ above, with the second term of (4.13) equal to

$$\mathbf{C}(\mathbf{Q}, \dot{\mathbf{Q}})\,\dot{\mathbf{Q}} = m_2\ell_1\ell_2\sin q_2 \begin{bmatrix} -2\dot{q}_1\dot{q}_2 - \dot{q}_2^2 \\ \dot{q}_1^2 \end{bmatrix},$$

and the third term equal to

$$\mathbf{K}(\mathbf{Q}) = \begin{bmatrix} -(m_1 + m_2)\,g\ell_1\sin q_1 - m_2 g\ell_2\sin(q_1 + q_2) \\ -m_2 g\ell_2\sin(q_1 + q_2) \end{bmatrix}.$$

State equations can be obtained by choosing the entries of $\mathbf{Q}$ and $\dot{\mathbf{Q}}$ as state variables.

| **3** | **Aggregation** |

The previous sections described models which were based on physical laws applicable over a large range of sizes, forces, and velocities. Models can also be based on empirical phenomena rather than on physical laws. In the biological and social sciences particularly, when behavioral models of large groups of individuals are constructed, it may be plausible to treat group population counts as continuous variables. Brief examples will be given.

Example 7
Population cohort model

The evolution of group populations is studied using *population cohort* models. For example, in a given geographic region let x_i, $i = 1, \cdots 9$ be the population of persons of age at least $10(i-1)$ but less than $10i$ years, and let x_{10} be the population 90 years old or more. Let t be the index of an interval of time, such as a year. Suppose that the survival rate of persons in age range i is given by coefficient a_i, $i = 1, \cdots 9$ such that

(4.24) $x_{i+1}(t + 1) = a_i x(t),$

and that for age group i the birthrate is b_i and the net immigration is u_i. Then the evolution of the group populations is given by the equation

(4.25a)
$$
\begin{bmatrix} x_1(t+1) \\ \vdots \\ x_{10}(t+1) \end{bmatrix} = \begin{bmatrix} b_1 & b_2 & \cdots & b_9 & b_{10} \\ a_1 & & & & \\ & a_2 & & & \\ & & \ddots & & \\ & & & a_9 & 0 \end{bmatrix} \begin{bmatrix} x_1(t) \\ \vdots \\ x_{10}(t) \end{bmatrix} + \begin{bmatrix} u_1(t) \\ \vdots \\ u_{10}(t) \end{bmatrix}.
$$

The coefficients a_i and b_i may be time-varying, for example, to account for periods of war or relative prosperity. Such models are used by demographers and actuaries.

Suppose the total population $y(t)$ is of particular interest. Then the output equation is

(4.25b) $y(t) = [\, 1, \cdots 1 \,] \begin{bmatrix} x_1(t) \\ \vdots \\ x_{10}(t) \end{bmatrix}.$

Example 8
Predator-prey model

Continuous-time equations may result if populations are aggregated but time is not. The classical *predator-prey* equations of population biology describe situations similar to the following. Suppose that in the abundance of food, the population x_1 of a group of small animals evolves according to $dx_1/dt = \alpha x_1$; that is, the population increases, potentially without limit. However, if there are

x_2 large animals that prey on the smaller animals, the number of small animals eaten is proportional to the product of the populations, so the previous equation becomes

(4.26a) $$\frac{dx_1}{dt} = \alpha x_1 - \beta x_1 x_2, \quad \alpha, \beta > 0.$$

In the absence of small animals, the predators starve and their population is described by $dx_2/dt = -\gamma x_2$; otherwise their population increases according to the rate of small animals consumed, giving the second equation

(4.26b) $$\frac{dx_1}{dt} = \delta x_1 x_2 - \gamma x_2, \quad \delta, \gamma > 0.$$

These equations have periodic solutions that describe the cyclic variation of the populations.

4   Operational diagrams: Digital filters

A digital filter is a clocked digital system or a software module, for which at any time $t \in \mathbb{Z}$, an input vector $\mathbf{U}(t)$, and an output vector $\mathbf{Y}(t)$ are defined. The system is normally designed so that prespecified properties of the infinite input sequence $\{\mathbf{U}(t)\}_{-\infty}^{\infty} = \{\cdots \mathbf{U}(-1), \mathbf{U}(0), \mathbf{U}(1), \cdots\}$ are enhanced in the output sequence $\{\mathbf{Y}(t)\}_{-\infty}^{\infty}$. Linear digital filters require the operations of addition, multiplication by a constant, and of delay by one clock time, and such a filter is illustrated in the operational diagram of Figure 4.5, in which the constant multipliers are shown with subscripted parameters a and b. All variables are real-valued and are defined at every clock time. The delay elements are register or memory locations for which the input at any clock time is the output at the next clock time. For example in the figure, the element for which the output at time t is $x_1(t)$ has input $x_1(t+1)$ at time t. Rules for writing the state-space equations can be given as follows:

1. Choose as state variables the outputs of the delay elements.
2. For each delay element, write the state-update equation of the form

(4.27) $x_i(t+1) =$ expression in the state and input variables.

3. Write the output equations in the form

(4.28) $y_i(t) =$ expression in the state and input variables.

Example 9
Digital circuit

For Figure 4.5, the equations in vector-matrix form are

(4.29a) $$\begin{bmatrix} x_1(t+1) \\ x_2(t+1) \\ x_3(t+1) \end{bmatrix} = \begin{bmatrix} a_{11} & a_{12} & a_{13} \\ a_{21} & a_{22} & 0 \\ a_{31} & 0 & 0 \end{bmatrix} \begin{bmatrix} x_1(t) \\ x_2(t) \\ x_3(t) \end{bmatrix} + \begin{bmatrix} b_1 \\ b_2 \\ 0 \end{bmatrix} u(t)$$

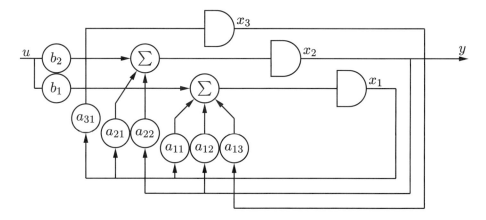

Fig. 4.5   A single-input, single-output digital filter.

$$(4.29b) \qquad y(t) = \begin{bmatrix} 0 & 1 & 0 \end{bmatrix} \begin{bmatrix} x_1(t) \\ x_2(t) \\ x_3(t) \end{bmatrix} + \begin{bmatrix} 0 \end{bmatrix} u(t).$$

The matrices in these equations have the same form as (4.7) for the circuit of Example 1, and may even have identical entries. This does not imply that the digital filter is an approximation of the continuous-time (analog) circuit.

4.1   Computer circuits

In the digital filter operational diagram Figure 4.5, let every variable (except time) take on only the values $\{0, 1\}$, let the summations be modulo-2, which in binary circuits is the "exclusive-or" operation, and let the values of the constant multipliers be likewise restricted to 0 or 1. Then equations (4.29) hold, but of course the arithmetic is modulo-2, and at any time t the vector $\begin{bmatrix} x_1(t) \\ x_2(t) \\ x_3(t) \end{bmatrix}$ is one of $2^3 = 8$ possible vectors, a finite number, and such a model is sometimes called a finite-state machine. In the models of previous sections, where x_1, x_2 and x_3 are real-valued, the set of possible state vectors is infinite.

Taking modulo-2 arithmetic into account, the state-space equations for this class of computer circuits can be written by using the same rules as for digital filters.

<table>
<tr><td>**5**</td><td>**Continuous-time operational diagrams**</td></tr>
</table>

Suppose the discrete-time operational diagram, Figure 4.5, is replaced by the continuous-time operational diagram of Figure 4.6. Such diagrams are often

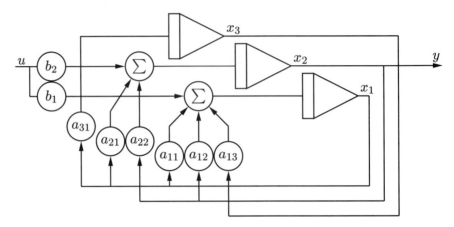

Fig. 4.6   Operational diagram of a continuous-time system.

used for developing models for simulation purposes. The variables are real, and the delay elements of the digital filter are here replaced by integrators, so that, for example, the element with output $x_1(t)$ has input dx_1/dt.

The rules for writing state-space equations for continuous-time operational diagrams are nearly identical to those for digital filter diagrams, as follows:

1. Choose the integrator outputs as state variables.

2. For each integrator, write the state-update equation of the form

(4.30)
$$\frac{dx_i(t)}{dt} = \text{expression in the state and input variables.}$$

3. Write the output equations

(4.31)
$$y_i(t) = \text{expression in the state and input variables.}$$

Example 10
Circuit analog

For the operational diagram of Figure 4.6, the equations obtained by choosing state variables as shown in the figure are

(4.32a)
$$\begin{bmatrix} dx_1/dt \\ dx_2/dt \\ dx_3/dt \end{bmatrix} = \begin{bmatrix} a_{11} & a_{12} & a_{13} \\ a_{21} & a_{22} & 0 \\ a_{31} & 0 & 0 \end{bmatrix} \begin{bmatrix} x_1(t) \\ x_2(t) \\ x_3(t) \end{bmatrix} + \begin{bmatrix} b_1 \\ b_2 \\ 0 \end{bmatrix} u(t)$$

(4.32b)
$$y(t) = \begin{bmatrix} 0 & 1 & 0 \end{bmatrix} \begin{bmatrix} x_1(t) \\ x_2(t) \\ x_3(t) \end{bmatrix} + \begin{bmatrix} 0 \end{bmatrix} u(t),$$

which are identical in form to (4.7) and (4.29). Therefore this operational diagram is an analog of the electric circuit provided the matrix entries have identical values.

---

6   High-order equations

An equation of order n in dependent variable v, of the form

(4.33) $f\left(\dfrac{d^n v}{dt^n}, \dfrac{d^{n-1} v}{dt^{n-1}}, \cdots v, \text{terms in other variables}\right) = 0,$

always can be replaced by a set of equations that are first-order in the state variables. Let

(4.34) $x_1 = v, \quad x_2 = \dfrac{dv}{dt}, \quad \cdots x_n = \dfrac{d^{n-1} v}{dt^{n-1}}.$

Then Equation (4.33) is rewritten as a set of n first-order equations:

(4.35a) $\dfrac{dx_1}{dt} = x_2, \quad \dfrac{dx_2}{dt} - x_3, \quad \cdots \dfrac{dx_{n-1}}{dt} = x_n,$

(4.35b) $f\left(\dfrac{dx_n}{dt}, x_{n-1}, \cdots x_1, \text{other terms as before}\right) = 0.$

Example 11
Second-order
equation

In the equation

$$\frac{d^2 y}{dt^2} + a_1 \frac{dy}{dt} + a_2 y = b_2 u$$

setting $x_1 = y$, $x_2 = dy/dt$ gives the two equations

$$\frac{dx_1}{dt} = x_2,$$

$$\frac{dx_2}{dt} + a_1 x_2 + a_2 x_1 = b_2 u,$$

which are first-order in x_1, x_2.

6.1   Direct realization of high-order linear equations

The previous technique does not result directly in state equations when derivatives of the inputs appear in the equation. In this case, suppose the highest derivative is of n-th order. Integrate (4.33) n times with respect to time, draw an operational diagram, and define the state variables to be the integrator outputs. Provided the system is linear it is then possible to solve for the outputs and derivatives of state variables as required by the state-space form.

Example 12
Input
derivatives

The previous example will be modified as shown:

$$\frac{d^2y}{dt^2} + a_1\frac{dy}{dt} + a_2y = b_0\frac{d^2u}{dt^2} + b_1\frac{du}{dt} + b_2u.$$

Then, setting $x_1 = y$, $x_2 = dy/dt$ gives the two equations

$$\frac{dx_1}{dt} = x_2,$$

$$\frac{dx_2}{dt} = -a_1x_2 - a_2x_1 + b_0\frac{d^2u}{dt^2} + b_1\frac{du}{dt} + b_2u,$$

which are first-order in x_1, x_2. Although in principle if $u(.)$ is a known function, du/dt and d^2u/dt^2 are also known, this equation is not in state-space form. Instead of using this substitution, integrate the original equation twice, giving

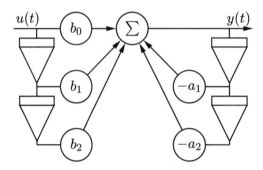

Fig. 4.7   Example realization.

$$y + a_1\int y\,dt + a_2\iint y\,dt\,dt = b_0u + b_1\int u\,dt + b_2\iint u\,dt\,dt,$$

which corresponds to Figure 4.7, so that the technique of Section 5 can be directly applied.

General case

The previous example also applies directly to high-order difference equations

(4.36) $y(t+n) + a_1y(t+n-1) + \cdots a_ny(t) = b_0u(t+n) + b_1u(t+n-1) + \cdots b_nu(t)$

or the corresponding $\mathcal{Z}$-transform transfer function

(4.37) $$\frac{y(z)}{u(z)} = \frac{b_0z^n + b_1z^{n-1} + \cdots b_n}{z^n + a_1z^{n-1} + \cdots a_n}.$$

If the time-domain equation is rewritten

(4.38) $y(t) = -a_1y(t-1) - \cdots a_ny(t-n) + b_0u(t) + b_1u(t-1) + \cdots b_nu(t-n),$

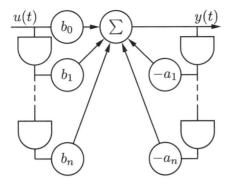

Fig. 4.8   Direct realization.

then the operational diagram of Figure 4.8 results, and if state variables are assigned to the $2n$ delay-element outputs, a set of state-space equations is obtained, in which the state vector has dimension $2n$. The form of (4.38) is often convenient for digital filtering because only the past values of the input and of the output need be stored, but it turns out to be easy to find state-space equations of order at most n for the same system or its continuous-time analog, as in the following sections.

Example 13
Direct
realization

Suppose $n = 2$, so that, choosing the state variables as outputs of the delays in Figure 4.8, and numbering them from the top, first on the left of the figure, and then on the right, the realization is

$$\mathbf{X}(t+1) = \begin{bmatrix} 0 & 0 & 0 & 0 \\ 1 & 0 & 0 & 0 \\ b_1 & b_2 & -a_1 & -a_2 \\ 0 & 0 & 1 & 0 \end{bmatrix} \mathbf{X}(t) + \begin{bmatrix} 1 \\ 0 \\ b_0 \\ 0 \end{bmatrix} u(t),$$

$$y(t) = \begin{bmatrix} b_1 & b_2 & -a_1 & -a_2 \end{bmatrix} \mathbf{X}(t) + \begin{bmatrix} b_0 \end{bmatrix} u(t).$$

7   Controllable and observable realizations

Two realizations which are very commonly used in theoretical contexts will be described. They both have the convenience of being related by inspection to scalar transfer functions. The adjectives *controllable* and *observable* have a meaning that will be discussed in Chapter 9. Continuous-time equations will be used, but the development applies equally to discrete-time systems, for which z replaces s in transform notation, and for which left-shift replaces differentiation in the time domain.

From the discussion of Section 1.3 of Chapter 3, any single-input, single-output system that has a state-space model has a rational, proper transfer function of the form

$$(4.39) \quad \frac{y(s)}{u(s)} = \frac{b_0 s^n + b_1 s^{n-1} + \cdots b_n}{s^n + a_1 s^{n-1} + \cdots a_n},$$

and from the properties of transforms, a time-domain equation corresponding to (4.39) can be written as

$$(4.40) \quad \frac{d^n y}{dt^n} + a_1 \frac{d^{n-1} y}{dt^{n-1}} + \cdots a_n y = b_0 \frac{d^n u}{dt^n} + b_1 \frac{d^{n-1} u}{dt^{n-1}} + \cdots b_n u.$$

If $b_0 \neq 0$, it is convenient to divide the denominator into the numerator to rewrite (4.39) as

$$(4.41) \quad \frac{y(s)}{u(s)} = b_0 + \frac{c_1 s^{n-1} + \cdots c_n}{s^n + a_1 s^{n-1} + \cdots a_n},$$

where

$$(4.42) \quad c_i = b_i - b_0 a_i, \quad i = 1, \cdots n.$$

Because (4.39) has one input and one output, a corresponding state-space model has dimensions $m = 1$, $p = 1$; that is, $\mathbf{B}$ must have one column, $\mathbf{C}$ must have one row, and $\mathbf{D}$ must be 1×1. The matrix $\mathbf{A}$ will be $n \times n$ where n is the degree of the denominator in (4.39), implying that $\mathbf{B}$ is $n \times 1$ and $\mathbf{C}$ is $1 \times n$.

Controllable form By copying the coefficients of (4.41) into matrices of the required dimensions, the state-space equations shown below, said to be in *controllable form,* can be written

$$(4.43a) \quad \frac{d}{dt} \begin{bmatrix} x_1 \\ x_2 \\ \vdots \\ x_n \end{bmatrix} = \begin{bmatrix} 0 & 1 & & \\ & & \ddots & \\ & & & 1 \\ -a_n & -a_{n-1} & \cdots & -a_1 \end{bmatrix} \begin{bmatrix} x_1 \\ x_2 \\ \vdots \\ x_n \end{bmatrix} + \begin{bmatrix} 0 \\ \vdots \\ 0 \\ 1 \end{bmatrix} u(t)$$

$$(4.43b) \quad y(t) = \begin{bmatrix} c_n & c_{n-1} & \cdots & c_1 \end{bmatrix} \begin{bmatrix} x_1 \\ x_2 \\ \vdots \\ x_n \end{bmatrix} + b_0 u(t).$$

The matrix $\mathbf{A}$ is called a *companion* matrix corresponding to the denominator polynomial of (4.41), which can be shown to be the characteristic polynomial of $\mathbf{A}$, as to be discussed in detail in Chapter 5, Section 5.

There are several ways to demonstrate that (4.43) has transfer function of the form of (4.39). One way is to obtain the Laplace transform of (4.43). The following reasoning can also be used.

Define variable $v(s)$ to satisfy

(4.44) $y(s) = (b_0 s^n + b_1 s^{n-1} + \cdots b_n)\dfrac{1}{s^n + a_1 s^{n-1} + \cdots a_n}u(s)$

$= (b_0 s^n + b_1 s^{n-1} + \cdots b_n)\, v(s),$

corresponding to time-domain equations

(4.45a) $\dfrac{d^n v}{dt^n} = -a_1 \dfrac{d^{n-1}v}{dt^{n-1}} - a_2 \dfrac{d^{n-2}v}{dt^{n-2}} \cdots - a_n v + u,$

(4.45b) $y = b_0 \dfrac{d^n v}{dt^n} + b_1 \dfrac{d^{n-1}v}{dt^{n-1}} + \cdots b_n v.$

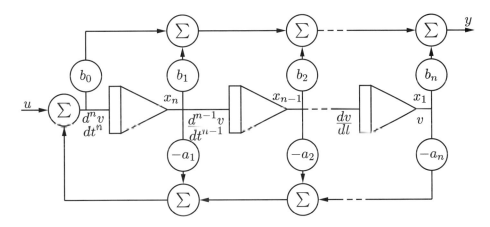

Fig. 4.9   Controllable realization.

Combining the operations of (4.45) in one diagram gives Figure 4.9, and with the choice of state variables shown in the figure, (4.43) is obtained.

Example 14
Controllable
realization

Given the transfer function $\hat{h}(s)$ shown below, expanded by long division,

$$\hat{h}(s) = \frac{s^2 + 2s - 7/2}{s^2 + (3/2)s + 1/2} = 1 + \frac{(1/2)s - 4}{s^2 + (3/2)s + 1/2},$$

the controllable realization is, by inspection,

$$\mathbf{A} = \begin{bmatrix} 0 & 1 \\ -1/2 & -3/2 \end{bmatrix}, \quad \mathbf{B} = \begin{bmatrix} 0 \\ 1 \end{bmatrix}, \quad \mathbf{C} = [\,-4 \quad 1/2\,], \quad \mathbf{D} = 1.$$

Observable
form

The state-space equations shown below, said to be in *observable form,* can be written by copying the coefficients from (4.41):

(4.46a)
$$\frac{d}{dt}\begin{bmatrix} x_1 \\ x_2 \\ \vdots \\ x_n \end{bmatrix} = \begin{bmatrix} 0 & & & -a_n \\ 1 & & & -a_{n-1} \\ & \ddots & & \vdots \\ & & 1 & -a_1 \end{bmatrix}\begin{bmatrix} x_1 \\ x_2 \\ \vdots \\ x_n \end{bmatrix} + \begin{bmatrix} c_n \\ c_{n-1} \\ \vdots \\ c_1 \end{bmatrix} u$$

(4.46b)
$$y = \begin{bmatrix} 0 & \cdots & 0 & 1 \end{bmatrix}\begin{bmatrix} x_1 \\ x_2 \\ \vdots \\ x_n \end{bmatrix} + b_0 u.$$

The matrix **A** above is another form of companion matrix.

To derive (4.46), integrate both sides of (4.40) n times and write the result as

(4.47)
$$y = b_0 u + \int \left(b_1 u - a_1 y + \int \left(b_2 u - a_2 y + \cdots \int (b_n u - a_n y)\, dt \cdots \right) dt \right) dt,$$

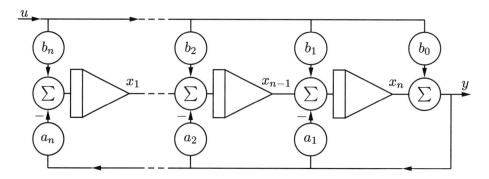

Fig. 4.10 Observable realization.

for which the operational diagram of Figure 4.10 can be drawn. Choosing the integrator outputs as state variables as in the diagram, we can write the equations below directly from the diagram:

(4.48a)
$$\frac{d}{dt}\begin{bmatrix} x_1 \\ x_2 \\ \vdots \\ x_n \end{bmatrix} = \begin{bmatrix} & & & 0 \\ 1 & & & \\ & \ddots & & \\ & & 1 & 0 \end{bmatrix}\begin{bmatrix} x_1 \\ x_2 \\ \vdots \\ x_n \end{bmatrix} - \begin{bmatrix} a_n \\ a_{n-1} \\ \vdots \\ a_1 \end{bmatrix} y + \begin{bmatrix} b_n \\ b_{n-1} \\ \vdots \\ b_1 \end{bmatrix} u$$

(4.48b)
$$0 = \begin{bmatrix} 0 & \cdots & 0 & 1 \end{bmatrix}\begin{bmatrix} x_1 \\ x_2 \\ \vdots \\ x_n \end{bmatrix} - 1y + b_0 u.$$

These equations are useful in the context of system identification, which is the determination of system parameters from input-output data. However, these equations are not in state-space form as required for simulation and other purposes because of the presence of y on the right-hand side. Solving for y from the bottom equation, substituting in the top, and using (4.42) gives (4.46).

Notes   1. The discrete-time equivalent of Figure 4.9 with integrators replaced by delays is a shift-register or tapped delay line. If the variables are binary and the arithmetic is modulo-2, then the result is a binary shift-register sequence generator, which in many applications would have no input, but which always would be started with nonzero initial state.

2. Although the controllable and observable forms are economical and convenient for theoretical realizations of transfer matrices, in the context of filter design or control design, other forms that are less sensitive to parameter uncertainty may be preferred.

Example 15
Observable
realization

The observable realization of Example 14 is

$$\mathbf{A} = \begin{bmatrix} 0 & -1/2 \\ 1 & -3/2 \end{bmatrix}, \quad \mathbf{B} = \begin{bmatrix} -4 \\ 1/2 \end{bmatrix}, \quad \mathbf{C} = [0 \quad 1], \quad \mathbf{D} = 1.$$

8   Factored realizations

Several other methods exist for finding state-space equations of single-input, single-output systems. The general method remains the same: find an operational diagram that represents the the initial description of the system, and write the state equations by using the outputs of the delays or integrators as state variables.

Two methods result from decomposing a transfer function either as a sum or as a product. The sum

(4.49) $$\frac{y(s)}{u(s)} = h(s) = d + h_1(s) + h_2(s) + \cdots h_r(s)$$

can be realized by combining in a summation element the outputs of $r+1$ blocks which realize the transfer functions $h_i(s)$ and the constant d. The $h_i(s)$ might be, for example, the partial fractions of $h(s)$.

Let $\mathbf{A}_i, \mathbf{B}_i, \mathbf{C}_i, 0$ be the state-space matrices for $h_i(s)$. Then, let

(4.50a) $$\mathbf{A} = \begin{bmatrix} \mathbf{A}_1 & & \\ & \ddots & \\ & & \mathbf{A}_r \end{bmatrix}, \quad \mathbf{B} = \begin{bmatrix} \mathbf{B}_1 \\ \vdots \\ \mathbf{B}_r \end{bmatrix}$$

(4.50b) $$\mathbf{C} = [\mathbf{C}_1, \cdots \mathbf{C}_r], \quad \mathbf{D} = d.$$

If the transfer function is factored as a product,

(4.51) $h(s) = h_t(s)\,h_{t-1}(s)\cdots h_1(s),$

then a realization is obtained by combining operational diagrams for the $h_i(s)$ in series.

Example 16
Partial fractions

Rewriting the transfer function of Example 14 as

$$\hat{h}(s) = 1 + \frac{-17/2}{s + 1/2} + \frac{9}{s + 1},$$

we see that the second term is realized by matrices

$$\mathbf{A}_1 = -1/2, \quad \mathbf{B}_1 = 1, \quad \mathbf{C}_1 = -17/2, \quad \mathbf{D}_1 = 0,$$

and the third term by

$$\mathbf{A}_2 = -1, \quad \mathbf{B}_2 = 1, \quad \mathbf{C}_2 = 9, \quad \mathbf{D}_2 = 0,$$

so that a realization of $\hat{h}(s)$ is

$$\mathbf{A} = \begin{bmatrix} -1/2 & 0 \\ 0 & -1 \end{bmatrix}, \quad \mathbf{B} = \begin{bmatrix} 1 \\ 1 \end{bmatrix}, \quad \mathbf{C} = [\,-17/2 \quad 9\,], \quad \mathbf{D} = 1.$$

Example 17
Product of factors

Factoring the numerator and denominator of the previous example, we can write the transfer function (nonuniquely) as the product

$$\hat{h}(s) = \left(\frac{s + 1 + 3/\sqrt{2}}{s + 1/2} \right)\left(\frac{s + 1 - 3/\sqrt{2}}{s + 1} \right)$$

$$= \left(1 + \frac{1/2 + 3/\sqrt{2}}{s + 1/2} \right)\left(1 + \frac{-3/\sqrt{2}}{s + 1} \right),$$

for which the operational diagrams are connected in series in Figure 4.11. From the diagram, the state-space equations are

$$\frac{d}{dt}\begin{bmatrix} x_1 \\ x_2 \end{bmatrix} = \begin{bmatrix} -1/2 & 0 \\ 1/2 + 3/\sqrt{2} & -1 \end{bmatrix}\begin{bmatrix} x_1 \\ x_2 \end{bmatrix} + \begin{bmatrix} 1 \\ 1 \end{bmatrix}u,$$

$$y = [\,1/2 + 3/\sqrt{2} \quad -3/\sqrt{2}\,]\begin{bmatrix} x_1 \\ x_2 \end{bmatrix} + [\,1\,]u.$$

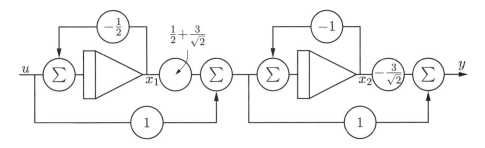

Fig. 4.11   Example product of two scalar terms.

9   Multi-input, multi-output (MIMO) transfer functions

Realizations for the transfer $p \times m$ matrix $\mathbf{H}(s) = [h_{ij}(s)]$ are available by inspection provided each scalar transfer function $h_{ij}(s)$ can be realized, using any of the previous methods to produce matrices $(\mathbf{A}_{ij}, \mathbf{B}_{ij}, \mathbf{C}_{ij}, d_{ij})$, as illustrated in Figure 4.12. If any $h_{ij}(s)$ is a constant d_{ij}, then the corresponding $\mathbf{A}_{ij}, \mathbf{B}_{ij}, \mathbf{C}_{ij}$ need not be defined.

Method

1. Let $\mathbf{A} = \operatorname{diag}[\mathbf{A}_{ij}]$, where the defined $\mathbf{A}_{ij}$ are listed in some fixed order, e.g., 11, 21, $\cdots$. Then $\mathbf{A}$ has dimension $n = \sum \dim \mathbf{A}_{ij}$, $\mathbf{B}$ is $n \times m$, and $\mathbf{C}$ is $p \times n$.

2. Construct $\mathbf{B}$ such that each defined $\mathbf{B}_{ij}$ is in column j of $\mathbf{B}$, and in the same rows as $\mathbf{A}_{ij}$ appears in $\mathbf{A}$.

3. Construct $\mathbf{C}$ such that each defined $\mathbf{C}_{ij}$ is in row i of $\mathbf{C}$, and in the same columns as $\mathbf{A}_{ij}$ appears in $\mathbf{A}$.

4. Finally, $\mathbf{D} = [d_{ij}]$.

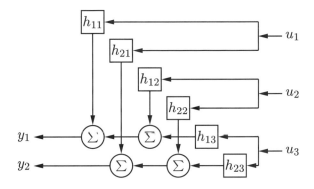

Fig. 4.12   Realization of a 2×3 transfer matrix.

Example 18
2 × 3 MIMO
system

Figure 4.12 illustrates the construction of a 2×3 transfer matrix, producing the matrices

(4.52a) $$\mathbf{A} = \begin{bmatrix} \mathbf{A}_{11} & & & & & \\ & \mathbf{A}_{21} & & & & \\ & & \mathbf{A}_{12} & & & \\ & & & \mathbf{A}_{22} & & \\ & & & & \mathbf{A}_{13} & \\ & & & & & \mathbf{A}_{23} \end{bmatrix}, \quad \mathbf{B} = \begin{bmatrix} \mathbf{B}_{11} \\ \mathbf{B}_{21} \\ & \mathbf{B}_{12} \\ & \mathbf{B}_{22} \\ & & \mathbf{B}_{13} \\ & & \mathbf{B}_{23} \end{bmatrix},$$

(4.52b) $$\mathbf{C} = \begin{bmatrix} \mathbf{C}_{11} & & \mathbf{C}_{12} & & \mathbf{C}_{13} & \\ & \mathbf{C}_{21} & & \mathbf{C}_{22} & & \mathbf{C}_{23} \end{bmatrix}, \quad \mathbf{D} = \begin{bmatrix} d_{11} & d_{12} & d_{13} \\ d_{21} & d_{22} & d_{23} \end{bmatrix}.$$

Example 19
MIMO system

In the transfer matrix

(4.53) $$\mathbf{H}(s) = \begin{bmatrix} 1 - \dfrac{5}{s+2} & \dfrac{s}{s^2+3} & 0 \\ 4 & \dfrac{7}{s+2} & \dfrac{9}{s+5} \end{bmatrix},$$

the entries $\hat{h}_{13}$ and $\hat{h}_{21}$ are constant, and do not require state variables to be defined, so that, by the above method, this transfer matrix can be realized by the matrices

(4.54a) $$\mathbf{A} = \begin{bmatrix} -2 & & & \\ & 0 & 1 & \\ & -3 & 0 & \\ & & & -2 & \\ & & & & -5 \end{bmatrix}, \quad \mathbf{B} = \begin{bmatrix} 1 \\ 0 \\ 1 \\ 1 \\ 1 \end{bmatrix},$$

(4.54b) $$\mathbf{C} = \begin{bmatrix} -5 & 0 & 1 & & \\ & & & 7 & 9 \end{bmatrix}, \quad \mathbf{D} = \begin{bmatrix} 1 & 0 & 0 \\ 4 & 0 & 0 \end{bmatrix}.$$

10   Further study

Although state-space models are clearly applicable to a broad class of physical and empirical laws, only a few examples are given in this chapter, rather than a definitive list of applications. More general methodologies are given in system-theoretic references such as Sandquist [47].

A set of engineering applications in the context of control systems is given by Friedland [16].

The use of state models of electric power systems can be found in Byerly and Kimbark [7].

Electric circuits provide a useful class of examples for state-space models, since most readers will know some elementary circuit theory, but the state-space form is typically not necessary for simulation, and has its greatest use for the qualitative analysis of nonlinear circuits; see Stern [49], or Chua [10]. However, analogous graph-based models are applicable in other specialized disciplines, such as mechanics, in which the state variables are typically positions and momenta (see Goldstein [18]), thermodynamics, and hydraulics (see for example, Koenig et al. [28] or Deo [13] or the more recent references by Dorny [15] or Rowell and Wormley [45]).

Readers intending to apply the results of Section 2 to robotic models might consult a reference such as Spong and Vidyasagar [48]. A good reference for classical mechanics is Goldstein [18].

The operational diagrams used here are familiar to many engineers, but alternatives for generating models are signal flow graphs, block diagrams, bond graphs, and the specialized languages for generating computer models of physical systems. Cellier [8] contains a survey. A reference for bond-graph modeling is Thoma [53].

A standard reference on classical mechanics, including the Euler-Lagrange method, is Goldstein [18]. Books on robotics often include these methods for modeling the dynamics of robot arms. Spong and Vidyasagar [48] and Paul [42] are two of many.

11   Problems

1   Write state-space equations for

(a) the linear circuit in Figure P4.1(a),

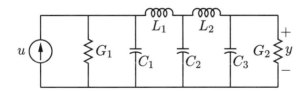

Fig. P4.1(a)   Linear circuit.

(b) the linear circuit in Figure P4.1(b),

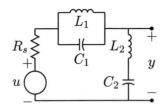

Fig. P4.1(b) Linear circuit.

(c) the nonlinear circuit in Figure P4.1(c),

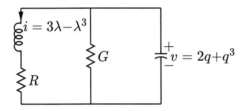

Fig. P4.1(c) Nonlinear circuit, with capacitor charge q, inductor flux-linkage λ.

(d) the binary circuit in Figure P4.1(d),

Fig. P4.1(d) Two-stage shift-register.

(e) the discrete-time real operational diagram in Figure P4.1(e),

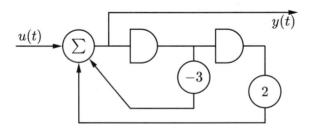

Fig. P4.1(e) Operational diagram of a real discrete-time system.

(f) the continuous-time real operational diagram in Figure P4.1(f).

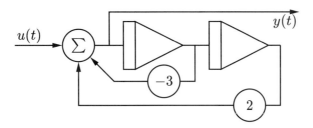

Fig. P4.1(f)   Operational diagram of a real continuous-time system.

2   Write the Euler-Lagrange equations for the inverted pendulum of Chapter 1, Example 19.

3   A tuned-mass damper is a damped vibratory mass designed to absorb energy at a predetermined frequency. Such devices may be used to reduce resonant vibrations of machines. Figure P4.3 illustrates a large tuned-mass damper used to reduce the swaying of tall buildings by absorbing energy at the resonant frequency of the structure. The pivot and base are attached to the building structure

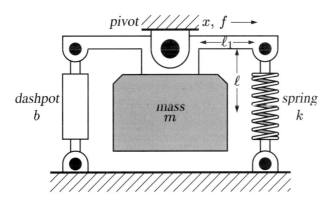

Fig. P4.3   Tuned-mass damper.

and move horizontally and simultaneously, with motion x and horizontal force f at the pivot. The mass m is at equivalent distance ℓ below the pivot. If the swing angle of the mass is θ, assume that the torque at the pivot produced by the spring is $k\ell_1^2\theta$, and that the damping torque is $b\ell_1^2\dot{\theta}$. Write the Euler-Lagrange equations for the tuned-mass damper.

4   Find a state-space realization of each of the systems

(a) $\hat{\mathbf{H}}(z) = \begin{bmatrix} z/(z+2) & 0 \\ 2\,z/(z+2) & 1/(z+1) \end{bmatrix}$,

(b) $\mathbf{H}(t) = [\, 4te^{-5t},\ e^{-5t}\,]$.

5   Write the state-space matrices $\mathbf{A}$, $\mathbf{B}$, $\mathbf{C}$, $\mathbf{D}$ for the following transfer matrices, where possible:

(a) $\hat{\mathbf{H}}(s) = 4 + \dfrac{7s^3 + 4s + 3}{s^4 + 6s^3 - 2s^2 + 8s + 5}$,   (b) $\hat{\mathbf{H}}(s) = \dfrac{2s+2}{s+4}$,

(c) $\hat{\mathbf{H}}(s) = \frac{1}{s}$,   (d) $\hat{\mathbf{H}}(z) = \frac{1}{z}$,   (e) $\hat{\mathbf{H}}(z) = \frac{z}{1}$,

(f) $\hat{\mathbf{H}}(s) = \begin{bmatrix} \dfrac{2s^2 + 2}{s^2 + s + 1} \\ 7 \\ 2/s \end{bmatrix}$.

6   A (p,m) convolutional encoder is a binary device or process, containing unit delays and mod-2 adders, that accepts a binary m-vector $\mathbf{U}(t)$ at time t and produces binary p-vector $\mathbf{Y}(t)$. The encoder is said to have *rate* m/p because p output bits are generated for every m input bits. Write state-space equations for the $(3,2)$ encoder of Figure P4.6.

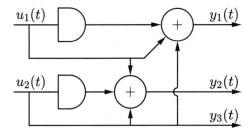

Fig. P4.6   Rate $2/3$ convolutional encoder.

7   If the transfer matrices $\hat{\mathbf{H}}_1(s)$, $\hat{\mathbf{H}}_2(s)$, are realized by system matrices $\mathbf{A}_1$, $\mathbf{B}_1$, $\mathbf{C}_1$, $\mathbf{D}_1$ and $\mathbf{A}_2$, $\mathbf{B}_2$, $\mathbf{C}_2$, $\mathbf{D}_2$ respectively, find matrices which realize

(a) the sum $\hat{\mathbf{H}}(s) = \hat{\mathbf{H}}_1(s) + \hat{\mathbf{H}}_2(s)$ (see Section 8),

(b) the product $\hat{\mathbf{H}}(s) = \hat{\mathbf{H}}_2(s)\,\hat{\mathbf{H}}_1(s)$.

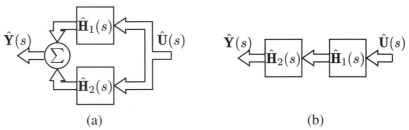

(a)   (b)

Fig. P4.7   Binary operations on transfer matrices: (a) addition (parallel connection), and
(b) multiplication (series connection).

---

8   Obtain a linearized state-space model of Figure P4.1(c) for the operating point
$\lambda = q = 0$.

---

9   It is sometimes necessary to approximate the delay function $y(t) = u(t-\tau)$
by using analog components. Expand this function as a Taylor series, to get

$$u(t) = y(t+\tau) = y(t) + \frac{\tau}{1!}\frac{dy}{dt} + \frac{\tau^2}{2!}\frac{d^2y}{dt^2} + \cdots.$$

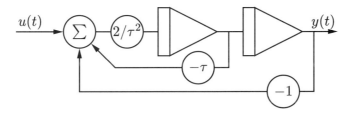

Fig. P4.9   Analog system.

(a) Show that Figure P4.9 is a second-order approximation to a delay of τ
seconds.

(b) Write the transfer function for the approximation, and note that the numerator has degree 0, and the denominator has degree 2. This is called a Padé
approximation of order $(0, 2)$ of the delay function.

(c) A pure delay has transfer function $e^{-s\tau}$. Calculate the coefficients for the
Padé approximation of order $(2, 2)$, which is of the form

$$b(s)/a(s) = (b_0 s^2 + b_1 s + b_2)/(s^2 + a_1 s + a_2),$$

by expanding $e^{-s\tau}$ as a series, writing

$$\frac{b(s)}{a(s)} \simeq 1 - \frac{s\tau}{1!} + \frac{s^2\tau^2}{2!} - \frac{s^3\tau^3}{3!} + \frac{s^4\tau^4}{4!} - \cdots,$$

and then by multiplying both sides by $a(s)$ and equating terms of degree 0 to 4 on the left and right.

10   In digital signal processing applications it is sometimes necessary to approximate the integral $y(t) = \int_0^t u(\tau)\, d\tau$, where $u(t)$ is sampled every T seconds to produce the sequence $\{y(k)\}$. One approximation method is Euler's forward rule shown in Figure P4.10, as in Example 11 of Chapter 1, in which $h\mathbf{F}(\mathbf{X}(k), kh))$ becomes simply $Tu(t)$.

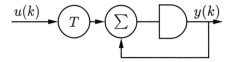

Fig. P4.10   Operational diagram for Euler's forward method.

Draw an operational diagram for the trapezoidal rule,

$$y(k+1) = y(k) + \frac{T}{2}(u(k) + u(k+1)).$$

11   A "T" flip-flop has one input and one output, and is described by the mod-2 equation $y(k+1) = y(k) + u(k)$.

(a) Draw an operational diagram for such a flip-flop.

(b) Draw an operational diagram that realizes a pure delay, or "D" flip-flop, using only "T" flip-flops and mod-2 adders.

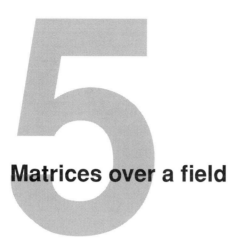

Matrices over a field

The purpose of this chapter is to recall basic definitions and facts about fields, so that the applicability of the state-space analysis can be checked in other contexts. Matrix arithmetic has been used in previous chapters, so the emphasis will be on summarizing basic results, and on introducing further material to be used later.

Using the simple and important concepts of elementary matrix operations, the normal form of a matrix under equivalence operations will be derived, with suitable warnings about sensitivity to errors in data. The normal form is a key to understanding linear equations and to the theory of vector-spaces, and leads directly to an important applied result in system realization and identification.

1   Basic definitions

This material is intended to be a brief summary of what may be found in a first course on linear algebra, with emphasis on the requirements of state-space analysis.

1.1   Field axioms

Model-building requires the ordinary arithmetic of a field.

A field is a set of scalars, together with the definition of two operations on elements of the set, multiplication $(\cdot)$, and addition, $(+)$, with the dot omitted when juxtaposition of elements conveys the same meaning. Some often-used fields are

1. The field of real numbers $\mathbb{R}$, with ordinary addition and multiplication.

2. The field of complex numbers $\mathbb{C}$ with complex arithmetic.

3. The field of rational functions with real or complex coefficients, typically the result of taking the Laplace or $\mathcal{Z}$ transform, with ordinary arithmetic.

4. The field of binary numbers, or more technically, the set $\{0, 1\}$ with modulo-2 arithmetic. This field is denoted $\mathbb{Z}_2$.

To check whether or not a set F of elements over which two operations are given is a field requires verification of the following axioms. In the following, let a, b, c, $\cdots$ be any elements of F.

A$_1$ *Uniqueness of addition:* $a + b$ is an element of F, uniquely determined by the pair (a, b).

A$_2$ *Commutative law of addition:* $a + b = b + a$.

A$_3$ *Associative law of addition:* $a + (b + c) = (a + b) + c$.

A$_4$ *Additive identity:* For every element a in F there exists an element 0 in F such that $a + 0 = 0 + a = a$.

A$_5$ *Additive inverse:* For each element a in F there exists a unique element $-a$ in F such that $a + (-a) = 0$.

M$_1$ *Uniqueness of multiplication:* $a \cdot b$ is an element of F, uniquely determined by the pair (a, b).

M$_2$ *Commutative law of multiplication:* $a \cdot b = b \cdot a$.

M$_3$ *Associative law of multiplication:* $a \cdot (b \cdot c) = (a \cdot b) \cdot c$.

M$_4$ *Multiplicative identity:* For every element a in F there exists an element $1 \neq 0$ in F such that $a \cdot 1 = 1 \cdot a = a$.

M$_5$ *Multiplicative inverse:* For each element $a \neq 0$ in F there exists a unique element a^{-1} in F such that $a \cdot a^{-1} = a^{-1} \cdot a = 1$.

D$_1$ *Distributive law:* $a(b + c) = ab + ac$.

D$_2$ *Distributive law:* $(a + b)c = ac + bc$.

Example 1
The integers
are not a field

The set $\mathbb{Z}$ of integers with ordinary arithmetic is not a field because axiom M$_5$ is not satisfied, but is an example of another kind of algebraic set with two defined operations, called a *ring*.

Example 2
Rational
numbers

The set of rational numbers m/n with ordinary arithmetic, where m and n are integers and nonunit common factors have been eliminated, satisfies the field axioms, with additive identity $0/1$ and multiplicative identity $1/1$.

Example 3
Rational
functions

The set of polynomials $\mathbb{R}[s]$ in indeterminate s, with real (or complex) coefficients, fails axiom M$_5$, as did the integers. However, the rational functions $\mathbb{R}(s)$, consisting of ratios of polynomials for which common factors have been eliminated, is a field, and obeys the same axioms as the rational numbers.

Example 4
Finite fields

Sets such as $\mathbb{Z}_2$ may contain a finite number of elements, in which case all possible additions and multiplications can be listed in tables, which can be inspected to check the axioms. Thus, the addition and multiplication tables for binary arithmetic are as shown:

$$
\begin{array}{c|cc}
+ & 0 & 1 \\
\hline
0 & 0 & 1 \\
1 & 1 & 0
\end{array},
\qquad
\begin{array}{c|cc}
\cdot & 0 & 1 \\
\hline
0 & 0 & 0 \\
1 & 0 & 1
\end{array}.
$$

By inspection of the tables, the existence and uniqueness of the additive and multiplicative identities can be checked. The commutative laws correspond to symmetry of the tables.

Example 5
$\mathbb{Z}_4$ is not a field

The set $\mathbb{Z}_4$ of integers mod 4 is the set $\{0, 1, 2, 3\}$, with addition defined such that $x + y = r$, where integer r is the remainder obtained on dividing 4 into the ordinary sum $(x + y)$, with $0 \leq r < 4$. Similarly, the result of multiplication is the remainder on division of the ordinary product by 4. Thus, all possible multiplications and divisions give values in the set $\mathbb{Z}_4$, and the addition and multiplication tables can be written as shown:

$$
\begin{array}{c|cccc}
+ & 0 & 1 & 2 & 3 \\
\hline
0 & 0 & 1 & 2 & 3 \\
1 & 1 & 2 & 3 & 0 \\
2 & 2 & 3 & 0 & 1 \\
3 & 3 & 0 & 1 & 2
\end{array},
\qquad
\begin{array}{c|cccc}
\cdot & 0 & 1 & 2 & 3 \\
\hline
0 & 0 & 0 & 0 & 0 \\
1 & 0 & 1 & 2 & 3 \\
2 & 0 & 2 & 0 & 2 \\
3 & 0 & 3 & 2 & 1
\end{array}.
$$

The tables show that $\mathbb{Z}_4$ is not a field since, by inspection of the multiplication table, axiom M_5 is not satisfied. In particular, the value 2, which is a member of $\mathbb{Z}_4$ and is not the zero member, does not have a multiplicative inverse.

Example 6
Digital filters

For reasons of efficiency, certain digital filters perform arithmetic over $\mathbb{Z}_p$, where p is chosen so that $\mathbb{Z}_p$ is a field, as for example, when p is a prime number.

1.2   **Matrix definitions and operations**

A *matrix* is a rectangular array of elements, with the subscript notation illustrated below:

(5.1) $$\mathbf{A} = [a_{ij}] = \begin{bmatrix} a_{11} & a_{12} & \cdots & a_{1n} \\ a_{21} & a_{22} & \cdots & a_{2n} \\ \cdots & & & \\ a_{m1} & a_{m2} & \cdots & a_{mn} \end{bmatrix}.$$

Thus the above matrix $\mathbf{A}$ has m rows and n columns, and when the elements are members of a field F, we write $\mathbf{A} \in F^{m \times n}$. If $m = n$ the matrix is *square,* or for emphasis, n-square.

If $\mathbf{A} \in F^{m \times n}$ is a matrix and $c \in F$, then $c\mathbf{A}$ is a matrix with every ij-element equal to $c\,a_{ij}$.

Two matrices $\mathbf{A}$ and $\mathbf{B}$ are equal only if they have the same dimensions, and for every element ij, $a_{ij} = b_{ij}$.

Addition of matrices is defined only for matrices of identical dimensions. If $\mathbf{A}, \mathbf{B} \in F^{m \times n}$, then $\mathbf{C} \in F^{m \times n}$ is defined such that $c_{ij} = a_{ij} + b_{ij}$.

The product $\mathbf{AB}$ of two matrices $\mathbf{A} \in F^{m \times n}$ and $\mathbf{B} \in F^{p \times r}$ is defined only if $n = p$, in which case $\mathbf{AB} = \mathbf{C} \in F^{m \times r}$ where the ij entry of $\mathbf{C}$ is $c_{ij} = \sum_{k=1}^{n} a_{ik} b_{kj}$.

Matrices of correct dimension for addition or multiplication to be defined are said to be *conformable* for the required operation.

A *diagonal* matrix is square and has zero elements except for at most the elements $11, 22, \cdots$.

The *zero* matrix is a matrix of zeros, denoted as $\mathbf{0}$ or often as 0, of dimension appropriate for the context.

The *identity* matrix $\mathbf{I}$ is a square diagonal matrix with all diagonal entries equal to 1. When the dimension is to be emphasized, $\mathbf{I}_n$ denotes an n-square identity matrix.

If $\mathbf{A} \in F^{m \times n}$ and there exists a matrix $\mathbf{B} \in F^{n \times m}$ such that $\mathbf{AB} = \mathbf{I}_m$, then $\mathbf{B}$ is called a *right inverse* of $\mathbf{A}$, sometimes written $\mathbf{A}^R$.

If $\mathbf{A} \in F^{m \times n}$ and there exists a matrix $\mathbf{B} \in F^{n \times m}$ such that $\mathbf{BA} = \mathbf{I}_n$, then $\mathbf{B}$ is called a *left inverse* of $\mathbf{A}$, sometimes written $\mathbf{A}^L$. Left and right inverses are illustrated in Figure 5.1.

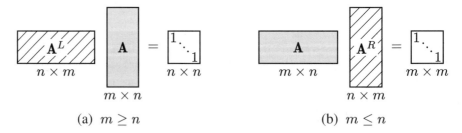

(a) $m \geq n$ (b) $m \leq n$

Fig. 5.1 Illustrating left and right inverses: (a) left inverse, (b) right inverse. In each case if the inverse exists it has the dimensions of $\mathbf{A}^T$.

If $\mathbf{AB} = \mathbf{BA} = \mathbf{I}$, then $\mathbf{B}$ is called the *inverse* of $\mathbf{A}$, and is written $\mathbf{A}^{-1}$.

If $\mathbf{A}$ has an inverse, the inverse is a unique matrix. If the inverse exists, the matrix is said to be *nonsingular;* otherwise it is *singular.* The inverse exists if and only if $\mathbf{A}$ is square, and its determinant $\det(\mathbf{A}) \neq 0$. Formulas for the inverse will be given later, in Section 4.

If $\mathbf{A}$ and $\mathbf{B}$ are nonsingular n-square matrices, then $(\mathbf{AB})^{-1} = \mathbf{B}^{-1}\mathbf{A}^{-1}$.

A square nonsingular matrix $\mathbf{A}$ for which $\mathbf{A}^{-1} = \mathbf{A}^T$ is called *orthogonal*.

If $\mathbf{A} \in F^{m \times n}$, then $\mathbf{A}^T$ denotes the $n \times m$ matrix obtained by interchanging the rows and columns of $\mathbf{A}$; that is, the ij entry of $\mathbf{A}^T$ is a_{ji}.

If $\mathbf{A}$ is $m \times n$ and $\mathbf{B}$ is $n \times p$, then $(\mathbf{AB})^T = \mathbf{B}^T\mathbf{A}^T$.

If $\mathbf{A} \in \mathbb{R}^{m \times n}$, then the nonnegative scalar $\|\mathbf{A}\|_E = +\sqrt{\sum_{i=1}^{m}\sum_{j=1}^{n} a_{ij}^2}$ is called the Euclidean norm of $\mathbf{A}$, and the sum of the squares of the elements of $\mathbf{A}$ equals $\|\mathbf{A}\|_E^2$. If $\mathbf{A}$ has one column or one row, then the Euclidean matrix norm is identical to the Euclidean vector norm. For matrices, the Euclidean norm is also called the Frobenius norm. If $\mathbf{A} \in \mathbb{C}^{m \times n}$, the formula is $\|\mathbf{A}\|_E = +\sqrt{\sum_{i=1}^{m}\sum_{j=1}^{n} |a_{ij}|^2}$.

Example 7
Algebra of
matrices

The algebra of matrices, or even of square matrices, is substantially different from the algebra of real or complex numbers, since in general the square matrices do not constitute a field. First, multiplication is not commutative in general. For example, if $\mathbf{A} = \begin{bmatrix} 1 & 1 \\ 0 & 1 \end{bmatrix}$, $\mathbf{B} = \begin{bmatrix} 1 & 0 \\ 0 & 2 \end{bmatrix}$, then $\mathbf{AB} = \begin{bmatrix} 1 & 2 \\ 0 & 2 \end{bmatrix}$, but $\mathbf{BA} = \begin{bmatrix} 1 & 1 \\ 0 & 2 \end{bmatrix}$. Second, a nonzero square matrix does not necessarily have an inverse. For example, $\mathbf{A} = \begin{bmatrix} 0 & 1 \\ 0 & 0 \end{bmatrix}$ does not have an inverse.

Example 8
A field
containing
matrices

The set of 2×2 matrices $\begin{bmatrix} a & -b \\ b & a \end{bmatrix}$, with $a, b \in \mathbb{R}$, is a field, with the unit elements of addition and multiplication respectively $\begin{bmatrix} 0 & 0 \\ 0 & 0 \end{bmatrix}$, $\begin{bmatrix} 1 & 0 \\ 0 & 1 \end{bmatrix}$, and the inverse given by the formula $\begin{bmatrix} a & -b \\ b & a \end{bmatrix}^{-1} = \begin{bmatrix} a/(a^2+b^2) & b/(a^2+b^2) \\ -b/(a^2+b^2) & a/(a^2+b^2) \end{bmatrix}$.

Example 9
Nonuniqueness
of $\mathbf{A}^L$ and $\mathbf{A}^R$

Left and right inverses are not unique in general. Thus, if $\mathbf{A} = [1, -2]$, a matrix $\mathbf{A}^R = \begin{bmatrix} \alpha \\ (\alpha-1)/2 \end{bmatrix}$ is a right inverse for any α, because $\mathbf{AA}^R = 1$.

Similarly, $\mathbf{A}^L\mathbf{A} = \left(\begin{bmatrix} 0 & 1 & 0 \\ 0 & 0 & 1/2 \end{bmatrix} + \begin{bmatrix} \alpha \\ \beta \end{bmatrix} [1 \quad -1 \quad 1] \right) \begin{bmatrix} 1 & -2 \\ 1 & 0 \\ 0 & 2 \end{bmatrix} = \begin{bmatrix} 1 & 0 \\ 0 & 1 \end{bmatrix}$,

so the matrix in parentheses is a left inverse of $\mathbf{A}$ for any α and β.

Example 10
Uniqueness of
$\mathbf{A}^{-1}$

If a left inverse $\mathbf{A}^L$ and a right inverse $\mathbf{A}^R$ both exist for matrix $\mathbf{A}$, then these inverses are identical, since by assumption, $\mathbf{A}\mathbf{A}^R = \mathbf{I}$, and $\mathbf{A}^L\mathbf{A} = \mathbf{I}$, so that $\mathbf{A}^L\mathbf{A}\mathbf{A}^R = (\mathbf{A}^L\mathbf{A})\mathbf{A}^R = \mathbf{A}^L(\mathbf{A}\mathbf{A}^R)$, from which $\mathbf{A}^R = \mathbf{A}^L$, written as $\mathbf{A}^{-1}$. Furthermore, $\mathbf{A}^{-1}$ is unique, for, supposing that $\mathbf{A}^R$ and $\mathbf{A}_2^R$ are distinct right inverses, then $\mathbf{A}\mathbf{A}^R = \mathbf{I}$, and $\mathbf{A}\mathbf{A}_2^R = \mathbf{I}$, so that, subtracting, $\mathbf{A}(\mathbf{A}^R - \mathbf{A}_2^R) = 0$, which becomes $\mathbf{A}^L\mathbf{A}(\mathbf{A}^R - \mathbf{A}_2^R) = \mathbf{A}^L 0$ on premultiplication by $\mathbf{A}^L$. The left-hand side of this equation is now $\mathbf{I}(\mathbf{A}^R - \mathbf{A}_2^R) = \mathbf{A}^R - \mathbf{A}_2^R$, but the right-hand side is $\mathbf{A}^L 0 = 0$, a contradiction.

Example 11
Inverse of
matrix product

To prove that $(\mathbf{A}\mathbf{B})^{-1} = \mathbf{B}^{-1}\mathbf{A}^{-1}$ when these inverses exist, observe that

$$(\mathbf{B}^{-1}\mathbf{A}^{-1})\mathbf{A}\mathbf{B} = \mathbf{B}^{-1}(\mathbf{A}^{-1}\mathbf{A})\mathbf{B} = \mathbf{B}^{-1}\mathbf{B} = \mathbf{I},$$

and that

$$\mathbf{A}\mathbf{B}(\mathbf{B}^{-1}\mathbf{A}^{-1}) = \mathbf{A}(\mathbf{B}\mathbf{B}^{-1})\mathbf{A}^{-1} = \mathbf{A}\mathbf{A}^{-1} = \mathbf{I},$$

showing that $\mathbf{B}^{-1}\mathbf{A}^{-1}$ is both a left inverse and a right inverse of $\mathbf{A}\mathbf{B}$. But from the previous example, this inverse is unique and equal to $(\mathbf{A}\mathbf{B})^{-1}$.

Example 12
Transpose of
matrix product

To prove that $(\mathbf{A}\mathbf{B})^T = \mathbf{B}^T\mathbf{A}^T$, let $\mathbf{A} \in F^{m \times n}$ and $\mathbf{B} \in F^{n \times p}$. Then the ij element of $\mathbf{C} = \mathbf{A}\mathbf{B}$ is $c_{ij} = \sum_{k=1}^{n} a_{ik}b_{kj}$, which is element ji of $(\mathbf{A}\mathbf{B})^T$. But element ji of $\mathbf{B}^T\mathbf{A}^T$ is

$$[\,\text{row } j \text{ of } \mathbf{B}^T\,]\,[\,\text{col } i \text{ of } \mathbf{A}^T\,] = [\,\text{col } j \text{ of } \mathbf{B}\,]^T\,[\,\text{row } i \text{ of } \mathbf{A}\,]^T$$

$$= \sum_{k=1}^{n} b_{kj}a_{ik} = c_{ij}.$$

Example 13
Euclidean norm
and norm
axioms

One can easily verify that the Euclidean norm $\|\cdot\|_E$ for real matrices satisfies the properties

1. $\|\mathbf{A}\| \geq 0$ for every matrix $\mathbf{A} \in \mathbb{R}^{m \times n}$, and $\|\mathbf{A}\| = 0$ if and only if every element of $\mathbf{A}$ is 0.

2. $\|\alpha\mathbf{A}\| = |\alpha|\|\mathbf{A}\|$ for every matrix $\mathbf{A} \in \mathbb{R}^{m \times n}$ and every scalar $\alpha \in \mathbb{R}$.

3. $\|\mathbf{A} + \mathbf{B}\| \leq \|\mathbf{A}\| + \|\mathbf{B}\|$ for every matrix $\mathbf{A}$ and $\mathbf{B} \in \mathbb{R}^{m \times n}$ (the triangle inequality).

These are the axioms of norms, which are scalars used as measures of the size of nonscalar objects, such as matrices or vectors, or of the distance between two such objects.

Matrices of identical norm are not necessarily identical. For example, the matrices

$$\begin{bmatrix} 0 & 1 \\ 0 & 0 \end{bmatrix}, \quad \begin{bmatrix} 1 & 0 \\ 0 & 0 \end{bmatrix}, \quad \begin{bmatrix} 1/\sqrt{2} & 0 \\ 0 & 1/\sqrt{2} \end{bmatrix}, \quad \begin{bmatrix} 1/2 & 1/2 \\ 1/2 & 1/2 \end{bmatrix}$$

all have Euclidean norm equal to 1.

Example 14
Vector and
matrix norms

Other norms than the Euclidean norm are often used, in particular for vectors $\mathbf{X} \in \mathbb{R}^n$, which are matrices with one row or column. The Euclidean norm is a special case of the class

$$\|\mathbf{X}\|_p = \left(\sum_{i=1}^{n} |x_i|^p \right)^{1/p}, \quad 1 \le p \le \infty,$$

where $\|\mathbf{X}\|_\infty = \max\{|x_i| : i = 1, \cdots n\}$.

A vector norm $\| \cdot \|_v$ and a matrix norm $\| \cdot \|_m$ which satisfy

$$\|\mathbf{AX}\|_v \le \|\mathbf{A}\|_m \|\mathbf{X}\|_v$$

for all matrices $\mathbf{A} \subset \mathbb{R}^{m \times n}$ and all vectors $\mathbf{X} \in \mathbb{R}^n$ are said to be compatible. Given a vector norm $\| \cdot \|_v$, the scalar

$$\|\mathbf{A}\|_v = \sup_{\mathbf{X}} \frac{\|\mathbf{AX}\|_v}{\|\mathbf{X}\|_v}$$

is a matrix norm said to be subordinate to or induced by $\| \cdot \|_v$. The notation sup means the supremum, that is, the largest value.

---

2   Determinants

The determinant is a function, of which the domain is the collection of square matrices, and the range is the field of scalars over which the matrices are defined.

The determinant of a 1×1 matrix α is defined to be α, and for any $n \times n$ matrix $\mathbf{A} = [a_{ij}]$ with $n > 1$, the function can be given in several ways:

1. *Permutation.* $\det(\mathbf{A}) = \sum \pm a_{1i} a_{2j} \cdots a_{nk}$, which is a sum of signed products of entries of $\mathbf{A}$. Each product contains exactly one entry from each of the n rows of $\mathbf{A}$, and exactly one entry from each of the n columns of $\mathbf{A}$. The list of n subscripts $i, j, \cdots k$ in this formula is a permutation of the integers $1, 2, \cdots n$. There are $n!$ such permutations, and $n!$ corresponding products in the sum. The sign is negative for odd permutations, and positive for even permutations. A permutation is even if it is obtained from $1, 2, \cdots n$ by exchanging pairs of numbers an even number of times, otherwise the permutation is odd.

2. *Expansion along the j-th column.* $\det(\mathbf{A}) = \sum_{i=1}^{n} a_{ij}(-1)^{i+j}\det(\mathbf{A}_{ij})$, where $\mathbf{A}_{ij}$ is $\mathbf{A}$ with row i and column j deleted.

3. *Expansion along the i-th row.* $\det(\mathbf{A}) = \sum_{j=1}^{n} a_{ij}(-1)^{i+j}\det(\mathbf{A}_{ij})$.

4. *Laplace expansion.* The previous two expansions are special cases of the Laplace expansion, which in general is taken along a set of $m \geq 1$ rows (or columns). Let the m rows of the expansion be chosen. Then all $m \times m$ matrices obtainable from these rows by deleting $n-m$ columns are formed. For each such square submatrix $\mathbf{S}$ of $\mathbf{A}$, the complementary $(n-m) \times (n-m)$ submatrix $\bar{\mathbf{S}}$ is constructed from $\mathbf{A}$ by deleting the m rows and m columns of $\mathbf{A}$ containing entries of $\mathbf{S}$. Also for each $\mathbf{S}$, let $\gamma(\mathbf{S})$ be the sum of the m row indices in $\mathbf{A}$ of the rows of $\mathbf{S}$, plus the m column indices in $\mathbf{A}$ of the columns of $\mathbf{S}$. Then $\det(\mathbf{A}) = \sum (-1)^{\gamma(\mathbf{S})}\det(\mathbf{S})\det(\bar{\mathbf{S}})$, where the sum is taken over all constructed submatrices $\mathbf{S}$ and corresponding $\gamma(\mathbf{S})$ and $\bar{\mathbf{S}}$.

A proof that the above definitions are equivalent is beyond the scope of this chapter, which will concentrate on system-theoretic applications.

Comment None of the above formulas for the determinant is to be recommended for computation with general matrices and floating-point arithmetic. In fact, the computation of numerical determinants is rarely necessary in the analysis of state-space systems. The determinant is often better treated as a theoretical quantity that is fundamental for analysis, rather than as a numerical tool.

Example 15
Expansion
along a row

Computing $\det(\mathbf{A})$ of Example 16 by expanding along (for example) row 4 gives

$$\det(\mathbf{A}) = (-1)^{4+1}2(2) + 0 + (-1)^{4+3}1(-1) + 0 = -4 + 0 + 1 + 0 = -3.$$

Example 16
Determinant by
permutation

The determinant of the matrix

$$\mathbf{A} = \begin{bmatrix} 1 & 0 & 2 & 0 \\ 2 & 3 & -1 & 1 \\ 4 & 1 & 0 & 0 \\ 2 & 0 & 1 & 0 \end{bmatrix}$$

will be constructed by the permutation method. The table below shows the permutations of the set of integers $1, 2, 3, 4$, the order of each permutation (the number of exchanges from $1, 2, 3, 4$), the term corresponding to the permutation, and its numerical value.

Perm.	Order	Term	Value
$1, 2, 3, 4$	0	$+a_{11}a_{22}a_{33}a_{44}$	0
$1, 2, 4, 3$	1	$-a_{11}a_{22}a_{34}a_{43}$	0
$1, 4, 2, 3$	2	$+a_{11}a_{24}a_{32}a_{43}$	1
$1, 4, 3, 2$	3	$-a_{11}a_{24}a_{33}a_{42}$	0
$1, 3, 4, 2$	4	$+a_{11}a_{23}a_{34}a_{42}$	0
$1, 3, 2, 4$	5	$-a_{11}a_{23}a_{32}a_{44}$	0
$3, 1, 2, 4$	6	$+a_{13}a_{21}a_{32}a_{44}$	0
$3, 1, 4, 2$	7	$-a_{13}a_{21}a_{34}a_{42}$	0
$3, 4, 1, 2$	8	$+a_{13}a_{24}a_{31}a_{42}$	0
$4, 3, 1, 2$	9	$-a_{14}a_{23}a_{31}a_{42}$	0
$4, 1, 3, 2$	10	$+a_{14}a_{21}a_{33}a_{42}$	0
$4, 1, 2, 3$	11	$-a_{14}a_{21}a_{32}a_{43}$	0
$4, 2, 1, 3$	12	$+a_{14}a_{22}a_{31}a_{43}$	0
$2, 4, 1, 3$	13	$-a_{12}a_{24}a_{31}a_{43}$	0
$2, 1, 4, 3$	14	$+a_{12}a_{21}a_{34}a_{43}$	0
$2, 1, 3, 4$	15	$-a_{12}a_{21}a_{33}a_{44}$	0
$2, 3, 1, 4$	16	$+a_{12}a_{23}a_{31}a_{44}$	0
$2, 3, 4, 1$	17	$-a_{12}a_{23}a_{34}a_{41}$	0
$2, 4, 3, 1$	18	$+a_{12}a_{24}a_{33}a_{41}$	0
$4, 2, 3, 1$	19	$-a_{14}a_{22}a_{33}a_{41}$	0
$4, 3, 2, 1$	20	$+a_{14}a_{23}a_{32}a_{41}$	0
$3, 4, 2, 1$	21	$-a_{13}a_{24}a_{32}a_{41}$	-4
$3, 2, 4, 1$	22	$+a_{13}a_{22}a_{34}a_{41}$	0
$3, 2, 1, 4$	23	$-a_{13}a_{22}a_{31}a_{44}$	0

Summing the value columns gives $\det(\mathbf{A}) = -3$.

Example 17
Expansion
along a column

Computing $\det(\mathbf{A})$ of Example 16 by expanding along the third column, for example, gives

$$\det(\mathbf{A}) = (-1)^{1+3}2(-2) + (-1)^{2+3}(-1)(0) + 0 + (-1)^{4+3}1(-1)$$
$$= -4 + 0 + 0 + 1 = -3.$$

Example 18
Expansion
along two rows

Use the Laplace expansion along rows 1 and 2 of the matrix of Example 16:

$$\det(\mathbf{A}) = (-1)^{1+2+1+2} \det\begin{bmatrix} 1 & 0 \\ 2 & 3 \end{bmatrix} \det\begin{bmatrix} 0 & 0 \\ 1 & 0 \end{bmatrix}$$

$$+ (-1)^{1+2+1+3} \det\begin{bmatrix} 1 & 2 \\ 2 & -1 \end{bmatrix} \det\begin{bmatrix} 1 & 0 \\ 0 & 0 \end{bmatrix}$$

$$+ (-1)^{1+2+1+4} \det\begin{bmatrix} 1 & 0 \\ 2 & 1 \end{bmatrix} \det\begin{bmatrix} 1 & 0 \\ 0 & 1 \end{bmatrix}$$

$$+(-1)^{1+2+2+3} \det \begin{bmatrix} 0 & 2 \\ 3 & -1 \end{bmatrix} \det \begin{bmatrix} 4 & 0 \\ 2 & 0 \end{bmatrix}$$

$$+(-1)^{1+2+2+4} \det \begin{bmatrix} 0 & 0 \\ 3 & 1 \end{bmatrix} \det \begin{bmatrix} 4 & 0 \\ 2 & 1 \end{bmatrix}$$

$$+(-1)^{1+2+3+4} \det \begin{bmatrix} 2 & 0 \\ -1 & 1 \end{bmatrix} \det \begin{bmatrix} 4 & 1 \\ 2 & 0 \end{bmatrix}$$

$$= +(3)(0) - (-5)(0) + (1)(1) + (-6)(0) - (0)(4) + (2)(-2)$$

$$= -3.$$

Example 19
Block-diagonal
matrix

The Laplace expansion is occasionally very useful for demonstrating properties of special matrices. Let $\mathbf{A} = \begin{bmatrix} \mathbf{P} & \mathbf{Q} \\ \mathbf{0} & \mathbf{R} \end{bmatrix}$, where $\mathbf{P}$ and $\mathbf{R}$ are square. Then by using the Laplace expansion along the rows containing $\mathbf{P}$ and $\mathbf{Q}$, one pair of submatrices in the expansion is $\mathbf{S} = \mathbf{P}$, $\bar{\mathbf{S}} = \mathbf{R}$, and all other submatrices $\bar{\mathbf{S}}$ will contain at least one column of zeros and hence will have determinant equal to 0, as is to be discussed in Section 2.1. Therefore, if α is the sum of the row indices of $\mathbf{P}$,

$$\det(\mathbf{A}) = (-1)^{2\alpha} \det(\mathbf{P}) \det(\mathbf{R}) = \det(\mathbf{P}) \det(\mathbf{R}).$$

2.1   Properties of determinants

Let $\mathbf{A}$, $\mathbf{B}$ be n-square matrices. The previous definitions of the determinant function allow the following conclusions.

1. $\det(\mathbf{A}^T) = \det(\mathbf{A})$.

2. If *all* elements of a row or a column of matrix $\mathbf{A}$ are zero, then $\det(\mathbf{A}) = 0$.

3. If $\mathbf{B}$ is obtained from $\mathbf{A}$ by multiplying all elements of one row or one column by a scalar α, then $\det(\mathbf{B}) = \alpha \det(\mathbf{A})$.

4. Let $\mathbf{A}$ and $\mathbf{B}$ be n-square matrices that are identical except possibly for row k. Then if $\mathbf{C}$ is identical to $\mathbf{A}$ except that the k-th row of $\mathbf{C}$ is the sum of the k-th row of $\mathbf{A}$ and the k-th row of $\mathbf{B}$, that is, $c_{kj} = a_{kj} + b_{kj}$, $j = 1, \cdots n$, then $\det(\mathbf{C}) = \det(\mathbf{A}) + \det(\mathbf{B})$.

5. If $\mathbf{B}$ is obtained from $\mathbf{A}$ by interchanging two rows or two columns, then $\det(\mathbf{B}) = -\det(\mathbf{A})$.

6. If $\mathbf{A}$ contains two identical rows or columns, then $\det \mathbf{A} = 0$.

7. If $\mathbf{B}$ is obtained from $\mathbf{A}$ by adding $\alpha \times$ row j to row i, $i \neq j$, then $\det(\mathbf{B}) = \det(\mathbf{A})$.

8. $\det(\mathbf{AB}) = \det(\mathbf{A}) \det(\mathbf{B})$.

Example 20
Properties of
determinants

It is instructive to demonstrate some of the above properties by using the definitions. Property 1 follows by expanding $\det(\mathbf{A}^T)$ along row i, and $\det(\mathbf{A})$ along column i. Property 2 is demonstrated by expansion along the zero row or column, properties 3 and 4 by expansion along the changed row or column. Properties 5 and 6 are demonstrated by computing the determinant, using permutations. Property 7 follows from properties 4 and 6.

Example 21
Determinant of
matrix product

To demonstrate Property 8, let $\mathbf{A}$ and $\mathbf{B}$ be $n \times n$, and write $\mathbf{AB} = \mathbf{C} = [c_{ij}] = [\sum_{k=1}^{n} a_{ik} b_{kj}]$. Then from Example 19, $\det(\mathbf{A}) \det(\mathbf{B}) = \det \begin{bmatrix} \mathbf{B} & -\mathbf{I} \\ 0 & \mathbf{A} \end{bmatrix}$, and to the $n{+}1$-th row, add a_{11} times the first row, a_{12} times the second, $\cdots$, a_{1n} times the n-th. Similarly to row $n{+}2$, add a_{21} times the first row, $\cdots$ a_{2n} times the n-th, and similarly modify the remaining rows. By Property 7, all of these operations leave the determinant unchanged, but give $\det(\mathbf{A}) \det(\mathbf{B}) = \det \begin{bmatrix} \mathbf{B} & -\mathbf{I} \\ \mathbf{C} & 0 \end{bmatrix}$, which, again applying the Laplace expansion, is

$$\det(\mathbf{A}) \ \det(\mathbf{B}) = (-1)^{(1+2+\cdots n)+(n+1+n+2\cdots 2n)} \det(\ \mathbf{I}) \det(\mathbf{C})$$
$$= (-1)^{2(1+2+\cdots n)+n^2}(-1)^n \det(\mathbf{C})$$
$$= (-1)^{n(n+1)+n^2+n} \det(\mathbf{C}) = \det(\mathbf{C}).$$

3   Rank, elementary transformations, and equivalence

Let $\mathbf{A} \in F^{m \times n}$ be an arbitrary rectangular matrix. The *rank* of $\mathbf{A}$ is the dimension of the largest square submatrix that has nonzero determinant and that is obtainable from $\mathbf{A}$ by deleting rows or columns or both.

Let the i-th row of $\mathbf{A} \in F^{m \times n}$ be written as $\mathbf{A}_i$. Given any set of distinct rows $\{\mathbf{A}_i, \mathbf{A}_j, \cdots \mathbf{A}_k\}$, not necessarily adjacent in $\mathbf{A}$, then if there is no set $\{\alpha_i, \alpha_j, \cdots \alpha_k\}$ of coefficients in F containing at least one nonzero entry and for which $\alpha_i \mathbf{A}_i + \alpha_j \mathbf{A}_j + \cdots \alpha_k \mathbf{A}_k$ is a row of zeros, then the rows $\mathbf{A}_i, \mathbf{A}_j, \cdots \mathbf{A}_k$ are said to be *linearly independent*. The rank of $\mathbf{A}$ will be shown to be the number of linearly independent rows, or, similarly, columns, contained in the matrix.

If $\mathbf{A} \in F^{m \times n}$ and $\mathrm{rank}(\mathbf{A}) = m$, the matrix is said to have full row rank; if $\mathrm{rank}(\mathbf{A}) = n$, the matrix has full column rank.

3.1   Elementary transformations

The following elementary row and column transformations, and their inverses, are of considerable conceptual use in matrix algebra. They are modified for floating-point computation. They do not change either the dimensions or the rank.

1. Let H_{ij} be the interchange of rows i and j. The inverse transformation is $H_{ij}^{-1} = H_{ij}$.

2. Let $H_i(\alpha)$ be the multiplication of all elements of row i by scalar $\alpha \neq 0$. The inverse transformation is $H_i^{-1}(\alpha) = H_i(1/\alpha)$.

3. Let $H_{ij}(\alpha)$ be the addition of the product of row j and scalar α to row i. The inverse transformation is $H_{ij}^{-1}(\alpha) = H_{ij}(-\alpha)$.

From the properties of determinants, transformation H_{ij} changes the sign of the determinant of a matrix, transformation $H_i(\alpha)$ multiplies the determinant by α, and transformation $H_{ij}(\alpha)$ leaves the determinant unchanged.

The above transformations are often called elementary row *operations*. Similarly, the following elementary column operations and their inverses are defined:

1. Let K_{ij} be the interchange of columns i and j. The inverse transformation is $K_{ij}^{-1} = K_{ij}$.

2. Let $K_i(\alpha)$ be the multiplication of all elements of column i by scalar $\alpha \neq 0$. The inverse transformation is $K_i^{-1}(\alpha) = K_i(1/\alpha)$.

3. Let $K_{ij}(\alpha)$ be the addition of the product of column j and scalar α to column i. The inverse transformation is $K_{ij}^{-1}(\alpha) = K_{ij}(-\alpha)$.

If matrix $\mathbf{B}$ can be obtained from $\mathbf{A}$ by a sequence of elementary row and column transformations, $\mathbf{B}$ is said to be *equivalent* to $\mathbf{A}$, written $\mathbf{B} \sim \mathbf{A}$. Since each elementary transformation has an inverse, $\mathbf{B} \sim \mathbf{A}$ implies $\mathbf{A} \sim \mathbf{B}$.

Example 22
Row operations:
row echelon
form

The following is a sequence of three elementary row operations, starting with matrix $\mathbf{A}$ as shown:

$$\mathbf{A} = \begin{bmatrix} -3 & 9 & 1 & -5 \\ -2 & 6 & 0 & -8 \\ -4 & 12 & 1 & -9 \end{bmatrix} \xrightarrow{H_{21}(-2/3)} \begin{bmatrix} -3 & 9 & 1 & -5 \\ 0 & 0 & -2/3 & -14/3 \\ -4 & 12 & 1 & -9 \end{bmatrix}$$

$$\xrightarrow{H_{31}(-4/3)} \begin{bmatrix} -3 & 9 & 1 & -5 \\ 0 & 0 & -2/3 & -14/3 \\ 0 & 0 & -1/3 & -7/3 \end{bmatrix}$$

$$\xrightarrow{H_{32}(-1/2)} \begin{bmatrix} -3 & 9 & 1 & -5 \\ 0 & 0 & -2/3 & -14/3 \\ 0 & 0 & 0 & 0 \end{bmatrix}.$$

Example 23
Column
operations:
column echelon
form

An example sequence of column operations, starting with $\mathbf{A}$ of the previous example, is

$$\mathbf{A} = \begin{bmatrix} -3 & 9 & 1 & -5 \\ -2 & 6 & 0 & -8 \\ -4 & 12 & 1 & -9 \end{bmatrix} \xrightarrow{K_{21}(3)} \begin{bmatrix} -3 & 0 & 1 & -5 \\ -2 & 0 & 0 & -8 \\ -4 & 0 & 1 & -9 \end{bmatrix}$$

$$\xrightarrow{K_{31}(1/3)} \begin{bmatrix} -3 & 0 & 0 & -5 \\ -2 & 0 & -2/3 & -8 \\ -4 & 0 & -1/3 & -9 \end{bmatrix} \xrightarrow{K_{41}(-5/3)} \begin{bmatrix} -3 & 0 & 0 & 0 \\ -2 & 0 & -2/3 & -14/3 \\ -4 & 0 & -1/3 & -7/3 \end{bmatrix}$$

$$\xrightarrow{K_{23}} \begin{bmatrix} -3 & 0 & 0 & 0 \\ -2 & -2/3 & 0 & -14/3 \\ -4 & -1/3 & 0 & -7/3 \end{bmatrix} \xrightarrow{K_{42}(-7)} \begin{bmatrix} -3 & 0 & 0 & 0 \\ -2 & -2/3 & 0 & 0 \\ -4 & -1/3 & 0 & 0 \end{bmatrix}.$$

3.2   Elementary matrices

An extremely useful analytical and computational fact about the above elementary transformations is that each transformation corresponds to multiplication by an *elementary matrix*. There are three kinds of elementary row matrices:

1. Denote by $\mathbf{H}_{ij}$ the identity matrix $\mathbf{I}_m$ to which transformation H_{ij} has been applied. $\mathbf{H}_{ij}$ has the form

$$\mathbf{H}_{ij} = \begin{bmatrix} 1 \\ & \ddots \\ & & 1 \\ & & & 0 & & 1 \\ & & & & 1 \\ & & & & & \ddots \\ & & & & & & 1 \\ & & & 1 & & 0 \\ & & & & & & & 1 \\ & & & & & & & & \ddots \\ & & & & & & & & & 1 \end{bmatrix} \begin{matrix} \\ \\ \\ \leftarrow \text{row } i \\ \\ \\ \\ \leftarrow \text{row } j \\ \\ \\ \end{matrix} \quad ,$$

   where all entries not explicitly shown are zero. Its inverse is $\mathbf{H}_{ij}^{-1} = \mathbf{H}_{ij}$.

2. Denote by $\mathbf{H}_i(\alpha)$ the identity matrix $\mathbf{I}_m$ to which transformation $H_i(\alpha)$ has been applied. $\mathbf{H}_i(\alpha)$ has the form

$$\mathbf{H}_i(\alpha) = \begin{bmatrix} 1 \\ & \ddots \\ & & 1 \\ & & & \alpha \\ & & & & 1 \\ & & & & & \ddots \\ & & & & & & 1 \end{bmatrix} \begin{matrix} \\ \\ \\ \leftarrow \text{row } i, \\ \\ \\ \end{matrix}$$

   and its inverse is $\mathbf{H}_i^{-1}(\alpha) = \mathbf{H}_i(\alpha^{-1})$.

3. Denote by $\mathbf{H}_{ij}(\alpha)$ the identity matrix $\mathbf{I}_m$ to which transformation $H_{ij}(\alpha)$ has been applied. $\mathbf{H}_{ij}(\alpha)$ has the form

$$\mathbf{H}_{ij}(\alpha) = \begin{bmatrix} 1 \\ & \ddots \\ & & 1 & & \alpha \\ & & & \ddots \\ & & & & 1 \\ & & & & & \ddots \\ & & & & & & 1 \end{bmatrix} \begin{matrix} \\ \\ \leftarrow \text{row } i \\ \\ \leftarrow \text{row } j \\ \\ \end{matrix} \quad ,$$

with inverse $\mathbf{H}_{ij}^{-1}(\alpha) = \mathbf{H}_{ij}(-\alpha)$.

To effect a given elementary row transformation on $m \times n$ matrix $\mathbf{A}$, perform the transformation on $\mathbf{I}_m$, and then premultiply $\mathbf{A}$ by the result.

Elementary column matrices $\mathbf{K}_{ij}$, $\mathbf{K}_i(\alpha)$, and $\mathbf{K}_{ij}(\alpha)$ are defined for the column operations. Each elementary column matrix results from applying an elementary column transformation to $\mathbf{I}_n$, and has the form of the transpose of the corresponding elementary row matrix. To effect a given elementary column transformation on $m \times n$ matrix $\mathbf{A}$, perform the transformation on $\mathbf{I}_n$, and then postmultiply $\mathbf{A}$ by the result.

Each of the elementary row and column matrices defined above is nonsingular.

Example 24
Elementary row
matrices

The elementary matrices corresponding to the row operations in Example 22 are

$$\mathbf{H}_{21}(-2/3) = \begin{bmatrix} 1 & 0 & 0 \\ -2/3 & 1 & 0 \\ 0 & 0 & 1 \end{bmatrix}, \quad \mathbf{H}_{31}(-4/3) = \begin{bmatrix} 1 & 0 & 0 \\ 0 & 1 & 0 \\ -4/3 & 0 & 1 \end{bmatrix},$$

$$\mathbf{H}_{32}(-1/2) = \begin{bmatrix} 1 & 0 & 0 \\ 0 & 1 & 0 \\ 0 & -1/2 & 1 \end{bmatrix}.$$

Example 25
Elementary
column
matrices

The elementary matrices in the previous example were 3×3 because $\mathbf{A}$ had 3 rows, but because $\mathbf{A}$ has 4 columns, the elementary column matrices corresponding to Example 23 are 4×4, as follows:

$$\mathbf{K}_{21}(3) = \begin{bmatrix} 1 & 3 & 0 & 0 \\ 0 & 1 & 0 & 0 \\ 0 & 0 & 1 & 0 \\ 0 & 0 & 0 & 1 \end{bmatrix}, \quad \mathbf{K}_{31}(1/3) = \begin{bmatrix} 1 & 0 & 1/3 & 0 \\ 0 & 1 & 0 & 0 \\ 0 & 0 & 1 & 0 \\ 0 & 0 & 0 & 1 \end{bmatrix},$$

$$\mathbf{K}_{23} = \begin{bmatrix} 1 & 0 & 0 & 0 \\ 0 & 0 & 1 & 0 \\ 0 & 1 & 0 & 0 \\ 0 & 0 & 0 & 1 \end{bmatrix}, \quad \mathbf{K}_{42}(-7) = \begin{bmatrix} 1 & 0 & 0 & 0 \\ 0 & 1 & 0 & -7 \\ 0 & 0 & 1 & 0 \\ 0 & 0 & 0 & 1 \end{bmatrix}.$$

Example 26
Operation pairs

If $\mathbf{A}$ is square, then the dimension of its elementary row matrices is the same as the dimension of its elementary column matrices, and each elementary row matrix is a possible elementary column matrix. In certain applications, elementary operations are applied in row-operation, column-operation pairs. For example, if $\mathbf{A}$ is square, then the elementary row matrix $\mathbf{H}_{ij}(\alpha)$ has inverse $\mathbf{H}_{ij}(-\alpha)$, which is identical to $\mathbf{K}_{ji}(-\alpha)$, and an operation pair that is useful in some algorithms is $(H_{ij}(\alpha), K_{ji}(-\alpha))$, equivalent to the multiplication $\mathbf{A} \rightarrow \mathbf{H}_{ij}(\alpha)\mathbf{A}\mathbf{K}_{ji}(-\alpha)$.

3.3   Echelon forms

A matrix $\mathbf{U} \in F^{m \times n}$ of which the nonzero rows are linearly independent and above all zero rows, and that is row-equivalent to a given matrix $\mathbf{A}$, is called an *upper row compression* of $\mathbf{A}$, and is illustrated in Figure 5.2.

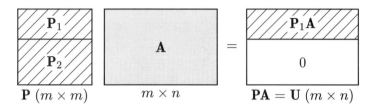

Fig. 5.2   Upper row compression. Since $\mathbf{A}$ and $\mathbf{U}$ are row-equivalent, $\mathbf{U} = \mathbf{P}\mathbf{A}$ where $\mathbf{P}$ is nonsingular, and $\mathbf{P}_1\mathbf{A}$ has $r = \text{rank}(\mathbf{A})$ rows.

One such upper compression is the *row echelon form,* or more descriptively, the *upper-right* row echelon form, which has the following properties, and is illustrated in Figure 5.3.

1. All zero rows are below the nonzero rows.

2. The leftmost nonzero entry of any row, called the *pivot* entry, is to the right of pivots in superior rows.

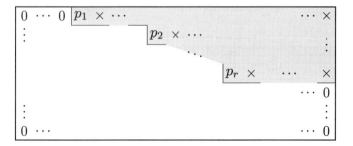

Fig. 5.3   The upper-right row echelon form. The pivot entries $p_1, \cdots p_r$ are nonzero, and the entries marked $\times$ are nonzero in general.

The *reduced* upper-right row echelon form has the additional properties, illustrated in Figure 5.4:

3. All pivots are equal to 1.

4. Any column containing a pivot is otherwise zero.

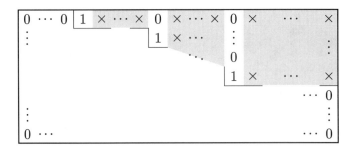

Fig. 5.4 Reduced upper-right row echelon form. The pivot columns are zero except for the pivots, which are 1.

The above form is called "upper-right" since the nonzero entries are above and to the right of the identically zero entries. Similarly, lower-right, and upper- and lower-left row echelon forms are defined, and upper-, lower-, left- and right-column echelon forms, or *column compressions,* are also defined.

The upper-right row-echelon form is illustrated in Example 22, and Example 23 results in a lower-left column echelon form.

Gaussian elimination Given an arbitrary matrix $\mathbf{A} \in F^{m \times n}$, an echelon form that is row-equivalent to $\mathbf{A}$ can be constructed by a sequence of elementary row transformations, the well-known Gaussian elimination algorithm, as follows:

1. Interchange rows to bring a leftmost nonzero entry in the matrix to the top row. This entry is now the pivot in the top row.

2. Add multiples of the (current) top row to lower rows to create zeros in the pivot column below the pivot.

3. Repeat the above operations on the submatrix currently below and to the right of the most recently created pivot, until all rows contain pivots or the current submatrix contains only zero entries.

Computer programs for solving linear equations often contain computations similar to the above scheme, with modifications for improved handling of data uncertainty and floating-point round-off error (see Problem 29). The method is also simply modified to produce the other row echelon forms, and when column operations are prescribed, to produce the column echelon forms.

Reduced form To produce the reduced row echelon form, append the following to step 2 above: Multiply the top row by the inverse of the pivot to make the pivot equal to 1, and add multiples of the current pivot row to superior rows, if any, to create zeros in the pivot column above the pivot.

Example 27
Reduced row
echelon form

Example 22 is modified to produce the reduced form shown:

$$\mathbf{A} = \begin{bmatrix} -3 & 9 & 1 & -5 \\ -2 & 6 & 0 & -8 \\ -4 & 12 & 1 & -9 \end{bmatrix} \xrightarrow[\longrightarrow]{\begin{array}{c} H_{21}(-2/3) \\ H_{31}(-4/3) \end{array}} \begin{bmatrix} -3 & 9 & 1 & -5 \\ 0 & 0 & -2/3 & -14/3 \\ 0 & 0 & -1/3 & -7/3 \end{bmatrix}$$

$$\xrightarrow[\longrightarrow]{H_1(-1/3)} \begin{bmatrix} 1 & -3 & -1/3 & 5/3 \\ 0 & 0 & -2/3 & -14/3 \\ 0 & 0 & -1/3 & -7/3 \end{bmatrix}$$

$$\xrightarrow[\longrightarrow]{H_{32}(-1/2)} \begin{bmatrix} 1 & -3 & -1/3 & 5/3 \\ 0 & 0 & -2/3 & -14/3 \\ 0 & 0 & 0 & 0 \end{bmatrix}$$

$$\xrightarrow[\longrightarrow]{H_2(-3/2)} \begin{bmatrix} 1 & -3 & -1/3 & 5/3 \\ 0 & 0 & 1 & 7 \\ 0 & 0 & 0 & 0 \end{bmatrix} \xrightarrow[\longrightarrow]{H_{12}(1/3)} \begin{bmatrix} 1 & -3 & 0 & 4 \\ 0 & 0 & 1 & 7 \\ 0 & 0 & 0 & 0 \end{bmatrix}.$$

3.4   Properties of echelon forms

Let $\mathbf{U}$ be an upper-right row echelon form that results on applying a sequence of elementary row operations to $\mathbf{A} \in F^{m \times n}$. Write the transformations as H_1, $H_2, \cdots H_t$, since the exact type of each operation is not known in advance. Then

(5.2) $\mathbf{U} = H_t(\cdots H_2(H_1\mathbf{A}) \cdots),$

and, since the application of an elementary row operation has the same effect as premultiplication by an elementary matrix,

(5.3) $\mathbf{U} = \mathbf{H}_t(\cdots \mathbf{H}_2(\mathbf{H}_1\mathbf{A}) \cdots) = (\mathbf{H}_t \cdots \mathbf{H}_2\mathbf{H}_1)\mathbf{A}$

for some set of elementary matrices $\mathbf{H}_1, \cdots \mathbf{H}_t$. Let $\mathbf{P} = \mathbf{H}_t \cdots \mathbf{H}_2\mathbf{H}_1$. By definition the $\mathbf{H}_i$ are nonsingular, and hence the product $\mathbf{P}$ is nonsingular. Then the following conclusions can be reached.

Multiplication by P   By construction using Gaussian elimination (or other algorithms), for any matrix $\mathbf{A} \in F^{m \times n}$, there exists a nonsingular matrix $\mathbf{P}$ such that $\mathbf{PA}$ is in upper-right row echelon form; that is,

(5.4) $\mathbf{U} = \mathbf{PA}.$

Rank   By inspection of the echelon form, the rank of $\mathbf{U}$ equals the number of nonzero rows. But $\mathbf{U}$ was obtained from $\mathbf{A}$ by a sequence of elementary transformations, which preserve rank. Therefore rank($\mathbf{A}$) also equals the number of nonzero rows of $\mathbf{U}$.

Computation of P

In order to find a matrix $\mathbf{P}$ for which (5.4) holds, create the matrix $[\mathbf{I}_m, \mathbf{A}]$, and perform row operations on it to change $\mathbf{A}$ to $\mathbf{U}$. Then, since the sequence of operations has the effect of premultiplying by $\mathbf{P}$,

(5.5) $\mathbf{P}[\mathbf{I}_m, \mathbf{A}] = [\mathbf{PI}_m, \mathbf{PA}] = [\mathbf{P}, \mathbf{U}],$

and at the conclusion of the algorithm the submatrix $\mathbf{I}_m$ has been transformed into $\mathbf{P}$.

Left multiplication

For any matrix $\mathbf{A} \in F^{m \times n}$ and any nonsingular $\mathbf{P}_1 \in F^{m \times m}$, the effect of premultiplication of $\mathbf{A}$ by $\mathbf{P}_1$ can be obtained by a sequence of elementary row transformations on $\mathbf{A}$. To prove this result, form the matrix $[\mathbf{I}_m, \mathbf{P}_1]$, and as above, perform row transformations on it to transform $\mathbf{P}_1$ to upper-right reduced row echelon form, which is $\mathbf{I}_m$ since $\mathbf{P}_1$ is square and nonsingular. The effect of the row transformations is identical to premultiplication by the product of elementary row matrices denoted $\mathbf{H}_t, \cdots \mathbf{H}_2, \mathbf{H}_1$. Thus,

(5.6) $\mathbf{H}_t \cdots \mathbf{H}_2\mathbf{H}_1 [\mathbf{I}_m, \mathbf{P}_1] = [\mathbf{P}_1^{-1}, \mathbf{I}_m].$

Therefore

(5.7) $\mathbf{H}_t \cdots \mathbf{H}_2\mathbf{H}_1 = \mathbf{P}_1^{-1},$

or

(5.8) $\mathbf{P}_1 = (\mathbf{H}_t \cdots \mathbf{H}_2\mathbf{H}_1)^{-1} = \mathbf{H}_1^{-1}\mathbf{H}_2^{-1} \cdots \mathbf{H}_t^{-1}.$

But from Section 3.2, each of the inverse matrices $\mathbf{H}_i^{-1}$ above is itself an elementary row matrix, corresponding to an elementary row operation, and premultiplication by $\mathbf{P}_1$ corresponds to application of elementary row transformations, beginning with H_t^{-1}, and ending with H_1^{-1}.

Example 28
Computation of $\mathbf{P}$

The matrix $\mathbf{P}$ that produces the result $\mathbf{U}$ for Example 22 is

$$\mathbf{P} = \mathbf{H}_{32}(-1/2)\mathbf{H}_{31}(-4/3)\mathbf{H}_{21}(-2/3) = \begin{bmatrix} 1 & 0 & 0 \\ -2/3 & 1 & 0 \\ -1 & -1/2 & 1 \end{bmatrix}.$$

Rather than calculating the elementary row matrices and multiplying to find $\mathbf{P}$, applying the operations to the array $[\mathbf{A}, \mathbf{I}]$ to produce $[\mathbf{U}, \mathbf{P}]$ results in

$$\begin{bmatrix} -3 & 9 & 1 & -5 & | & 1 & 0 & 0 \\ -2 & 6 & 0 & -8 & | & 0 & 1 & 0 \\ -4 & 12 & 1 & -9 & | & 0 & 0 & 1 \end{bmatrix} \longrightarrow \begin{bmatrix} -3 & 9 & 1 & -5 & | & 1 & 0 & 0 \\ 0 & 0 & -2/3 & -14/3 & | & -2/3 & 1 & 0 \\ 0 & 0 & 0 & 0 & | & -1 & -1/2 & 1 \end{bmatrix}.$$

Example 29
Rank from
echelon form

The echelon form $\mathbf{U}$ for $\mathbf{A}$ of Example 22 has two nonzero rows; consequently its largest submatrix with nonzero determinant has size 2, so that $\mathbf{U}$ has rank 2, and because rank is preserved under elementary row operations, $\mathrm{rank}(\mathbf{A}) = 2$.

Example 30
Nonuniqueness
of $\mathbf{P}$

In general, the matrix $\mathbf{P}$ which transforms $\mathbf{A}$ to row echelon form is not unique, and consequently the corresponding row transformations are not unique. Let the upper-right echelon form be

$$(5.9) \qquad \begin{bmatrix} \mathbf{U} \\ \mathbf{0} \end{bmatrix} = \mathbf{PA} = \begin{bmatrix} \mathbf{P}_1 \\ \mathbf{P}_2 \end{bmatrix} \mathbf{A},$$

where $\mathbf{U}$ has r nonzero rows, and where $\mathbf{P}$ has been partitioned as shown, with r rows in $\mathbf{P}_1$. Further addition of multiples of the bottom $m-r$ rows to the upper rows leaves $\mathbf{PA}$ unchanged, but corresponds to multiplication by a nonsingular matrix of the form $\begin{bmatrix} \mathbf{I}_r & \mathbf{L} \\ \mathbf{0} & \mathbf{I}_{m-r} \end{bmatrix}$, for some $\mathbf{L}$, so the complete set of row operations resulting in $\mathbf{U}$ corresponds to premultiplication of $\mathbf{A}$ by $\begin{bmatrix} \mathbf{P}_1 + \mathbf{LP}_2 \\ \mathbf{P}_2 \end{bmatrix}$, which is not unique since $\mathbf{L}$ is arbitrary.

3.5   The normal form

In the previous section it was shown how to create row echelon and reduced row echelon forms by row transformations, or equivalently by premultiplication with a nonsingular matrix. Similarly, column forms can be produced by column transformations, or equivalently by postmultiplication with a nonsingular matrix. The following result uses both row and column operations, and is central to both applied and theoretical linear algebra.

For any matrix $\mathbf{A} \in F^{m \times n}$ of rank $r \leq \min(m, n)$ there exist nonsingular matrices $\mathbf{P} \in F^{m \times m}$, $\mathbf{Q} \in F^{n \times n}$ such that

$$(5.10) \qquad \mathbf{PAQ} = \mathbf{N} = \begin{bmatrix} \mathbf{I}_r & \mathbf{0} \\ \mathbf{0} & \mathbf{0} \end{bmatrix},$$

where by convention, not all of the blocks of zeros in $\mathbf{N}$ need exist, depending on r; that is,

$$\mathbf{N} = \mathbf{I}_n, \quad \text{or } \mathbf{N} = \begin{bmatrix} \mathbf{I}_n \\ \mathbf{0} \end{bmatrix}, \quad \text{or } \mathbf{N} = [\mathbf{I}_m, \mathbf{0}], \quad \text{or } \mathbf{N} = \begin{bmatrix} \mathbf{I}_r & \mathbf{0} \\ \mathbf{0} & \mathbf{0} \end{bmatrix}.$$

The matrix $\mathbf{N}$ above is called the *normal form* of $\mathbf{A}$ under equivalence operations. The matrices $\mathbf{P}$, $\mathbf{Q}$ required to produce $\mathbf{N}$ are not unique.

Constructing N

One possibility for constructing $\mathbf{N}$ is first to do row transformations on $\mathbf{A}$ to produce the upper-right echelon form $\mathbf{U}$ as previously, and then to perform column transformations to obtain the lower-left reduced column echelon form of $\mathbf{U}$, which is $\mathbf{N}$. Expressed symbolically,

$$\mathbf{A} \overset{\text{row}}{\sim} \mathbf{U} \overset{\text{col}}{\sim} \mathbf{N},$$

so that, to construct $\mathbf{P}$ and $\mathbf{Q}$ as well as $\mathbf{N}$, the following array is formed:

$$\mathbf{I}_m \quad \mathbf{A}$$
$$\mathbf{I}_n$$

and the row transformations are performed on the upper blocks, and the column transformations are performed on the rightmost blocks, giving, in order,

$$\begin{bmatrix} \mathbf{I}_m & \mathbf{A} \\ & \mathbf{I}_n \end{bmatrix} \overset{\text{row}}{\sim} \begin{bmatrix} \mathbf{PI}_m & \mathbf{PA} \\ & \mathbf{I}_n \end{bmatrix} = \begin{bmatrix} \mathbf{P} & \mathbf{U} \\ & \mathbf{I}_n \end{bmatrix} \overset{\text{col}}{\sim} \begin{bmatrix} \mathbf{P} & \mathbf{UQ} \\ & \mathbf{I}_n\mathbf{Q} \end{bmatrix} = \begin{bmatrix} \mathbf{P} & \mathbf{N} \\ & \mathbf{Q} \end{bmatrix}.$$

This result will be applied in discussing vector spaces and linear operators over vector spaces, in the solution of general linear equations, and in Section 6, which contains an important result in system identification.

Example 31
Normal form

Constructing the required array for the matrix $\mathbf{A}$ of Example 22, performing row operations to produce the upper-right row echelon form, and then column operations to produce the lower-left reduced column echelon form gives the result

$$\begin{bmatrix} \mathbf{I}_3 & \mathbf{A} \\ & \mathbf{I}_4 \end{bmatrix} \overset{\text{row}}{\sim} \left[\begin{array}{ccc|cccc} 1 & 0 & 0 & -3 & 9 & 1 & -5 \\ -2/3 & 1 & 0 & 0 & 0 & -2/3 & -14/3 \\ -1 & -1/2 & 1 & 0 & 0 & 0 & 0 \\ \hline & & & 1 & 0 & 0 & 0 \\ & & & 0 & 1 & 0 & 0 \\ & & & 0 & 0 & 1 & 0 \\ & & & 0 & 0 & 0 & 1 \end{array} \right]$$

$$\overset{\text{col}}{\sim} \left[\begin{array}{ccc|cccc} 1 & 0 & 0 & 1 & 0 & 0 & 0 \\ -2/3 & 1 & 0 & 0 & 1 & 0 & 0 \\ -1 & -1/2 & 1 & 0 & 0 & 0 & 0 \\ \hline & & & -1/3 & -1/2 & 3 & -4 \\ & & & 0 & 0 & 1 & 0 \\ & & & 0 & -3/2 & 0 & -7 \\ & & & 0 & 0 & 0 & 1 \end{array} \right] = \begin{bmatrix} \mathbf{P} & \mathbf{N} \\ & \mathbf{Q} \end{bmatrix},$$

where the sequence of row operations is

$$H_{21}(-2/3), H_{31}(-4/3), H_{32}(-1/2),$$

and the column-operation sequence is

$$K_{54}(3), K_{64}(1/3), K_{74}(-5/3), K_4(-1/3), K_{56}, K_{75}(-7), K_5(-3/2).$$

3.6   **The Singular-Value Decomposition (SVD)**

A very useful modification of the normal form is known as the *singular-value decomposition,* which has good numerical robustness properties and is widely used for least-squares problems. Here it is assumed that the working field is $\mathbb{R}$, although extension to $\mathbb{C}$ is possible.

From Section 3.5, the matrices $\mathbf{P}$, $\mathbf{Q}$ that reduce a matrix $\mathbf{A} \in \mathbb{R}^{m \times n}$ to normal form are not unique. By generalizing (5.10), it can be shown that there exist matrices $\mathbf{U}$, $\mathbf{V}$ that are nonsingular and orthogonal; that is, $\mathbf{U}^{-1} = \mathbf{U}^T$, $\mathbf{V}^{-1} = \mathbf{V}^T$, such that

$$(5.11) \quad \mathbf{U}^T \mathbf{A} \mathbf{V} = \Sigma = \begin{bmatrix} \Sigma_r & 0 \\ 0 & 0 \end{bmatrix},$$

where Σ is $m \times n$ and zero except for the main diagonal, which has nonnegative entries $\sigma_1, \sigma_2, \cdots$, called the *singular values* of $\mathbf{A}$, ordered largest to smallest from the upper left. With exact arithmetic, Σ has the form shown above, where $r = \mathrm{rank}(\mathbf{A})$, and Σ_r is $r \times r$, diagonal, and nonsingular. With floating-point computation, all entries of the diagonal of Σ are typically nonzero, but those at the lower right may be small, of a size attributable to arithmetic rounding error. If we partition $\mathbf{U} = [\mathbf{U}_1, \mathbf{U}_2]$ and $\mathbf{V} = [\mathbf{V}_1, \mathbf{V}_2]$ in (5.11), where $\mathbf{U}_1$ and $\mathbf{V}_1$ have r columns, then $\mathbf{A}$ can be factored as

$$(5.12) \quad \mathbf{A} = \mathbf{U} \Sigma \mathbf{V}^T = [\mathbf{U}_1, \mathbf{U}_2] \begin{bmatrix} \Sigma_r & 0 \\ 0 & 0 \end{bmatrix} \begin{bmatrix} \mathbf{V}_1^T \\ \mathbf{V}_2^T \end{bmatrix} = \mathbf{U}_1 \Sigma_r \mathbf{V}_1^T.$$

To relate the SVD to the normal form, matrices $\mathbf{P}$, $\mathbf{Q}$ that satisfy (5.10) are, for example, $\mathbf{P} = \mathbf{U}^T$, and $\mathbf{Q} = [\mathbf{V}_1 \Sigma_r^{-1}, \mathbf{V}_2]$, which is shown as follows:

$$(5.13) \quad \mathbf{P} \mathbf{A} \mathbf{Q} = \mathbf{U}^T (\mathbf{U} \Sigma \mathbf{V}^T) \mathbf{V} \begin{bmatrix} \Sigma_r^{-1} & 0 \\ 0 & \mathbf{I} \end{bmatrix} = \begin{bmatrix} \Sigma_r & 0 \\ 0 & 0 \end{bmatrix} \begin{bmatrix} \Sigma_r^{-1} & 0 \\ 0 & \mathbf{I} \end{bmatrix} = \mathbf{N}.$$

Another useful property of the singular values is that the sum of their squares equals the sum of squares of the entries of $\mathbf{A}$ in (5.11). To show this, we will show first that the sum of squares of the entries of $\mathbf{A}$ is invariant under multiplication by an orthogonal matrix. Let $\mathbf{A} \in \mathbb{R}^{m \times n}$ be written by columns as $\mathbf{A} = [\mathbf{A}_1, \cdots \mathbf{A}_n]$, and let $\mathbf{U} \in \mathbb{R}^{m \times m}$ be orthogonal. The sum of squares of the entries of a matrix is the sum of squares of the entries of its columns, so

$$(5.14) \quad \|\mathbf{U}\mathbf{A}\|_E^2 = \| [\mathbf{U}\mathbf{A}_1, \mathbf{U}\mathbf{A}_2, \cdots \mathbf{U}\mathbf{A}_n] \|_E^2 = \sum_{i=1}^{n} \mathbf{A}_i^T \mathbf{U}^T \mathbf{U} \mathbf{A}_i = \sum_{i=1}^{n} \mathbf{A}_i^T \mathbf{A}_i$$

$$= \| [\mathbf{A}_1, \mathbf{A}_2, \cdots \mathbf{A}_n] \|_E^2 = \|\mathbf{A}\|_E^2.$$

A similar argument applies for postmultiplication by an orthogonal matrix. Therefore, using (5.11), the sum of squares of the entries of $\mathbf{A}$ equals the sum of squares of the entries of Σ, which is the sum $\sum_i \sigma_i^2$ of the squared singular values.

Numerical illustrations and some basic applications of the SVD will be given in examples.

Example 32
Numerical
example

A numerical calculation of the SVD for $\mathbf{A}$ of Example 22 gives, to four figures,

$$\mathbf{U} = \begin{bmatrix} -0.4998 & 0.5529 & -0.6667 \\ -0.4636 & -0.8210 & -0.3333 \\ -0.7316 & 0.1424 & 0.6667 \end{bmatrix}, \quad \Sigma = \begin{bmatrix} 21.25 & 0 & 0 & 0 \\ 0 & 3.21 & 0 & 0 \\ 0 & 0 & 5.61e{-}16 & 0 \end{bmatrix},$$

$$\mathbf{V} = \begin{bmatrix} 0.2519 & -0.1829 & 0.8735 & -0.3742 \\ -0.7556 & 0.5488 & 0.3541 & 0.0499 \\ -0.0579 & 0.2169 & -0.3306 & -0.9167 \\ 0.6019 & 0.7864 & 0.0472 & 0.1310 \end{bmatrix}.$$

The rank of $\mathbf{A}$ is known to be 2, implying that the magnitude of entry $3, 3$ is attribuable to floating-point error, and an exact computation would have produced a zero in this location. In the absence of such prior knowledge, treating this value as zero without further information would be an *assumption,* but the value shown is representative of floating-point computation on many computers.

Example 33
Least squares
using SVD

Fitting a straight line to measured data is often done by minimizing the squared errors of the fitting equation. Suppose the line is described by $y = ax + b$, and the values of a and b that give the best least-squares fit to a set of m data points (x_i, y_i) is to be found, defining the error of the i-th data point as $z_i = ax_i + b - y_i$. Then writing the equations together as

$$(5.15) \qquad \begin{bmatrix} x_1 & 1 \\ x_2 & 1 \\ \vdots & \vdots \\ x_m & 1 \end{bmatrix} \begin{bmatrix} a \\ b \end{bmatrix} - \begin{bmatrix} y_1 \\ y_2 \\ \vdots \\ y_m \end{bmatrix} = \begin{bmatrix} z_1 \\ z_2 \\ \vdots \\ z_m \end{bmatrix}$$

gives one example in which the size of the error vector $\mathbf{Z}$ in an equation of the form

$$(5.16) \qquad \mathbf{AX} - \mathbf{B} = \mathbf{Z}$$

is to be minimized with respect to $\mathbf{X}$. A small example is shown in Figure 5.5. In more general applications of Equation (5.16), typically the matrices $\mathbf{A} \in \mathbb{R}^{m \times n}$, $\mathbf{B} \in \mathbb{R}^m$ contain measured quantities, such that $m > n$, with $\mathbf{A}$ of full column rank n. With these assumptions, the SVD of $\mathbf{A}$ is

$$(5.17) \qquad \mathbf{A} = \mathbf{U}\Sigma\mathbf{V}^T = [\,\mathbf{U}_1, \mathbf{U}_2\,] \begin{bmatrix} \Sigma_n \\ 0 \end{bmatrix} \mathbf{V}^T = \mathbf{U}_1 \Sigma_n \mathbf{V}^T,$$

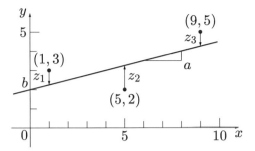

Fig. 5.5   Fitting a straight line $y = ax + b$ to three points, showing the components of error vector $\mathbf{Z} = [z_1,\ z_2,\ z_3]^T$.

where Σ_n is diagonal and nonsingular, and $\mathbf{U}$ has been partitioned as shown, with n columns in $\mathbf{U}_1$.

If the size of $\mathbf{Z}$ is defined to be its Euclidean norm $\|\mathbf{Z}\|_E$, then minimizing the error corresponds to minimizing the sum of squares $\|\mathbf{Z}\|_E^2 = \mathbf{Z}^T\mathbf{Z}$.

Calculate the SVD of $\mathbf{A}$, let $\mathbf{X} = \mathbf{VY}$, and multiply (5.16) by $\mathbf{U}^T$ to give

$$(5.18) \quad \mathbf{U}^T(\mathbf{U}\Sigma\mathbf{V}^T)\mathbf{VY} - \mathbf{U}^T\mathbf{B} = \Sigma\mathbf{Y} - \mathbf{U}^T\mathbf{B} = \mathbf{U}^T\mathbf{Z}.$$

Then the sum of squares is

$$(5.19) \quad \begin{aligned} \mathbf{Z}^T\mathbf{Z} &= \mathbf{Z}^T\mathbf{U}\mathbf{U}^T\mathbf{Z} = (\mathbf{U}^T\mathbf{Z})^T(\mathbf{U}^T\mathbf{Z}) \\ &= (\Sigma\mathbf{Y} - \mathbf{U}^T\mathbf{B})^T(\Sigma\mathbf{Y} - \mathbf{U}^T\mathbf{B}) = \|\Sigma\mathbf{Y} - \mathbf{U}^T\mathbf{B}\|_E^2 \\ &= \left\| \begin{bmatrix} \Sigma_n\mathbf{Y} \\ 0 \end{bmatrix} - \begin{bmatrix} \mathbf{U}_1^T\mathbf{B} \\ \mathbf{U}_2^T\mathbf{B} \end{bmatrix} \right\|_E^2 = \|\Sigma_n\mathbf{Y} - \mathbf{U}_1^T\mathbf{B}\|_E^2 + \|\mathbf{U}_2^T\mathbf{B}\|_E^2, \end{aligned}$$

which is minimized with respect to $\mathbf{Y}$ when $\mathbf{Y} = \Sigma_n^{-1}\mathbf{U}_1^T\mathbf{B}$, or, solving for $\mathbf{X}$,

$$(5.20) \quad \mathbf{X} = \mathbf{VY} = \mathbf{V}\Sigma_n^{-1}\mathbf{U}_1^T\mathbf{B}.$$

To use the artificially small example of Figure 5.5, the matrices $\mathbf{A}$, $\mathbf{B}$ are shown below, together with $\mathbf{U}$, Σ, $\mathbf{V}$, calculated by computer:

$$\mathbf{A} = \begin{bmatrix} 1 & 1 \\ 5 & 1 \\ 9 & 1 \end{bmatrix}, \quad \mathbf{B} = \begin{bmatrix} 3 \\ 2 \\ 5 \end{bmatrix},$$

$$\mathbf{U} = [\,\mathbf{U}_1,\ \mathbf{U}_2\,] = \begin{bmatrix} 0.1082 & 0.9064 & 0.4082 \\ 0.4873 & 0.3096 & -0.8165 \\ 0.8665 & -0.2873 & 0.4082 \end{bmatrix},$$

$$\Sigma = \begin{bmatrix} \Sigma_n \\ 0 \end{bmatrix} = \begin{bmatrix} 10.45 & 0 \\ 0 & 0.9380 \\ \hline 0 & 0 \end{bmatrix}, \quad \mathbf{V} = \begin{bmatrix} 0.9902 & -0.1400 \\ 0.1400 & 0.9902 \end{bmatrix},$$

from which the best fit is

$$\begin{bmatrix} a \\ b \end{bmatrix} = \mathbf{X} = \mathbf{V}\Sigma_n^{-1}\mathbf{U}_1^T\mathbf{B} = \begin{bmatrix} 0.2500 \\ 2.0833 \end{bmatrix}.$$

Example 34
Estimating the
rank of an
inexact matrix

If the entries of a given matrix $\mathbf{A}$ are inexact, as they are when derived from measured data, then the rank of $\mathbf{A}$ may differ from the rank of the "true" matrix which would be obtained if the values were exact. Since the true matrix is unknown, its rank cannot be calculated. In some circumstances, the true rank may be postulated as the smallest n for which there is a matrix $\mathbf{K}$ of rank n "near" enough to $\mathbf{A}$ so that differences can be accounted for by the data imprecision.

Using the Euclidean norm, the distance from $\mathbf{A}$ to $\mathbf{K}$ is $\|\mathbf{A} - \mathbf{K}\|_E = (\sum_{ij}(a_{ij} - k_{ij})^2)^{1/2}$. Finding the nearest $\mathbf{K}$ to $\mathbf{A}$ by minimizing this norm corresponds to minimizing the sum of squares $\|\mathbf{A} - \mathbf{K}\|_E^2$; therefore we wish to find $\mathbf{K}$ of prespecified rank which minimizes this sum of squares. The nearest rank-r matrix $\mathbf{K}$ to $\mathbf{A}$ is found by computing the SVD of $\mathbf{A}$,

$$(5.21) \quad \mathbf{A} = \mathbf{U}\Sigma\mathbf{V}^T = [\,\mathbf{U}_1, \mathbf{U}_2\,] \begin{bmatrix} \Sigma_r & 0 \\ 0 & \Sigma_2 \end{bmatrix} \begin{bmatrix} \mathbf{V}_1^T \\ \mathbf{V}_2^T \end{bmatrix},$$

where Σ has been partitioned so that Σ_r contains the r largest singular values, $\mathbf{U}$, $\mathbf{V}$ have been partitioned conformably, and Σ_2 can contain zeros on its diagonal, as when $\mathbf{A}$ has less than full rank or is nonsquare. Then the required $\mathbf{K}$ is

$$(5.22) \quad \mathbf{K} = \mathbf{U}_1\Sigma_r\mathbf{V}_1^T,$$

and the squared distance from $\mathbf{K}$ to $\mathbf{A}$ is $\|\Sigma_2\|_E^2 = \sum_{i>r}\sigma_i^2$, which is immediately available from (5.21), so the distances from $\mathbf{A}$ to the closest matrices of different ranks are easily calculated by choosing different partitions of Σ into Σ_r and Σ_2.

The above formula can be verified as follows. Assuming that $\mathbf{K}$ of rank r minimizes $\|\mathbf{A} - \mathbf{K}\|_E^2$, let the SVD of $\mathbf{K}$ (rather than of $\mathbf{A}$ as above) be

$$(5.23) \quad \mathbf{K} = \mathbf{U}\Sigma\mathbf{V}^T = [\,\mathbf{U}_1, \mathbf{U}_2\,] \begin{bmatrix} \Sigma_r & 0 \\ 0 & 0 \end{bmatrix} \begin{bmatrix} \mathbf{V}_1^T \\ \mathbf{V}_2^T \end{bmatrix} = \mathbf{U}_1\Sigma_r\mathbf{V}_1^T,$$

where Σ_r is $r \times r$, and $\mathbf{U}$, $\mathbf{V}$ have been partitioned conformably. Then the sum of squares is

$$(5.24) \quad \|\mathbf{A} - \mathbf{K}\|_E^2 = \|\mathbf{U}^T(\mathbf{A} - \mathbf{K})\mathbf{V}\|_E^2 = \left\| \begin{bmatrix} \mathbf{U}_1^T\mathbf{A}\mathbf{V}_1 & \mathbf{U}_1^T\mathbf{A}\mathbf{V}_2 \\ \mathbf{U}_2^T\mathbf{A}\mathbf{V}_1 & \mathbf{U}_2^T\mathbf{A}\mathbf{V}_2 \end{bmatrix} - \begin{bmatrix} \Sigma_r & 0 \\ 0 & 0 \end{bmatrix} \right\|_E^2.$$

In (5.24) the submatrix $\mathbf{U}_1^T \mathbf{A} \mathbf{V}_2$ is zero, for if it were not, then

(5.25) $\mathbf{K'} = \mathbf{U} \begin{bmatrix} \Sigma_r & \mathbf{U}_1^T \mathbf{A} \mathbf{V}_2 \\ 0 & 0 \end{bmatrix} \mathbf{V}^T$

would be a rank-r matrix for which

(5.26) $\|\mathbf{A} - \mathbf{K'}\|_E^2 = \|\mathbf{U}^T (\mathbf{A} - \mathbf{K'}) \mathbf{V}\|_E^2$

$$= \left\| \begin{bmatrix} \mathbf{U}_1^T \mathbf{A} \mathbf{V}_1 & \mathbf{U}_1^T \mathbf{A} \mathbf{V}_2 \\ \mathbf{U}_2^T \mathbf{A} \mathbf{V}_1 & \mathbf{U}_2^T \mathbf{A} \mathbf{V}_2 \end{bmatrix} - \begin{bmatrix} \Sigma_r & \mathbf{U}_1^T \mathbf{A} \mathbf{V}_2 \\ 0 & 0 \end{bmatrix} \right\|_E^2 < \|\mathbf{A} - \mathbf{K}\|_E^2,$$

so that $\mathbf{K}$ would not minimize $\|\mathbf{A} - \mathbf{K}\|_E^2$, a contradiction. Similar arguments imply that $\mathbf{U}_2^T \mathbf{A} \mathbf{V}_1 = 0$ and that $\mathbf{U}_1^T \mathbf{A} \mathbf{V}_1 = \Sigma_r$. Therefore, the minimal sum of squares is $\|\mathbf{A} - \mathbf{K}\|_E^2 = \|\mathbf{U}_2^T \mathbf{A} \mathbf{V}_2\|_E^2$. Now because $\mathbf{U}_1^T \mathbf{A} \mathbf{V}_1 = \Sigma_r$ in (5.24) is diagonal, its r diagonal entries are singular values of $\mathbf{A}$, and the sum of squares of the entries of $\mathbf{U}_2^T \mathbf{A} \mathbf{V}_2$ must equal the sum of squares of the remaining singular values of $\mathbf{A}$, which is minimized when the largest r singular values are in Σ_r.

Example 35
Distance to matrices of given rank

A numerical illustration of the previous example will be given. The matrix $\mathbf{A}$ of Example 22 has rank 2, but suppose that due to measurement error, the matrix $\mathbf{A}_m$ of measured values of the entries of $\mathbf{A}$ is as shown, with resulting Σ of the SVD,

$$\mathbf{A}_m = \begin{bmatrix} -2.98 & 9.03 & 1.00 & -4.96 \\ -1.98 & 6.05 & 0 & -8.01 \\ -4.01 & 11.97 & 0.99 & -9.03 \end{bmatrix}, \quad \Sigma = \begin{bmatrix} 21.27 & 0 & 0 & 0 \\ 0 & 3.213 & 0 & 0 \\ 0 & 0 & 0.044 & 0 \end{bmatrix}.$$

Then since rank $\Sigma = 3$, the measured $\mathbf{A}_m$ has rank 3. The distance from $\mathbf{A}_m$ to the least-squares nearest matrix $\mathbf{K}$ of rank 2 is $\sigma_3 = 0.044$, and to the nearest matrix of rank 1 the distance is $\sqrt{\sigma_2^2 + \sigma_3^2} = \sqrt{3.213^2 + 0.044^2} = 3.213$. Consequently, the hypothesis that the measured values represent a matrix of true rank of 2 is reasonable provided measurement error is known to be of the order of σ_3, which is small compared to the magnitude of the nonzero measured values. The hypothesis that the true rank is 1 would require measurement errors of approximately 3, which is of the order of each of the measurements themselves. These conclusions crucially depend on the distance measure chosen, which weights all matrix entries equally, and which may be invalid if there is a known structure for the matrix or the measurement errors.

For this example, the rank-2 matrix $\mathbf{K}$ nearest to $\mathbf{A}_m$ is calculated from Equation (5.23) as

$$\mathbf{K} = \mathbf{U}_1 \Sigma_2 \mathbf{V}_1^T = \begin{bmatrix} -3.007 & 9.020 & 1.005 & -4.961 \\ -1.994 & 6.045 & 0.003 & -8.010 \\ -3.983 & 11.98 & 0.985 & -9.029 \end{bmatrix},$$

where $\mathbf{U}_1$ and $\mathbf{V}_1$ contain the first two columns of the computed $\mathbf{U}$ and $\mathbf{V}$ respectively, and Σ_2 is the top-left 2×2 block of Σ. Then this matrix is the least-squares nearest matrix of rank 2 to $\mathbf{A}_m$, but it does not exactly equal the rank-2 matrix $\mathbf{A}$ of Example 22.

4   Matrix inverses

The matrix normal form allows the computation of matrix inverses.

4.1   Left inverse

Let $\mathbf{A}$ be $m \times n$, with rank $\mathbf{A} = r$. Let $\mathbf{P} = \begin{bmatrix} \mathbf{P}_1 \\ \mathbf{P}_2 \end{bmatrix}$ be such that $\mathbf{PA}$ is in reduced upper-right row echelon form, with $\mathbf{P}_1$ containing r rows. If $r = n$, then the echelon form is

$$(5.27) \quad \begin{bmatrix} \mathbf{P}_1 \\ \mathbf{P}_2 \end{bmatrix} \mathbf{A} = \begin{bmatrix} \mathbf{P}_1 \mathbf{A} \\ \mathbf{P}_2 \mathbf{A} \end{bmatrix} = \begin{bmatrix} \mathbf{I}_n \\ 0 \end{bmatrix},$$

showing that $\mathbf{P}_1$ is a left inverse $\mathbf{A}^L$ of $\mathbf{A}$. From Example 30, any matrix $\mathbf{P}_1 + \mathbf{LP}_2$ is also a left inverse, where $\mathbf{L}$ is arbitrary. Conversely, if rank $\mathbf{A} < n$, then rank $\mathbf{P}_1\mathbf{A} \leq r < n$ for all $\mathbf{P}_1$, so there can be no matrix $\mathbf{P}_1$ for which $\mathbf{P}_1\mathbf{A} = \mathbf{I}_n$.

Example 36
Left inverse
calculation

Transforming the matrix $\mathbf{A}$ shown to reduced row echelon form, using methods of the previous sections gives, nonuniquely,

$$[\mathbf{I}, \mathbf{A}] = \begin{bmatrix} 1 & 0 & 0 & 2 & -1 \\ 0 & 1 & 0 & -5 & 3 \\ 0 & 0 & 1 & 1 & 0 \end{bmatrix} \overset{\text{row}}{\sim} \begin{bmatrix} 3 & 1 & 0 & 1 & 0 \\ 5 & 2 & 0 & 0 & 1 \\ -3 & -1 & 1 & 0 & 0 \end{bmatrix} = \begin{bmatrix} \mathbf{P}_1 & \mathbf{I}_2 \\ \mathbf{P}_2 & 0 \end{bmatrix},$$

so that left inverses can be constructed as

$$\mathbf{A}^L = \begin{bmatrix} 3 & 1 & 0 \\ 5 & 2 & 0 \end{bmatrix} + \mathbf{L}[-3 \quad -1 \quad 1],$$

for arbitrary matrix $\mathbf{L}$ of dimension 2×1.

4.2   Right inverse

Similarly, $\mathbf{A}$ has a right inverse if and only if rank $\mathbf{A} = m$, in which case, let $\mathbf{Q} = [\mathbf{Q}_1, \mathbf{Q}_2]$ be such that $\mathbf{AQ}$ is in lower-left reduced column echelon form, with $\mathbf{Q}_1$ containing r columns. Then, if rank $\mathbf{A} = m$,

$$(5.28) \quad \mathbf{A}[\mathbf{Q}_1, \mathbf{Q}_2] = [\mathbf{I}_m, 0]$$

so that $\mathbf{Q}_1$ (or in general, $\mathbf{Q}_1 + \mathbf{Q}_2\mathbf{K}$ for arbitrary $\mathbf{K}$) is a right inverse $\mathbf{A}^R$ of $\mathbf{A}$.

Example 37
Right-inverse
calculation

By column operations, the matrix $\mathbf{A}$ shown can be put into reduced column echelon form,

$$
\begin{bmatrix} \mathbf{A} \\ \mathbf{I} \end{bmatrix} = \begin{bmatrix} 2 & 1 \\ 1 & 0 \\ 0 & 1 \end{bmatrix} \overset{\text{col}}{\sim} \begin{bmatrix} 1 & 0 \\ 0 & 1 \\ 1 & -2 \end{bmatrix} = \begin{bmatrix} \mathbf{I}_1 & 0 \\ \mathbf{Q}_1 & \mathbf{Q}_2 \end{bmatrix},
$$

so that right inverses can be computed as

$$
\mathbf{A} = \begin{bmatrix} 0 \\ 1 \end{bmatrix} + \begin{bmatrix} 1 \\ -2 \end{bmatrix} k,
$$

for arbitrary k.

4.3   Inverse

Let $\mathbf{A}$ be $n \times n$ and rank $\mathbf{A} = n$. Then the normal form of $\mathbf{A}$ is $\mathbf{I}_n$, and inverting both sides of (5.10) gives

(5.29) $\mathbf{Q}^{-1}\mathbf{A}^{-1}\mathbf{P}^{-1} = \mathbf{I}_n,$

so that, premultiplying by $\mathbf{Q}$ and postmultiplying by $\mathbf{P}$,

(5.30) $\mathbf{A}^{-1} = \mathbf{QP},$

which gives the inverse of $\mathbf{A}$ as a formula that is useful in computation, for several choices of $\mathbf{P}$ and $\mathbf{Q}$. But since the reduced row echelon form of $\mathbf{A}$ is $\mathbf{I}_n$, in principle the matrix $\mathbf{P}$ can be chosen to produce this form, with $\mathbf{Q} = \mathbf{I}_n$ in (5.30); that is, the inverse can be computed by using row operations exclusively. Similarly, the reduced column echelon form is also $\mathbf{I}_n$, so choosing $\mathbf{P} = \mathbf{I}_n$, the inverse can be computed by using column operations exclusively.

Example 38
LU-factorization

Matrix inverses, like determinants, are of great conceptual value, but for reasons related to floating-point numerical analysis, computing them is often considered to be bad practice when solving for $\mathbf{X}$ in equations of the form $\mathbf{AX} = \mathbf{B}$. A common technique is to perform row and column operations on $\mathbf{A}$, confining the column operations exclusively to interchanges, such that

$$\mathbf{L}^{-1}\mathbf{AK} = \mathbf{U},$$

where $\mathbf{U}$ is in upper-right row echelon form, where $\mathbf{L}$ (and $\mathbf{L}^{-1}$) is in lower-left row echelon form with 1's on the diagonal, and where $\mathbf{K}$ is a product of column permutation matrices $\mathbf{K}_{ij}$. Then, symbolically,

$$\mathbf{A} = \mathbf{LUK}^{-1},$$

which is known as the *LU*-factorization of **A** with complete pivoting. Then the equation

$$\mathbf{AX} = \mathbf{LUK}^{-1}\mathbf{X} = \mathbf{B}$$

is solved in three steps:

1. First $\mathbf{LY} = \mathbf{B}$ is solved for $\mathbf{Y}$, a simple and rapid computation because $\mathbf{L}$ is triangular.

2. Next $\mathbf{UZ} = \mathbf{Y}$ is solved for $\mathbf{Z}$, taking advantage of triangular $\mathbf{U}$.

3. Then $\mathbf{X} = \mathbf{KZ}$, which is a permutation of the entries of $\mathbf{Z}$.

Adjoint formula   A formula for the inverse that is of considerable theoretical use can be derived from the properties of determinants. Given $\mathbf{A} \in F^{n \times n}$, let the *adjoint matrix* be the $n \times n$ matrix adj($\mathbf{A}$) for which each entry ij is the *cofactor* $(-1)^{i+j} \det(\mathbf{A}_{ji})$, where $\mathbf{A}_{ij}$ denotes $\mathbf{A}$ with row i and column j removed.

First, although the details are beyond the scope of this chapter, using the properties of determinants it is possible to show that for any $\mathbf{A} \in F^{n \times n}$,

(5.31) $\det(\mathbf{A})\,\mathbf{I} = \mathbf{A}\,\text{adj}(\mathbf{A})$,

so that, if $\det(\mathbf{A}) \neq 0$, dividing by $\det(\mathbf{A})$ and premultiplying by $\mathbf{A}^{-1}$ yields

(5.32) $\mathbf{A}^{-1} = \dfrac{1}{\det(\mathbf{A})}\,\text{adj}(\mathbf{A})$.

Example 39
Adjoint formula

Applying the adjoint formula to a simple example,

$$\begin{bmatrix} 0 & 1 \\ -2 & -3 \end{bmatrix}^{-1} = \frac{1}{2}\begin{bmatrix} (-1)^{1+1}(-3) & (-1)^{1+2}(-2) \\ (-1)^{2+1}(1) & (-1)^{2+2}(0) \end{bmatrix}^{T} = \begin{bmatrix} -3/2 & -1/2 \\ 1 & 0 \end{bmatrix}.$$

5   The characteristic equation

Let λ be an algebraic variable, and consider the matrix $(\lambda\mathbf{I}-\mathbf{A})$, which has considerable importance in stability analysis, transform analysis as in Chapter 3, and diagonal decompositions. The entries of this matrix are polynomials of degree one or zero. The set of polynomials with ordinary addition and multiplication is not a field, since there is no multiplicative inverse for all nonzero members of the set. However, by writing each polynomial as a fraction with denominator 1, the polynomials are a subset of the rational functions, that is, the set of fractions with polynomial numerator and denominator, which is a field. Therefore, within the field of rational functions of λ with coefficients in the base field of constants, usually the real or complex numbers or the binary numbers, the matrix $(\lambda\mathbf{I}-\mathbf{A})$ has an inverse, given in this case, from (5.32), as

(5.33) $(\lambda\mathbf{I}-\mathbf{A})^{-1} = \dfrac{1}{\det(\lambda\mathbf{I}-\mathbf{A})}\mathrm{adj}(\lambda\mathbf{I}-\mathbf{A}),$

where, as in Section 1.3, the determinant in the denominator is a polynomial $\phi(\lambda)$ of the form

(5.34) $\det(\lambda\mathbf{I}-\mathbf{A}) = \phi(\lambda) = \lambda^n + a_1\lambda^{n-1} + \cdots a_{n-1}\lambda + a_n,$

which is called the *characteristic polynomial* of $\mathbf{A}$. Equating this polynomial to zero gives the *characteristic equation*

(5.35) $\phi(\lambda) = 0,$

which has central importance in the analysis of dynamical systems. Its roots determine the stability of the system, as will be shown, but the theorem in the following section provides a very useful algebraic result about general matrices over a field.

Example 40
Characteristic
polynomial

The characteristic polynomial of $\mathbf{A}$ of Example 39 is

$\phi(\lambda) = \det \begin{bmatrix} \lambda & -1 \\ 2 & \lambda+3 \end{bmatrix} = \lambda^2 + 3\lambda + 2.$

5.1   The Cayley-Hamilton theorem

As shown in Equations (2.4) and (2.17), high powers of $\mathbf{A}$ appear in the solution of LTI state-space equations. The Cayley-Hamilton result shows that only a finite number of powers of $\mathbf{A}$ are of central importance, or in the system-theoretic sense, that the character of system responses to initial conditions or to inputs is determined by a finite number of matrices.

Let $\phi(\lambda) = \det(\lambda\mathbf{I}-\mathbf{A})$ as in (5.34). The Cayley-Hamilton theorem states

(5.36) $\phi(\mathbf{A}) = 0;$

that is, writing $\mathbf{A}$ for λ in (5.34), a matrix $\mathbf{A}$ satisfies its own characteristic equation (5.35). Equivalently,

(5.37) $\mathbf{A}^n = -a_1\mathbf{A}^{n-1} \cdots - a_{n-1}\mathbf{A} - a_n\mathbf{I},$

where $a_1, \cdots a_n$ are the coefficients in (5.34), and where the term $a_n = a_n \lambda^0$ becomes $a_n \mathbf{A}^0 = a_n \mathbf{I}$.

Equation (5.36) is derived as follows. The adjoint matrix contains polynomial elements of degree $n-1$ at most, so it can be written as a polynomial with constant matrix coefficients:

(5.38) $\text{adj}(\lambda \mathbf{I} - \mathbf{A}) = \mathbf{B}_1 \lambda^{n-1} + \mathbf{B}_2 \lambda^{n-2} + \cdots \mathbf{B}_n.$

Then, by multiplying both sides of (5.33) by $\phi(\lambda)(\lambda \mathbf{I} - \mathbf{A})$,

(5.39) $(\lambda^n + a_1 \lambda^{n-1} + \cdots a_{n-1} \lambda + a_n) \mathbf{I} = (\mathbf{B}_1 \lambda^{n-1} + \mathbf{B}_2 \lambda^{n-2} + \cdots \mathbf{B}_n)(\lambda \mathbf{I} - \mathbf{A}),$

and by comparing terms with equal powers of λ on the left and right sides,

(5.40)

$$
\begin{aligned}
\mathbf{I} &= \mathbf{B}_1 & \Rightarrow \quad \mathbf{B}_1 &= \mathbf{I} \\
a_1 \mathbf{I} &= \mathbf{B}_2 - \mathbf{B}_1 \mathbf{A} & \Rightarrow \quad \mathbf{B}_2 &= a_1 \mathbf{I} + \mathbf{B}_1 \mathbf{A} \\
&\ \ \vdots \\
a_{n-2} \mathbf{I} &= \mathbf{B}_{n-1} - \mathbf{B}_{n-2} \mathbf{A} & \Rightarrow \quad \mathbf{B}_{n-1} &= a_{n-2} \mathbf{I} + \mathbf{B}_{n-2} \mathbf{A} \\
a_{n-1} \mathbf{I} &= \mathbf{B}_n - \mathbf{B}_{n-1} \mathbf{A} & \Rightarrow \quad \mathbf{B}_n &= a_{n-1} \mathbf{I} + \mathbf{B}_{n-1} \mathbf{A} \\
a_n \mathbf{I} &= 0 - \mathbf{B}_n \mathbf{A} & \Rightarrow \quad 0 &= a_n \mathbf{I} + \mathbf{B}_n \mathbf{A}.
\end{aligned}
$$

In the last row, substituting for $\mathbf{B}_n$, using the preceding row, gives

(5.41) $0 = a_n \mathbf{I} + (a_{n-1} \mathbf{I} + \mathbf{B}_{n-1} \mathbf{A})\mathbf{A} = a_n \mathbf{I} + a_{n-1} \mathbf{A} + \mathbf{B}_{n-1} \mathbf{A}^2,$

in which $\mathbf{B}_{n-1}$ can be replaced using the preceding row, and so on, finally giving

(5.42) $0 = a_n \mathbf{I} + a_{n-1} \mathbf{A} + \cdots a_1 \mathbf{A}^{n-1} + \mathbf{B}_1 \mathbf{A}^n = a_n \mathbf{I} + a_{n-1} \mathbf{A} + \cdots a_1 \mathbf{A}^{n-1} + \mathbf{A}^n,$

which is (5.36).

The Cayley-Hamilton theorem will be used in the proof of an important result for system identification in the next section. In later chapters the theorem will also have central importance in finding the smallest system that is externally equivalent to a given system.

Minimal polynomial   Although the details will not be given here, for each matrix $\mathbf{A} \in \mathbb{R}^{n \times n}$, a nonzero *minimal polynomial* $m(\lambda)$ of smallest degree $l \leq n$ such that $m(\mathbf{A}) = 0$ can be defined, so that $m(\lambda)$ sometimes may be used in proofs instead of the characteristic polynomial.

Example 41
Cayley-Hamilton   Substituting the matrix $\mathbf{A}$ of Example 39 into its characteristic polynomial gives

$$
\mathbf{A}^2 + 3\mathbf{A} + 2\mathbf{I} = \begin{bmatrix} -2 & -3 \\ 6 & 7 \end{bmatrix} + 3 \begin{bmatrix} 0 & 1 \\ -2 & -3 \end{bmatrix} + 2 \begin{bmatrix} 1 & 0 \\ 0 & 1 \end{bmatrix} = \begin{bmatrix} 0 & 0 \\ 0 & 0 \end{bmatrix}.
$$

6 | The Ho algorithm

An application of the matrix normal form discussed in Section 3.5 and of the Cayley-Hamilton theorem will be described. This important calculation solves the *inverse* of the problem of generating from $(\mathbf{A}, \mathbf{B}, \mathbf{C}, \mathbf{D})$ the Markov sequence

$$(5.43) \quad \{\mathbf{H}_k\}_0^\infty = \left\{ \begin{array}{ll} \mathbf{D}, & k = 0 \\ \mathbf{C}\mathbf{A}^{k-1}\mathbf{B}, & k > 0 \end{array} \right\}.$$

That is, given $\{\mathbf{H}_k\}_0^\infty$ and the knowledge or assumption that this sequence can be generated by an LTI system, it is required to find at least one set of matrices $(\mathbf{A}, \mathbf{B}, \mathbf{C}, \mathbf{D})$ satisfying the above formula. Finding a state-space model from input-output information such as the Markov sequence is one kind of *system identification*.

Minimal order   The system with matrices $(\mathbf{A}, \mathbf{B}, \mathbf{C}, \mathbf{D})$ produced by the Ho method has *minimal order* n, where $\mathbf{A} \in \mathbb{R}^{n \times n}$, in the class of LTI systems satisfying (5.43).

First some circumstances in which the $\{\mathbf{H}_k\}$ are obtained will be given, then the algorithm and its derivation, followed by examples.

6.1 | The context

The sequence $\{\mathbf{H}_k\}$ is obtained in the following situations, and others:

1. $\{\mathbf{H}_k\}$ is the impulse-response sequence of an unknown discrete-time system, as in Section 1.5 of Chapter 2.

2. The rational proper discrete-time transfer matrix $\mathbf{H}(z)$ is known, and can be expanded (by long division!) as $\mathbf{H}(z) = \mathbf{H}_0 + \mathbf{H}_1 z^{-1} + \mathbf{H}_2 z^{-2} + \cdots$, as for continuous-time systems in Equation (3.25).

3. The transfer matrix $\mathbf{H}(s)$ of a continuous-time system is known and can be expanded as for the discrete-time system above into $\mathbf{H}(s) = \mathbf{H}_0 + \mathbf{H}_1 s^{-1} + \mathbf{H}_2 s^{-2} + \cdots$.

4. Given the continuous-time impulse response matrix $\mathbf{H}(t)$, the $\mathbf{H}_k$ can be obtained, using (2.33), as

$$\mathbf{H}_1 = \frac{d^0}{dt^0} \mathbf{H}(t)|_{t=0+}$$
$$\mathbf{H}_2 = \frac{d^1}{dt^1} \mathbf{H}(t)|_{t=0+}$$
$$\vdots$$
$$\mathbf{H}_k = \frac{d^{k-1}}{dt^{k-1}} \mathbf{H}(t)|_{t=0+},$$

with the zeroth term expressed consistently with the others, using the convention

$$\mathbf{H}_0 = \int_{0-}^{0+} \mathbf{H}(t)\, dt = \frac{d^{-1}}{dt^{-1}} \mathbf{H}(t) .$$

5. A realization $(\mathbf{A}, \mathbf{B}, \mathbf{C}, \mathbf{D})$ of $\{\mathbf{H}_k\}$ is known, but may not be of minimal order, for example, if the realization has been found by inspecting the transfer matrix, using the methods of Chapter 4. Then the $\mathbf{H}_k$ can be calculated directly, using (5.43).

6.2   Constructive solution

The solution to this problem is given by the following, known as the B. L. Ho algorithm:

Step 0   First, by definition,

(5.44) $\mathbf{D} = \mathbf{H}_0.$

Step 1   For r "large enough," construct the $pr \times mr$ matrix

(5.45) $\mathbf{S}_r = \begin{bmatrix} \mathbf{H}_1 & \mathbf{H}_2 & \cdots & \mathbf{H}_r \\ \mathbf{H}_2 & \mathbf{H}_3 & \cdots & \mathbf{H}_{r+1} \\ \cdots & & & \\ \mathbf{H}_r & \mathbf{H}_{r+1} & \cdots & \mathbf{H}_{2r-1} \end{bmatrix}.$

A matrix with the above structure is called a Hankel matrix. Find nonsingular $\mathbf{P}, \mathbf{Q}$ such that

(5.46) $\mathbf{P}\mathbf{S}_r\mathbf{Q} = \begin{bmatrix} \mathbf{I}_n & 0 \\ 0 & 0 \end{bmatrix} = \mathbf{N},$

where $\mathbf{N}$ is the normal form of $\mathbf{S}_r$, and n is the rank of $\mathbf{S}_r$. The required value of r will become clear later in the discussion. As illustrated in Figure 5.6, partition $\mathbf{P}, \mathbf{Q}$ into

(5.47) $\mathbf{P} = \begin{bmatrix} \mathbf{P}_1 \\ \mathbf{P}_2 \end{bmatrix}, \quad \mathbf{Q} = [\mathbf{Q}_1, \mathbf{Q}_2],$

where $\mathbf{P}_1$ has n rows and $\mathbf{Q}_1$ has n columns.

Step 2   As illustrated in Figure 5.7, calculate the matrices

(5.48a) $\mathbf{A} = \mathbf{P}_1 \begin{bmatrix} \mathbf{H}_2 & \mathbf{H}_3 & \cdots & \mathbf{H}_{r+1} \\ \mathbf{H}_3 & \mathbf{H}_4 & \cdots & \mathbf{H}_{r+2} \\ \cdots & & & \\ \mathbf{H}_{r+1} & \mathbf{H}_{r+2} & \cdots & \mathbf{H}_{2r} \end{bmatrix} \mathbf{Q}_1, \quad \mathbf{B} = \mathbf{P}_1 \begin{bmatrix} \mathbf{H}_1 \\ \mathbf{H}_2 \\ \vdots \\ \mathbf{H}_r \end{bmatrix},$

(5.48b) $\mathbf{C} = [\mathbf{H}_1, \mathbf{H}_2, \cdots \mathbf{H}_r] \mathbf{Q}_1.$

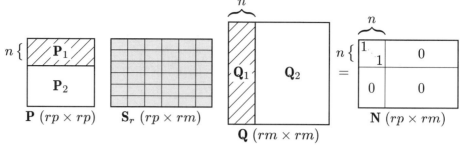

Fig. 5.6   Construction of **P**, **Q**, and **N**, showing matrix dimensions.

| **6.3** | **Development of the algorithm** |

The proof that the previous construction produces a minimal system generating $\{\mathbf{H}_k\}_0^\infty$ rests on the following results.

Proposition 1   *If there is a realization of finite order n, then rank $\mathbf{S}_r \leq n$ for all $r = 1, 2, \cdots$.*

Proof: Factor $\mathbf{S}_r$ as the product of matrices $\mathscr{O}\mathscr{C}$ as shown:

$$(5.49) \quad \mathbf{S}_r = \mathscr{O}\mathscr{C} = \begin{bmatrix} \mathbf{C} \\ \mathbf{CA} \\ \vdots \\ \mathbf{CA}^{r-1} \end{bmatrix} [\mathbf{B}, \mathbf{AB}, \cdots \mathbf{A}^{r-1}\mathbf{B}],$$

where $\mathbf{C}$, $\mathbf{A}$, $\mathbf{B}$ are matrices of the finite-order realization. Because $\mathscr{O}$ has n columns and $\mathscr{C}$ has n rows, rank $\mathscr{O}\mathscr{C} \leq \min\{\text{rank } \mathscr{O}, \text{rank } \mathscr{C}\} \leq n$. $\qquad\square$

Proposition 2   *If there is a realization of finite order then there exist constants $\alpha_1, \cdots \alpha_r$ such that, for any $k > 0$,*

$$(5.50) \quad \mathbf{H}_{k+r} = \alpha_1 \mathbf{H}_{k+r-1} + \alpha_2 \mathbf{H}_{k+r-2} + \cdots \alpha_r \mathbf{H}_k .$$

Proof: Let $\mathbf{A}$ be the $n \times n$ state-vector coefficient matrix of a realization of order n, with characteristic polynomial

$$(5.51) \quad \det(\lambda \mathbf{I} - \mathbf{A}) = \lambda^n + a_1 \lambda^{n-1} + \cdots a_n .$$

Then by the Cayley-Hamilton theorem,

$$(5.52) \quad \mathbf{A}^n = -a_1 \mathbf{A}^{n-1} \cdots - a_n \mathbf{I}_n,$$

so that

$$(5.53) \quad \begin{aligned} \mathbf{H}_{k+n} &= \mathbf{CA}^{k+n-1}\mathbf{B} = \mathbf{CA}^{k-1}(\mathbf{A}^n)\mathbf{B} \\ &= \mathbf{CA}^{k-1}(-a_1 \mathbf{A}^{n-1} - a_2 \mathbf{A}^{n-2} - \cdots a_n \mathbf{I}_n)\mathbf{B} \\ &= -a_1 \mathbf{H}_{k+n-1} - a_2 \mathbf{H}_{k+n-2} - \cdots a_n \mathbf{H}_k. \end{aligned}$$

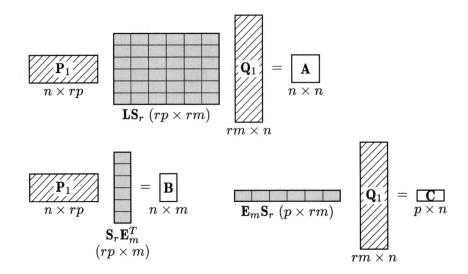

Fig. 5.7 Construction of **A**, **B**, and **C**.

Thus a suitable set of α_i is $\alpha_i = -a_i, i = 1, \cdots n$.   □

Proposition 3   *If coefficients α_i that satisfy (5.50) exist, then there is a realization of finite order.*

Proof: Define the matrices

$$(5.54) \quad \mathbf{L} = \begin{bmatrix} 0 & \mathbf{I}_p & & \\ & & \ddots & \\ & & & \mathbf{I}_p \\ \alpha_r\mathbf{I}_p & \alpha_{r-1}\mathbf{I}_p & \cdots & \alpha_1\mathbf{I}_p \end{bmatrix}, \quad \mathbf{R} = \begin{bmatrix} 0 & & & \alpha_r\mathbf{I}_m \\ \mathbf{I}_m & & & \alpha_{r-1}\mathbf{I}_m \\ & \ddots & & \vdots \\ & & \mathbf{I}_m & \alpha_1\mathbf{I}_m \end{bmatrix},$$

and note that the effect of premultiplying $\mathbf{S}_r$ by $\mathbf{L}$ is to shift the block rows of $\mathbf{S}_r$ up one place and to insert a bottom block row in which each entry satisfies (5.50). Furthermore the effect of premultiplying $\mathbf{S}_r$ by $\mathbf{L}$ is identical to postmultiplying by $\mathbf{R}$:

$$(5.55) \quad \mathbf{L}^k\mathbf{S}_r = \begin{bmatrix} \mathbf{H}_{k+1} & \mathbf{H}_{k+2} & \cdots & \mathbf{H}_{k+r} \\ \mathbf{H}_{k+2} & \mathbf{H}_{k+3} & \cdots & \mathbf{H}_{k+r+1} \\ \cdots & & & \\ \mathbf{H}_{k+r} & \mathbf{H}_{k+r+1} & \cdots & \mathbf{H}_{k+2r-1} \end{bmatrix} = \mathbf{S}_r\mathbf{R}^k, \quad k \geq 0.$$

In addition, define the matrices

$$(5.56) \quad \mathbf{E}_p = [\,\mathbf{I}_p, 0, \cdots 0\,]_{p\times pr}, \quad \mathbf{E}_m = [\,\mathbf{I}_m, 0, \cdots 0\,]_{m\times mr},$$

so that the terms of $\{\mathbf{H}_k\}$ can be extracted from $\mathbf{S}_r$ by either of the multiplications shown:

(5.57) $\mathbf{H}_k = \mathbf{E}_p \mathbf{L}^{k-1} \mathbf{S}_r \mathbf{E}_m^T = \mathbf{E}_p \mathbf{S}_r \mathbf{R}^{k-1} \mathbf{E}_m^T.$

Thus, by comparing these two products to (5.43), two different possible realizations $(\mathbf{A}, \mathbf{B}, \mathbf{C}, \mathbf{D})$ can be written as $(\mathbf{L}, \mathbf{S}_r \mathbf{E}_m^T, \mathbf{E}_p, \mathbf{H}_0)$, and $(\mathbf{R}, \mathbf{E}_m^T, \mathbf{E}_p \mathbf{S}_r, \mathbf{H}_0)$, of order pr and mr respectively. $\square$

Proposition 4   *The Ho algorithm results in a realization of order n.*

Proof: The matrix $\mathbf{A}$ in Step 2 of the algorithm is $n \times n$, but it remains to show that the given $\{\mathbf{H}_k\}$ are generated by the resulting $(\mathbf{A}, \mathbf{B}, \mathbf{C}, \mathbf{D})$.

Some identities will be used: first, the normal form $\mathbf{N}$ of $\mathbf{S}_r$ computed in (5.46) satisfies

(5.58a) $\mathbf{N} = \mathbf{N}\mathbf{N}^T\mathbf{N}$

(5.58b) $\mathbf{N}^T = \mathbf{N}^T\mathbf{N}\mathbf{N}^T$

(5.58c) $\mathbf{S}_r = \mathbf{S}_r (\mathbf{Q}\mathbf{N}^T\mathbf{P}) \mathbf{S}_r.$

The latter identity is demonstrated by multiplication by $\mathbf{P}$ and $\mathbf{Q}$:

(5.59) $\mathbf{P}\mathbf{S}_r(\mathbf{Q}\mathbf{N}^T\mathbf{P})\mathbf{S}_r\mathbf{Q} = (\mathbf{P}\mathbf{S}_r\mathbf{Q})\mathbf{N}^T(\mathbf{P}\mathbf{S}_r\mathbf{Q}) = \mathbf{N}\mathbf{N}^T\mathbf{N} = \mathbf{N} = \mathbf{P}\mathbf{S}_r\mathbf{Q} \ .$

We shall show that the $\mathbf{H}_k$ obtained from (5.57) as $\mathbf{H}_k = \mathbf{E}_p \mathbf{L}^{k-1} \mathbf{S}_r \mathbf{E}_m^T$ can also be rewritten as the simpler (but more complicated-looking at first glance) formula

(5.60) $\mathbf{H}_k = (\mathbf{E}_p \mathbf{S}_r \mathbf{Q}\mathbf{N}^T)(\mathbf{N}\mathbf{N}^T\mathbf{P}\mathbf{L}\mathbf{S}_r\mathbf{Q}\mathbf{N}^T)^{k-1}(\mathbf{N}\mathbf{N}^T\mathbf{P}\mathbf{S}_r\mathbf{E}_m^T).$

To derive this formula, first expand $\mathbf{H}_k$ as

(5.61) $\mathbf{E}_p \mathbf{L}^{k-1} \mathbf{S}_r \mathbf{E}_m^T = \mathbf{E}_p \mathbf{L}^{k-1} \mathbf{S}_r \mathbf{Q}\mathbf{N}^T\mathbf{P}\mathbf{S}_r \mathbf{E}_m^T$

$\qquad\qquad\qquad = \mathbf{E}_p \mathbf{S}_r \mathbf{R}^{k-1} \mathbf{Q}\mathbf{N}^T\mathbf{P}\mathbf{S}_r \mathbf{E}_m^T$

$\qquad\qquad\qquad = \mathbf{E}_p \mathbf{S}_r \mathbf{Q}\mathbf{N}^T\mathbf{P}\mathbf{S}_r \mathbf{R}^{k-1} \mathbf{Q}\mathbf{N}^T\mathbf{P}\mathbf{S}_r \mathbf{E}_m^T$

$\qquad\qquad\qquad = \mathbf{E}_p \mathbf{S}_r \mathbf{Q}\mathbf{N}^T\mathbf{P}\mathbf{L}^{k-1} \mathbf{S}_r \mathbf{Q}\mathbf{N}^T\mathbf{P}\mathbf{S}_r \mathbf{E}_m^T$

$\qquad\qquad\qquad = (\mathbf{E}_p \mathbf{S}_r \mathbf{Q}\mathbf{N}^T)(\mathbf{N}\mathbf{N}^T\mathbf{P}\mathbf{L}^{k-1} \mathbf{S}_r \mathbf{Q}\mathbf{N}^T)(\mathbf{N}\mathbf{N}^T\mathbf{P}\mathbf{S}_r \mathbf{E}_m^T),$

and now it will be shown that the middle factor can be changed to

(5.62) $\mathbf{N}\mathbf{N}^T\mathbf{P}\mathbf{L}^{k-1}\mathbf{S}_r\mathbf{Q}\mathbf{N}^T = (\mathbf{N}\mathbf{N}^T\mathbf{P}\mathbf{L}\mathbf{S}_r\mathbf{Q}\mathbf{N}^T)^{k-1} \ ,$

as follows, by induction:

(5.63) $(\mathbf{NN}^T\mathbf{PL}^\ell\mathbf{S}_r\mathbf{QN}^T)(\mathbf{NN}^T\mathbf{PL}^1\mathbf{S}_r\mathbf{QN}^T) = \mathbf{NN}^T\mathbf{PL}^\ell(\mathbf{S}_r\mathbf{Q}(\mathbf{N}^T\mathbf{NN}^T)\mathbf{PLS}_r)\mathbf{QN}^T$
$$= \mathbf{NN}^T\mathbf{PL}^\ell(\mathbf{S}_r\mathbf{QN}^T\mathbf{PLS}_r)\mathbf{QN}^T$$
$$= \mathbf{NN}^T\mathbf{PL}^\ell(\mathbf{S}_r\mathbf{QN}^T\mathbf{PS}_r\mathbf{R})\mathbf{QN}^T$$
$$= \mathbf{NN}^T\mathbf{PL}^\ell(\mathbf{S}_r\mathbf{R})\mathbf{QN}^T$$
$$= \mathbf{NN}^T\mathbf{PL}^{\ell+1}\mathbf{S}_r\mathbf{QN}^T.$$

The consequence of formula (5.60) is that there is a realization $(\bar{\mathbf{A}}, \bar{\mathbf{B}}, \bar{\mathbf{C}}, \mathbf{D})$ with $\mathbf{D} = \mathbf{H}_0$ and

(5.64) $\bar{\mathbf{A}} = \mathbf{NN}^T\mathbf{PLS}_r\mathbf{QN}^T = \begin{bmatrix} \mathbf{I}_n & 0 \\ 0 & 0 \end{bmatrix}\begin{bmatrix} \mathbf{P}_1 \\ \mathbf{P}_2 \end{bmatrix}\mathbf{LS}_r\,[\,\mathbf{Q}_1, \mathbf{Q}_2\,]\begin{bmatrix} \mathbf{I}_n & 0 \\ 0 & 0 \end{bmatrix}$
$$= \begin{bmatrix} \mathbf{P}_1 \\ 0 \end{bmatrix}\mathbf{LS}_r\,[\,\mathbf{Q}_1, 0\,] = \begin{bmatrix} \mathbf{P}_1(\mathbf{LS}_r)\mathbf{Q}_1 & 0 \\ 0 & 0 \end{bmatrix},$$

(5.65) $\bar{\mathbf{B}} = \mathbf{NN}^T\mathbf{PS}_r\mathbf{E}_m^T = \begin{bmatrix} \mathbf{I}_n & 0 \\ 0 & 0 \end{bmatrix}\begin{bmatrix} \mathbf{P}_1 \\ \mathbf{P}_2 \end{bmatrix}\mathbf{S}_r\begin{bmatrix} \mathbf{I}_m \\ 0 \end{bmatrix} = \begin{bmatrix} \mathbf{P}_1 \\ 0 \end{bmatrix}\begin{bmatrix} \mathbf{H}_1 \\ \vdots \\ \mathbf{H}_r \end{bmatrix},$

(5.66) $\bar{\mathbf{C}} = \mathbf{E}_p\mathbf{S}_r\mathbf{QN}^T = [\,\mathbf{I}_p, 0\,]\,\mathbf{S}_r\,[\,\mathbf{Q}_1, \mathbf{Q}_2\,]\begin{bmatrix} \mathbf{I}_n & 0 \\ 0 & 0 \end{bmatrix} = [\,\mathbf{H}_1, \cdots \mathbf{H}_r\,]\,[\,\mathbf{Q}_1, 0\,].$

This realization is of the form

(5.67) $(\bar{\mathbf{A}} = \begin{bmatrix} \mathbf{A} & 0 \\ 0 & 0 \end{bmatrix}, \bar{\mathbf{B}} = \begin{bmatrix} \mathbf{B} \\ 0 \end{bmatrix}, \bar{\mathbf{C}} = [\,\mathbf{C}, 0\,], \mathbf{D}),$

where $\mathbf{A}$ is $n \times n$, the partitions of $\bar{\mathbf{B}}$ and $\bar{\mathbf{C}}$ are conformal, and where $\mathbf{A}, \mathbf{B}, \mathbf{C}$ above are given by Equation (5.48). Because of the zeros in $\bar{\mathbf{A}}, \bar{\mathbf{B}}$, and $\bar{\mathbf{C}}$, $\mathbf{H}_k = \bar{\mathbf{C}}\bar{\mathbf{A}}^{k-1}\bar{\mathbf{B}} = \mathbf{CA}^{k-1}\mathbf{B}, k > 0$, and a realization of order n is $(\mathbf{A}, \mathbf{B}, \mathbf{C}, \mathbf{D})$, given by the Ho algorithm.   □

Proposition 5   *The realization obtained by the algorithm is of minimal order n.*

Proof: Assume, to the contrary, that the algorithm produces a realization of order $n = $ rank $\mathbf{S}_r$, and that there is a realization of order $n_1 < n$. Then by Proposition 1, rank $\mathbf{S}_r \le n_1$, a contradiction.   □

Notes   1. *Computation vs proof:* Although the α_i play a central role in the proofs, they are not calculated in practice. Steps 0–2 of the algorithm are followed instead.

2. *Algorithm modifications:* It is not necessary to calculate all of $\mathbf{P}$ and $\mathbf{Q}$ in Step 1, but only submatrices $\mathbf{P}_1$ and $\mathbf{Q}_1$, which are not unique. They can be chosen to minimize computations for example, or to produce system matrices of particular form.

3. *Bound on order:* If, perhaps from physics or from a known nonminimal realization, an upper bound n' is available for n, then from the proofs given above, this upper bound is a suitable value for r, and the $2n' + 1$ matrices $\mathbf{H}_0, \cdots \mathbf{H}_{2n'}$ enter into the algorithm. If such a bound is not available, then since the desired r satisfies $n = \text{rank}\,\mathbf{S}_r = \text{rank}\,\mathbf{S}_{r+1} = \text{rank}\,\mathbf{S}_{r+2} = \cdots$, an attempt may be made to compute the sequence $\{\mathbf{S}_1, \mathbf{S}_2, \cdots\}$ and $\{\text{rank}\,\mathbf{S}_1, \text{rank}\,\mathbf{S}_2, \cdots\}$ until there is reason to believe that the maximal rank has been reached.

4. *Noisy data:* In practice, if the $\mathbf{H}_k$ are not known exactly, the rank of $\mathbf{S}_r$ that would have been obtained from exact data is unknown. If the data are corrupted by independent random noise then, with probability one, $\mathbf{S}_r$ will have full rank for all r. One approximate approach to such a situation is to choose n, then to find the rank-n matrix $\mathbf{K}_n$ that is nearest $\mathbf{S}_r$, using a suitable measure of distance, and to substitute $\mathbf{K}_n$ for $\mathbf{S}_r$ in the B. L. Ho algorithm. One simple measure is the sum of squares of the difference. This quantity can be written $\|\mathbf{K}_n - \mathbf{S}_r\|_E^2$, since the sum of squares is the square of the Euclidean norm.

For each $\mathbf{K}_n$, $n = 1, 2 \cdots$ the matrices $\mathbf{A}, \mathbf{B}, \mathbf{C}$ and the error

(5.68) $$e_n = \left(\sum_{k=1}^{N} \|\mathbf{H}_k - \mathbf{C}\mathbf{A}^{k-1}\mathbf{B}\|_E^2 \right)^{1/2}$$

can be calculated for some large fixed N. Comparing the e_n allows the smallest n that gives an acceptable fit of the data to be determined. This technique has the advantage of simplicity, since it is easily done by using the singular-value decomposition of $\mathbf{S}_r$ discussed below, but *not* of optimality since the $\mathbf{K}_n$ do not have exact Hankel-matrix form.

One way of computing $\mathbf{K}_n$ to minimize the sum of squares is via the singular-value decomposition of $\mathbf{S}_r$ as in Section 3.6, producing matrices $\mathbf{U}, \mathbf{V}, \Sigma$ such that

(5.69) $$\mathbf{S}_r = \mathbf{U}\Sigma\mathbf{V}^T.$$

If $\mathbf{S}_r$ has rank n, then the n-square top-left block Σ_n of Σ is diagonal and invertible, and partitioning $\mathbf{V}$ as $\mathbf{V} = [\mathbf{V}_1, \mathbf{V}_2]$ where $\mathbf{V}_1$ has n columns, matrices $\mathbf{P}, \mathbf{Q}$ that satisfy (5.46) in the Ho algorithm are $\mathbf{P} = \mathbf{U}^T$, $\mathbf{Q} = [\mathbf{V}_1\Sigma_n^{-1}, \mathbf{V}_2]$, as for Equation (5.13), with $r = n$. If rank $\mathbf{S}_r > n$ as will be true for noisy data, then because $\mathbf{U} = [\mathbf{U}_1, \mathbf{U}_2]$, $\mathbf{V}$ are orthogonal (see Example 34), the best least-squares rank-n approximation to $\mathbf{S}_r$ is $\mathbf{K}_n = \mathbf{U}_1 \Sigma_n \mathbf{V}_1^T$, for which (5.46) becomes, as required,

(5.70)
$$\mathbf{PK}_n\mathbf{Q} = \begin{bmatrix} \mathbf{U}_1^T \\ \mathbf{U}_2^T \end{bmatrix} \mathbf{U}_1 \Sigma_n \mathbf{V}_1^T [\mathbf{V}_1 \Sigma_n^{-1}, \mathbf{V}_2] = \begin{bmatrix} \mathbf{I}_n & 0 \\ 0 & 0 \end{bmatrix}.$$

In fact, $\mathbf{K}_n$ need never be computed explicitly. Rather, the singular-value decomposition of $\mathbf{S}_r$ is first obtained, then for different values of n, the matrices $\mathbf{P}_1 = \mathbf{U}_1^T$, $\mathbf{Q}_1 = \mathbf{V}_1 \Sigma_n^{-1}$ are obtained; the matrices $\mathbf{A}$, $\mathbf{B}$, $\mathbf{C}$ are calculated from (5.48); and the sequence $\{\mathbf{CA}^{k-1}\mathbf{B}\}_1^N$ is computed and compared to the data as above.

An efficient alternative to the above procedure is to use the error measure $\|\mathbf{K}_n - \mathbf{S}_r\|_E = (\sum_{i>n} \sigma_i^2)^{1/2}$ as in Example 34. This quantity is equal to (5.68) with terms weighted by integer constants. Simpler still, it can be shown that the quantity σ_{n+1} is also a norm of $\mathbf{K}_n - \mathbf{S}_r$.

5. *System approximation:* It can be shown that the B. L. Ho algorithm can be used to produce an optimal realization $(\mathbf{A}_\mu, \mathbf{B}_\mu, \mathbf{C}_\mu, \mathbf{D})$ of given order μ for which

(5.71)
$$\sum_{k=1}^{\infty} \|\mathbf{H}_k - \mathbf{C}_\mu \mathbf{A}_\mu^{k-1} \mathbf{B}_\mu\|_E^2$$

is a minimum, but except in special cases there is no known construction of the required $\mathbf{P}_1$ and $\mathbf{Q}_1$ without use of nonlinear optimization techniques.

Example 42
Scalar system

As a simple illustration, the Ho algorithm will be used to find a minimal state-space realization of the sequence

$$\{\mathbf{H}_k\}_0^{\infty} = \{\frac{d^{k-1}}{dt^{k-1}} h(t)|_{t=0}\} = \{0, 0, 1, -6, 27, -108, 405, \cdots\},$$

for the impulse response

$$h(t) = 0\delta(t) + te^{-3t}.$$

Step 0   First, $\mathbf{D} = \mathbf{H}_0 = 0$.
Step 1   Computing $\mathbf{S}_r$ and its rank, for $r = 1, 2, \cdots$,

$$\mathbf{S}_1 = 0, \quad \text{rank}(\mathbf{S}_1) = 0, \quad \mathbf{S}_2 = \begin{bmatrix} 0 & 1 \\ 1 & -6 \end{bmatrix}, \quad \text{rank}(\mathbf{S}_2) = 2,$$

$$\mathbf{S}_3 = \begin{bmatrix} 0 & 1 & -6 \\ 1 & -6 & 27 \\ -6 & 27 & -108 \end{bmatrix}, \quad \text{rank}(\mathbf{S}_3) = 2, \cdots,$$

and further $\mathbf{S}_r$ have rank 2, as confirmed by observing that $h(t)$ is a function corresponding to a double pole at -3, that is, to a characteristic polynomial of degree 2. Hence, for convenience, the smallest $\mathbf{S}_r$ of rank 2 can be used in the Ho algorithm. Reducing $\mathbf{S}_2$ to normal form, first by row operations and then by column operations gives, for example,

$$\begin{bmatrix} \mathbf{I} & \mathbf{S}_2 \\ & \mathbf{I} \end{bmatrix} \underset{\text{row}}{\sim} \left[\begin{array}{cc|cc} 0 & 1 & 1 & -6 \\ 1 & 0 & 0 & 1 \\ \hline & & 1 & 0 \\ & & 0 & 1 \end{array}\right] = \begin{bmatrix} \mathbf{P}_1 & \mathbf{P}_1\mathbf{S}_2 \\ & \mathbf{I} \end{bmatrix}$$

$$\underset{\text{col}}{\sim} \left[\begin{array}{cc|cc} 0 & 1 & 1 & 0 \\ 1 & 0 & 0 & 1 \\ \hline & & 1 & 6 \\ & & 0 & 1 \end{array}\right] = \begin{bmatrix} \mathbf{P}_1 & \mathbf{N} \\ & \mathbf{Q}_1 \end{bmatrix}.$$

Step 2   Computing the system matrices yields

$$\mathbf{B} = \mathbf{P}_1 \begin{bmatrix} 0 \\ 1 \end{bmatrix} = \begin{bmatrix} 1 \\ 0 \end{bmatrix}, \quad \mathbf{C} = [0 \quad 1]\mathbf{Q}_1 = [0 \quad 1],$$

$$\mathbf{A} = \mathbf{P}_1 \begin{bmatrix} 1 & -6 \\ -6 & 27 \end{bmatrix} \mathbf{Q}_1 = \begin{bmatrix} -6 & -9 \\ 1 & 0 \end{bmatrix},$$

which are not unique, since $\mathbf{P}_1$ and $\mathbf{Q}_1$ are not unique.

Example 43
Sequence
generator

We wish to design a system that generates a sequence of output zeros up to time $t = 0$, after which time it produces the counting sequence $\{y(k)\}_0^\infty = \{0, 1, 2, \cdots\}$. Equating the output to the free response of a linear system,

$$y(k) = \begin{cases} 0, & k \le 0 \\ \mathbf{CA}^{k-1}\mathbf{X}(0), & k > 0 \end{cases},$$

which is the Markov sequence of a system with $\mathbf{B} = \mathbf{X}(0)$.

Perform the steps of the algorithm to calculate $\mathbf{A}, \mathbf{B}, \mathbf{C}, \mathbf{D}$:

Step 0 $\mathbf{D} = \mathbf{H}_0 = 0$.

Step 1   Construct the sequence of matrices $\mathbf{S}_r$ and their rank,

$$\mathbf{S}_1 = \mathbf{H}_1 = 1, \quad \text{rank}(\mathbf{S}_1) = 1,$$

$$\mathbf{S}_2 = \begin{bmatrix} \mathbf{H}_1 & \mathbf{H}_2 \\ \mathbf{H}_2 & \mathbf{H}_3 \end{bmatrix} = \begin{bmatrix} 1 & 2 \\ 2 & 3 \end{bmatrix}, \quad \text{rank}(\mathbf{S}_2) = 2,$$

$$\mathbf{S}_3 = \begin{bmatrix} \mathbf{H}_1 & \mathbf{H}_2 & \mathbf{H}_3 \\ \mathbf{H}_2 & \mathbf{H}_3 & \mathbf{H}_4 \\ \mathbf{H}_3 & \mathbf{H}_4 & \mathbf{H}_5 \end{bmatrix} = \begin{bmatrix} 1 & 2 & 3 \\ 2 & 3 & 4 \\ 3 & 4 & 5 \end{bmatrix}, \quad \text{rank}(\mathbf{S}_3) = 2,$$

at which point, since the rank has not exceeded 2, a system of order 2 may be tried, provided it will be verified that the required sequence is obtained. The smallest $\mathbf{S}_r$ with rank 2 is $\mathbf{S}_2$, and by computing the normal form,

$$
\begin{bmatrix} \mathbf{I}_2 & \mathbf{S}_2 \\ & \mathbf{I}_2 \end{bmatrix} = \left[\begin{array}{cc|cc} 1 & 0 & 1 & 2 \\ 0 & 1 & 2 & 3 \\ \hline & & 1 & 0 \\ & & 0 & 1 \end{array} \right] \overset{\text{row}}{\sim} \left[\begin{array}{cc|cc} -3 & 2 & 1 & 0 \\ 2 & -1 & 0 & 1 \\ \hline & & 1 & 0 \\ & & 0 & 1 \end{array} \right],
$$

so that

$$
\mathbf{P}_1 = \mathbf{P} = \begin{bmatrix} -3 & 2 \\ 2 & -1 \end{bmatrix}, \quad \text{and} \quad \mathbf{Q}_1 = \mathbf{Q} = \mathbf{I}_2.
$$

Step 2   The linear system is computed as

$$
\mathbf{B} = \mathbf{P}_1 \begin{bmatrix} \mathbf{H}_1 \\ \mathbf{H}_2 \end{bmatrix} = \begin{bmatrix} -3 & 2 \\ 2 & -1 \end{bmatrix} \begin{bmatrix} 1 \\ 2 \end{bmatrix} = \begin{bmatrix} 1 \\ 0 \end{bmatrix}, \quad \mathbf{C} = [\,\mathbf{H}_1, \mathbf{H}_2\,]\mathbf{Q}_1 = [\,1 \quad 2\,],
$$

$$
\mathbf{A} = \mathbf{P}_1 \begin{bmatrix} \mathbf{H}_2 & \mathbf{H}_3 \\ \mathbf{H}_3 & \mathbf{H}_4 \end{bmatrix} \mathbf{Q}_1 = \begin{bmatrix} -3 & 2 \\ 2 & -1 \end{bmatrix} \begin{bmatrix} 2 & 3 \\ 3 & 4 \end{bmatrix} \mathbf{I}_2 = \begin{bmatrix} 0 & -1 \\ 1 & 2 \end{bmatrix}.
$$

Since it was not verified that $\text{rank}(\mathbf{S}_r)$ never exceeds 2, it must be checked that the calculated system generates the correct Markov sequence. One way to check is by recursion. First, the generated sequence is correct up to $\mathbf{H}_{2n} = \mathbf{H}_4 = 4$, since these matrices were used explicitly in the computation. Now assume that the sequence up to term k is $\{0, 1, \cdots k{-}1, k\}$. The characteristic polynomial is $\det(\lambda\mathbf{I} - \mathbf{A}) = \lambda^2 - 2\lambda + 1$, so that suitable α_i of Proposition 3 are $\alpha_1 = -2$, $\alpha_2 = 1$. Calculating term $k{+}1$ of the sequence, $\mathbf{H}_{k+1} = -\alpha_1\mathbf{H}_k - \alpha_2\mathbf{H}_{k-1} = -(-2)(k) - (1)(k - 1) = k{+}1$, as required.

Example 44
Binary sequence generator

As another example of sequence generation, suppose it is required to generate the binary sequence $\{\mathbf{H}_k\} = \left\{ \begin{bmatrix} 0 \\ 0 \end{bmatrix}, \begin{bmatrix} 1 \\ 0 \end{bmatrix}, \begin{bmatrix} 1 \\ 1 \end{bmatrix}, \begin{bmatrix} 0 \\ 1 \end{bmatrix}, \begin{bmatrix} 0 \\ 0 \end{bmatrix}, \cdots \right\}$, which repeats with cycle length 4. The system must cycle through four states, not including the zero state, since, if at any time k the state $\mathbf{X}(k)$ is 0, then $\mathbf{Y}(\ell) = \mathbf{A}^{\ell-k}\mathbf{X}(k) = 0$, for $\ell \geq k$. A binary system of order n has 2^n possible state vectors, including the zero vector, so a linear system that generates the required sequence will have order 3, which determines r in the Ho algorithm.

The matrix $\mathbf{S}_3$ below has rank 3, and with $\mathbf{Q}_1 = \mathbf{I}_3$, by inspection a suitable matrix $\mathbf{P}_1$ picks out the third row from $\mathbf{S}_3$, the (mod 2) sum of the first and third, and the fifth, as shown:

$$
\mathbf{S}_3 = \begin{bmatrix} 1 & 1 & 0 \\ 0 & 1 & 1 \\ 1 & 0 & 0 \\ 1 & 1 & 0 \\ 0 & 0 & 1 \\ 1 & 0 & 0 \end{bmatrix}, \quad \mathbf{P}_1 = \begin{bmatrix} 0 & 0 & 1 & 0 & 0 & 0 \\ 1 & 0 & 1 & 0 & 0 & 0 \\ 0 & 0 & 0 & 0 & 1 & 0 \end{bmatrix};
$$

that is, $\mathbf{P}_1\mathbf{S}_3 = \mathbf{I}_3$. The matrices of the realization are then calculated as

$$
\mathbf{A} = \mathbf{P}_1 \begin{bmatrix} 1 & 0 & 0 \\ 1 & 1 & 0 \\ 0 & 0 & 1 \\ 1 & 0 & 0 \\ 0 & 1 & 1 \\ 0 & 0 & 1 \end{bmatrix} \mathbf{Q}_1 = \begin{bmatrix} 0 & 0 & 1 \\ 1 & 0 & 1 \\ 0 & 1 & 1 \end{bmatrix}, \quad \mathbf{B} = \mathbf{P}_1 \begin{bmatrix} 1 \\ 0 \\ 1 \\ 1 \\ 0 \\ 1 \end{bmatrix} = \begin{bmatrix} 1 \\ 0 \\ 0 \end{bmatrix},
$$

$$
\mathbf{C} = \begin{bmatrix} 1 & 1 & 0 \\ 0 & 1 & 1 \end{bmatrix} \mathbf{Q}_1 = \begin{bmatrix} 1 & 1 & 0 \\ 0 & 1 & 1 \end{bmatrix}, \quad \mathbf{D} = \begin{bmatrix} 0 \\ 0 \end{bmatrix}.
$$

Example 45
Approximate
realization

To illustrate the sub-optimal approximation of noisy measured data as discussed previously, suppose the following 10 measured $\mathbf{H}_k$ are available:

$$\{\mathbf{H}_k\}_0^{10} = \{0.0141,\ 0.0104,\ 1.0024,\ 0.7108,\ 0.7961,\ 0.4092,\ 0.4190,\ 0.1698,$$
$$0.2013,\ 0.0695,\ 0.0973\}.$$

Sufficient data are available to compute $\mathbf{S}_5$, for which the computed singular values are $\{\sigma_i\} = \{2.6096,\ 0.9271,\ 0.3207,\ 0.0093,\ 0.0015\}$, which shows that $\mathbf{S}_5$ has rank 5, so that there is a system of order 5 that matches the data exactly, but a computed system of order 4 corresponds to data a distance $\sigma_5 = 0.0015$ away from the measured data, one of order 3 is $\sqrt{\sigma_4^2 + \sigma_5^2} = 0.009$ away, and one of order 2 is $\sqrt{\sigma_3^2 + \sigma_4^2 + \sigma_5^2} = 0.3$ away, which would lead to a postulated minimal order of 3 for the noise-free data on the assumption that noise accounts for perturbations in the order of 1%, say.

Example 46
Impulse
response by
cross
correlation

It may be impractical to measure the impulse response, or the transfer matrix, to obtain the sequence $\{\mathbf{H}_k\}$. An indirect means of obtaining this sequence uses a zero-mean random or pseudorandom input signal $\{\mathbf{U}_k\}$, typically of low power, perhaps inserted as a perturbation of existing inputs to the system to be identified. The input sequence $\{\mathbf{U}_k\}$ is measured, together with the response $\{\mathbf{Y}_k\}$, over a long interval, and the cross-correlation matrices $\sum_{t=-\infty}^{\infty} \mathbf{Y}(t)\,\mathbf{U}^T(t-k)$ of the input and response are computed, to give $\{\mathbf{H}_k\}$. This process will be illustrated for a scalar system, for simplicity.

The response at time t to an input sequence $\{u(\ell)\}$ is the convolution sum

$$(5.72) \quad y(t) = \sum_{\ell=0}^{t} h(\ell)\,u(t-\ell),$$

and the required cross-correlation, with shift parameter k, is

$$(5.73) \quad \sum_{t=-\infty}^{\infty} y(t)\,u(t-k) = \sum_{t=-\infty}^{\infty} \sum_{\ell=0}^{t} h(\ell)\,u(t-\ell)\,u(t-k)$$

$$= \sum_{t=-\infty}^{\infty} \sum_{\ell=-\infty}^{t} h(\ell)\, u(t-\ell)\, u(t-k),$$

where the lower summation limit for ℓ has been changed, since $h(\ell) = 0$ for $\ell < 0$. Then exchanging the order of summation gives

$$(5.74) \qquad \sum_{\ell=-\infty}^{\infty} \sum_{t=\ell}^{\infty} h(\ell)\, u(t-\ell)\, u(t-k) = \sum_{\ell=-\infty}^{\infty} h(\ell) \left(\sum_{t=\ell}^{\infty} u(t-\ell)\, u(t-k) \right).$$

The input sequence $\{u(\ell)\}$ must be such that the sum in parentheses is zero when $\ell \neq k$, as when $\{u(\ell)\}$ is white noise, say. The sum must also have a known value q when $\ell = k$. Then the cross-correlation function is

$$(5.75) \qquad \sum_{t=-\infty}^{\infty} y(t)\, u(t-k) = h(k)\, q,$$

and is computed, for a range of values of k and finite but large limits of summation, and divided by q to find the Markov parameters $h(k)$.

7 | Solution of linear equations

Consider the equation

$$(5.76) \qquad \mathbf{AX} = \mathbf{B}$$

where $\mathbf{A} \in \mathbb{R}^{m \times n}$, $\mathbf{X} \in \mathbb{R}^{n \times q}$, $\mathbf{B} \in \mathbb{R}^{m \times q}$. With $\mathbf{A}$ and $\mathbf{B}$ known, we want to solve for $\mathbf{X}$. When $\mathbf{A}$ is singular, that is, $m \neq n$ or $\det \mathbf{A} = 0$, this equation is a generalization of the more familiar case in which $\mathbf{A}$ is square and nonsingular, and for which the solution is written symbolically as

$$(5.77) \qquad \mathbf{X} = \mathbf{A}^{-1}\mathbf{B}.$$

The solution of the general case of (5.76) is important for applications and for illustrating theoretical concepts in the study of vector spaces and linear operators.

7.1 | General method

Denote the rank of $\mathbf{A}$ in (5.76) by r, so that $r \leq m$ and $r \leq n$. A vector $\mathbf{X}$ that satisfies (5.76) can be obtained by the following steps:

Step 1   As in Section 3.5, find nonsingular matrices $\mathbf{P} = \begin{bmatrix} \mathbf{P}_1 \\ \mathbf{P}_2 \end{bmatrix}$ and $\mathbf{Q} = [\,\mathbf{Q}_1, \mathbf{Q}_2\,]$, where $\mathbf{P}_1$ has r rows and $\mathbf{Q}_1$ has r columns, such that

(5.78) $$\mathbf{PAQ} = \begin{bmatrix} \mathbf{P}_1\mathbf{AQ}_1 & \mathbf{P}_1\mathbf{AQ}_2 \\ \mathbf{P}_2\mathbf{AQ}_1 & \mathbf{P}_2\mathbf{AQ}_2 \end{bmatrix} = \begin{bmatrix} \mathbf{I}_r & 0 \\ 0 & 0 \end{bmatrix}$$

is the normal form of $\mathbf{A}$, where $\mathbf{I}_r$ is the $r{\times}r$ identity matrix, and not all zero blocks shown occur in every case, depending on the value of r.

Step 2   Premultiply $\mathbf{B}$ by $\mathbf{P}$ to calculate

(5.79) $$\mathbf{PB} = \begin{bmatrix} \mathbf{P}_1\mathbf{B} \\ \mathbf{P}_2\mathbf{B} \end{bmatrix}.$$

If $\mathbf{P}_2\mathbf{B} \neq 0$ then (5.76) is inconsistent, and no solution exists. This conclusion is demonstrated by premultiplying (5.76) by $\mathbf{P}$ and substituting

(5.80) $$\mathbf{X} = [\,\mathbf{Q}_1, \mathbf{Q}_2\,] \begin{bmatrix} \mathbf{Y}_1 \\ \mathbf{Y}_2 \end{bmatrix},$$

changing (5.76) to

(5.81) $$\begin{bmatrix} \mathbf{I}_r & 0 \\ 0 & 0 \end{bmatrix} \begin{bmatrix} \mathbf{Y}_1 \\ \mathbf{Y}_2 \end{bmatrix} = \begin{bmatrix} \mathbf{P}_1\mathbf{B} \\ \mathbf{P}_2\mathbf{B} \end{bmatrix}.$$

Because $\mathbf{P}$ and $\mathbf{Q}$ are nonsingular, (5.76) has a solution if and only if (5.81) has a solution. Consequently, if $\mathbf{P}_2\mathbf{B} \neq 0$, then (5.81), and hence (5.76), is inconsistent.

Step 3   If a solution exists, then from (5.81) and (5.80),

(5.82) $$\mathbf{X} = [\,\mathbf{Q}_1, \mathbf{Q}_2\,] \begin{bmatrix} \mathbf{P}_1\mathbf{B} \\ \mathbf{Y}_2 \end{bmatrix} - \mathbf{Q}_1\mathbf{P}_1\mathbf{B} \mid \mathbf{Q}_2\mathbf{Y}_2,$$

where $\mathbf{Q}_1\mathbf{P}_1\mathbf{B}$ is called a *particular solution,* and where $\mathbf{Q}_2\mathbf{Y}_2$ is called the *complimentary function,* and contains arbitrary parameters in $\mathbf{Y}_2$.

Example 47
General
solution

For the equation $\mathbf{AX} = \mathbf{B}$, below, with $\mathbf{A}$ as in Example 22,

$$\begin{bmatrix} -3 & 9 & 1 & -5 \\ -2 & 6 & 0 & -8 \\ -4 & 12 & 1 & -9 \end{bmatrix} \mathbf{X} = \begin{bmatrix} -8 \\ -4 \\ -10 \end{bmatrix},$$

matrices $\mathbf{P}$, $\mathbf{Q}$ that result in the normal form were computed in Example 31. Since $\mathbf{P}_2\mathbf{B} = 0$, the equation has solutions, and since $\mathbf{P}_1\mathbf{B} = \begin{bmatrix} -8 \\ 4/3 \end{bmatrix}$, the general solution is

$$\mathbf{X} = \mathbf{Q}_1 \mathbf{P}_1 \mathbf{B} + \mathbf{Q}_2 \mathbf{Y}_2$$

$$= \begin{bmatrix} -1/3 & -1/2 \\ 0 & 0 \\ 0 & -3/2 \\ 0 & 0 \end{bmatrix} \begin{bmatrix} -8 \\ 4/3 \end{bmatrix} + \begin{bmatrix} 3 & -4 \\ 1 & 0 \\ 0 & -7 \\ 0 & 1 \end{bmatrix} \mathbf{Y}_2 = \begin{bmatrix} 2 \\ 0 \\ -2 \\ 0 \end{bmatrix} + \begin{bmatrix} 3 & -4 \\ 1 & 0 \\ 0 & -7 \\ 0 & 1 \end{bmatrix} \mathbf{Y}_2,$$

where $\mathbf{Y}_2$ contains two arbitrary parameters.

Example 48
General
solution

The equation

$$\begin{bmatrix} 1 & 2 & 3 \\ -1 & -2 & -3 \end{bmatrix} \mathbf{X} = \begin{bmatrix} 3 \\ -3 \end{bmatrix}$$

can be solved as follows. The elementary operations $K_{12}(-2)$, $K_{13}(-3)$, $H_{12}(1)$ produce the transformation

$$\begin{array}{c} \mathbf{I}_2 \quad \mathbf{A} \quad \mathbf{B} \\ \mathbf{I}_3 \end{array} \longrightarrow \left[\begin{array}{cc|ccc|c} 1 & 0 & 1 & 0 & 0 & 3 \\ 1 & 1 & 0 & 0 & 0 & 0 \\ \hline & & 1 & -2 & -3 \\ & & 0 & 1 & 0 \\ & & 0 & 0 & 1 \end{array}\right] = \begin{array}{cc} \mathbf{P}_1 & \mathbf{I} & 0 & \mathbf{P}_1\mathbf{B} \\ \mathbf{P}_2 & 0 & 0 & \mathbf{P}_2\mathbf{B} \\ & \mathbf{Q}_1 & \mathbf{Q}_2 \end{array}.$$

Since rank $\mathbf{A} = 1$ and $\mathbf{P}_2\mathbf{B} = 0$, a solution exists, and since $\mathbf{P}_1\mathbf{B} = 3$, the complete solution is

$$\mathbf{X} = \begin{bmatrix} 1 \\ 0 \\ 0 \end{bmatrix} 3 + \begin{bmatrix} -2 & -3 \\ 1 & 0 \\ 0 & 1 \end{bmatrix} \mathbf{Y}_2,$$

where $\mathbf{Y}_2$ is a 2×1 matrix of arbitrary parameters.

7.2   Abbreviated method

A method which is often given descriptively in elementary textbooks will be shown to be a special case of the previous solution.

Not all of the blocks of $\mathbf{P}$ and $\mathbf{Q}$ are required in computing the solution, using the method of Section 7.1. In (5.81), only $\mathbf{P}_2\mathbf{B}$ is required to confirm that a solution exists, and $\mathbf{P}_2$ is not required in (5.81) or (5.82). In (5.82), $\mathbf{Q}_2$ is explicitly required, but only $\mathbf{P}_1\mathbf{B}$ is required, and not $\mathbf{P}_1$. Consequently, an alternative algorithm can be stated as follows:

Step 1   This step produces the effect of premultiplication by $\mathbf{P}$ in (5.78) without explicitly constructing $\mathbf{P}$. Perform elementary row operations on the array $[\mathbf{A}, \mathbf{B}]$ to put the submatrix $\mathbf{A}$ into reduced row echelon form, as shown,

(5.83) $[\mathbf{A}, \mathbf{B}] \rightarrow [\mathbf{PA}, \mathbf{PB}] = \begin{bmatrix} \mathbf{P}_1\mathbf{A} & \mathbf{P}_1\mathbf{B} \\ 0 & \mathbf{P}_2\mathbf{B} \end{bmatrix} = \begin{bmatrix} [\mathbf{I}_r, \mathbf{V}]\mathbf{J} & \mathbf{P}_1\mathbf{B} \\ 0 & \mathbf{P}_2\mathbf{B} \end{bmatrix}$, where rank $\mathbf{A} = r$, and $\mathbf{J}$ is some *permutation matrix,* an identity matrix $\mathbf{I}_n$ with permuted columns. Postmultiplication by $\mathbf{J}$ in the above equation is just a notational device to specify that the computed $\mathbf{P}_1\mathbf{A}$ is of the form $[\,\mathbf{I}_r, \mathbf{V}\,]$ except for column interchanges. This postmultiplication corresponds to column permutation, premultiplication by $\mathbf{J}^T$ corresponds to row permutation, and $\mathbf{J}$ has the property that $\mathbf{J}\mathbf{J}^T = \mathbf{I}$. If the first r columns of $\mathbf{A}$ are linearly independent, then $\mathbf{J} = \mathbf{I}_n$. The matrix $\mathbf{V}$ has no particular form, and does not exist if $r = n$.

Step 2   If $\mathbf{P}_2\mathbf{B} \neq 0$, then the equations are not consistent.

Step 3   In principle, $\mathbf{Q}$ in (5.78) can be constructed by inspection of (5.83) as

(5.84) $\mathbf{Q} = \mathbf{J}^T \begin{bmatrix} \mathbf{I}_r & -\mathbf{V} \\ 0 & \mathbf{I}_{n-r} \end{bmatrix}$,

since this $\mathbf{Q}$ produces the required form from $\mathbf{A}$ when applied with $\mathbf{P}$, as shown:

(5.85) $\mathbf{P}_1\mathbf{A}\mathbf{Q} = ([\,\mathbf{I}_r, \mathbf{V}\,]\,\mathbf{J}) \left(\mathbf{J}^T \begin{bmatrix} \mathbf{I}_r & -\mathbf{V} \\ 0 & \mathbf{I}_{n-r} \end{bmatrix} \right) = [\,\mathbf{I}_r, \mathbf{V}\,] \begin{bmatrix} \mathbf{I}_r & -\mathbf{V} \\ 0 & \mathbf{I}_{n-r} \end{bmatrix} = [\,\mathbf{I}_r, 0\,]$.

Consequently, the complete solution (5.82) is constructible from (5.83) by inspection as

(5.86) $\mathbf{X} = \mathbf{J}^T \left(\begin{bmatrix} \mathbf{P}_1\mathbf{B} \\ 0 \end{bmatrix} + \begin{bmatrix} -\mathbf{V} \\ \mathbf{I}_{n-r} \end{bmatrix} \mathbf{Y}_2 \right)$.

Example 49
Abbreviated method

For the data of the previous example, the matrix $[\,\mathbf{PA}, \mathbf{PB}\,]$ in reduced row echelon is

$$\begin{bmatrix} 1 & -3 & 0 & 4 & | & 2 \\ 0 & 0 & 1 & 7 & | & -2 \\ 0 & 0 & 0 & 0 & | & 0 \end{bmatrix}, \quad \text{with } \mathbf{V} = \begin{bmatrix} -3 & 4 \\ 0 & 7 \end{bmatrix}, \quad \mathbf{P}_1\mathbf{B} = \begin{bmatrix} 2 \\ -2 \end{bmatrix}, \quad \mathbf{P}_2\mathbf{B} = 0,$$

from which, by inspection, since the pivot columns are the first and third,

$$\begin{bmatrix} x_1 \\ x_3 \end{bmatrix} = \begin{bmatrix} 2 \\ -2 \end{bmatrix} + \begin{bmatrix} 3 \\ 0 \end{bmatrix} x_2 + \begin{bmatrix} -4 \\ -7 \end{bmatrix} x_4.$$

Equivalently, noting that postmultiplication by $\mathbf{J}$ to produce (5.83) has the effect of exchanging columns 2 and 3 in this case, the solution (5.86) becomes

$$\begin{bmatrix} x_1 \\ x_2 \\ x_3 \\ x_4 \end{bmatrix} = \mathbf{J}^T \left(\begin{bmatrix} 2 \\ -2 \\ 0 \\ 0 \end{bmatrix} + \begin{bmatrix} 3 & -4 \\ 0 & -7 \\ 1 & 0 \\ 0 & 1 \end{bmatrix} \begin{bmatrix} y_1 \\ y_2 \end{bmatrix} \right) = \begin{bmatrix} 2 + 3y_1 - 4y_2 \\ y_1 \\ -2 - 7y_2 \\ y_2 \end{bmatrix},$$

where y_1, y_2 are arbitrary parameters, and where premultiplication by $\mathbf{J}^T$ exchanges rows 2 and 3 within the parentheses on the right-hand side.

| 7.3 | **Uniqueness and generality of solutions** |

In general the matrices $\mathbf{P}$ and $\mathbf{Q}$ are not unique, as in Example 30, from which a dual example can be constructed for $\mathbf{Q}$. Nevertheless, certain conclusions can be reached, as follows.

Theorem 5.1   *If rank* $\mathbf{A} = n$, *then the solution* $\mathbf{X}$ *is unique.*

Proof: Let $\text{rank}(\mathbf{A}) = n$, and construct a particular solution $\mathbf{X}$ as in (5.82). Assume $\mathbf{X}' \neq \mathbf{X}$ also satisfies (5.76). Then

$$0 = \mathbf{AX} - \mathbf{AX}' = \mathbf{A}(\mathbf{X} - \mathbf{X}')$$
$$= \mathbf{A}_1(x_1 - x_1') + \mathbf{A}_2(x_2 - x_2') + \cdots \mathbf{A}_n(x_n - x_n'),$$

where the i-th column of $\mathbf{A}$ is written $\mathbf{A}_i$, and the i-th entries of $\mathbf{X}$ and $\mathbf{X}'$ are x_i and x_i' respectively. Therefore there exists a linear combination of the columns of $\mathbf{A}$ equal to zero, but in which the scalar coefficients are not all zero; that is $\text{rank}(\mathbf{A}) < n$, a contradiction. $\qquad\square$

It is important to ensure that to find all solutions (5.82), only one pair of matrices $\mathbf{P}, \mathbf{Q}$ is required.

Theorem 5.2   *Let particular matrices* $\mathbf{P}, \mathbf{Q}$ *satisfying (5.78) be given. Then all solutions of (5.76) are given by (5.82).*

Proof: Let $\mathbf{P}, \mathbf{Q}$ be fixed matrices satisfying (5.78), and let $\mathbf{X}'$ be any solution of (5.76). We shall prove that the equation

$$(5.87) \quad \mathbf{X}' = \mathbf{Q}_1\mathbf{P}_1\mathbf{B} + \mathbf{Q}_2\mathbf{Y}_2$$

is satisfied for some $\mathbf{Y}_2$.

First, because $\mathbf{Q}$ is nonsingular, the equation

$$(5.88) \quad \mathbf{X}' = [\,\mathbf{Q}_1, \mathbf{Q}_2\,] \begin{bmatrix} \mathbf{Y}_1' \\ \mathbf{Y}_2' \end{bmatrix}$$

has a unique solution $\begin{bmatrix} \mathbf{Y}_1' \\ \mathbf{Y}_2' \end{bmatrix}$. Now, by assumption, $\mathbf{AX}' = \mathbf{B}$, and premultiplying this equation by $\mathbf{P}$ yields

$$\mathbf{PAQ} \begin{bmatrix} \mathbf{Y}_1' \\ \mathbf{Y}_2' \end{bmatrix} = \begin{bmatrix} \mathbf{I}_r & 0 \\ 0 & 0 \end{bmatrix} \begin{bmatrix} \mathbf{Y}_1' \\ \mathbf{Y}_2' \end{bmatrix} = \begin{bmatrix} \mathbf{P}_1 \\ \mathbf{P}_2 \end{bmatrix} \mathbf{B},$$

from which $\mathbf{Y}_1' = \mathbf{P}_1\mathbf{B}$, and (5.87) becomes $\mathbf{X}' = \mathbf{Q}_1\mathbf{Y}_1' + \mathbf{Q}_2\mathbf{Y}_2$. By comparing this equation to (5.88), both (5.88) and (5.87) are seen to be satisfied if $\mathbf{Y}_2 = \mathbf{Y}_2'$. $\qquad\square$

| 7.4 | Special cases |

Several observations and simplifications can be made in special cases.

1. In (5.78), if $\mathbf{A}$ has full column rank $r = n$, then $\mathbf{Q} = \mathbf{Q}_1 = \mathbf{I}_n$ can be assumed, and only the particular solution appears in (5.82). In this case, augmenting $\mathbf{A}$ on the right with $\mathbf{B}$ and performing row operations to produce the normal form gives the result shown:

(5.89)
$$[\,\mathbf{A},\,\mathbf{B}\,] \rightarrow \begin{bmatrix} \mathbf{I}_n & \mathbf{P}_1\mathbf{B} \\ 0 & \mathbf{P}_2\mathbf{B} \end{bmatrix}.$$

Then provided $\mathbf{P}_2\mathbf{B} = 0$, the solution is $\mathbf{X} = \mathbf{P}_1\mathbf{B}$. Because, in this case, $\mathbf{P}_1\mathbf{A} = \mathbf{I}_n$, the matrix $\mathbf{P}_1$ is a left inverse of $\mathbf{A}$.

2. If $\mathbf{A}$ has full row rank $r = m$, then $\mathbf{P} = \mathbf{P}_1 = \mathbf{I}_m$ can be assumed. Here a matrix $\mathbf{Q}_1$ can be computed such that $\mathbf{A}\mathbf{Q}_1 = \mathbf{I}_m$, and $\mathbf{Q}_1$ is a right inverse of $\mathbf{A}$.

3. If $\mathrm{rank}(\mathbf{A}) < n$, then the solution is not unique, for by construction $\mathbf{Q}_2$ has $(n - \mathrm{rank}(A)) > 0$ nonzero columns, and the solution (5.82) has parameters in $\mathbf{Y}_2$.

4. If (5.76) is homogeneous, that is, $\mathbf{B} = 0$, then only the complementary function appears in (5.82), and the solution is obtained by column operations as shown:

(5.90)
$$\begin{bmatrix} \mathbf{A} \\ \mathbf{I}_n \end{bmatrix} \rightarrow \begin{bmatrix} \mathbf{A}\mathbf{Q}_1 & 0 \\ \mathbf{Q}_1 & \mathbf{Q}_2 \end{bmatrix},$$

where $\mathbf{A}\mathbf{Q}_1$ has full column rank r.

Example 50
Homogeneous
solution

Suppose we wish to find all solutions of the equation $\mathbf{A}\mathbf{X} = 0$ for $\mathbf{A}$ of Example 31. A set of elementary column operations for computation (5.90) corresponds to the matrix $\mathbf{Q}$ and the general solution shown,

$$\mathbf{Q} = [\,\mathbf{Q}_1,\mathbf{Q}_2\,] = \begin{bmatrix} 0 & 1 & 3 & -4 \\ 0 & 0 & 1 & 0 \\ 1 & 3 & 0 & -7 \\ 0 & 0 & 0 & 1 \end{bmatrix}, \quad \mathbf{X} = \begin{bmatrix} 3 & -4 \\ 1 & 0 \\ 0 & -7 \\ 0 & 1 \end{bmatrix}\mathbf{Y},$$

where $\mathbf{Y}$ contains arbitrary entries. The matrix $\mathbf{A}$ does not have full column rank, therefore $\mathbf{Q}$ is not unique.

Example 51
Homogeneous
solution from
SVD

Suppose that $\mathbf{A}\mathbf{X} = 0$ is to be solved for $\mathbf{X}$ using SVD, with $\mathbf{U}^T\mathbf{A}\mathbf{V} = \Sigma$. If $\mathbf{A}$ has less than full column rank, that is if $\Sigma = \begin{bmatrix} \Sigma_r & 0 \\ 0 & 0 \end{bmatrix}$ or $\Sigma = \begin{bmatrix} \Sigma_r & 0 \end{bmatrix}$, then the general solution is $\mathbf{X} = \mathbf{V}_2\mathbf{Y}$, where $\mathbf{V} = \begin{bmatrix} \mathbf{V}_1, & \mathbf{V}_2 \end{bmatrix}$ has been partitioned so that $\mathbf{V}_1$ contains $r = \mathrm{rank}(\mathbf{A})$ columns.

8   Further study

References to the algebra used here are Birkhoff and MacLane [5], Maclane and Birkhoff [34], or more recent books on linear algebra such as Strang [51]. A readable introduction and numerous examples are found in Ayres [4]. Implementing linear algebra reliably by using floating-point arithmetic requires knowledge of numerical analysis, for which introductions can be found in Stewart [50] or Golub and Van Loan [19]. Professional routines meant to be accurate and reliable when physical insight cannot be used directly in the computations can be found in scientific libraries such as LAPACK (Anderson et al. [1]), and routines of similar quality are built into commercial software packages such as MATLAB (The MathWorks, Inc. [35]) or MATRIX$_X$ (Integrated Systems Inc. [23]).

The Ho algorithm given here or with modifications can be used to estimate the order of a model of measured data, and as an initial estimate of the system; see, for example, Juang [24]. The algorithm is discussed in standard texts such as Chen [9], but the original article by Ho and Kalman [21] is still a useful reference.

The analysis of linear equations given in this chapter has direct relevance, via the SVD, to least-squares problems, as in Lawson and Hanson [31]. The normal-form computation for nonsingular square matrices is directly related to the LU factorization of Example 38; see Stewart [50].

9   Problems

1   Construct the multiplication table to show that $\mathbb{Z}_6$ cannot be a field, since the nonzero elements of the set do not all have multiplicative inverses, violating axiom M_5.

2   Analog sensors typically produce numerical values within a bounded range. Show that the set of bounded real numbers $\mathbb{R}_{[-a,a]} = \{x : |x| \leq a\}$, where $x, a \in \mathbb{R}$ and $a > 0$, is not a field.

3 Determine why each of the following is not a field: (a) the set of 4-byte signed integers, with computer integer arithmetic, (b) the set of 8-byte floating-point numbers, with floating-point computer arithmetic.

4 Let $\mathbb{R}[x]$ be the set of polynomials in x with real coefficients, and with ordinary arithmetic. This set satisfies the same axioms as the integers $\mathbb{Z}$. Like the integers, $\mathbb{R}[x]$ contains some polynomials that are products of other polynomials in $\mathbb{R}[x]$, whereas other polynomials, having no such factors, are called irreducible, or *prime*. One prime polynomial is $x^2 + 1$. By long division, any polynomial $p(x)$ can be written $p(x) = q(x)(x^2 + 1) + r(x)$, where $q(x)$ is the quotient and $r(x) = ax + b$ is the remainder, a polynomial of degree less than 2, with a and b real quantities. Let $\mathbb{R}_{x^2+1}[x]$ be the set of polynomials modulo $(x^2 + 1)$, that is, the set of remainders $r(x)$ on division by $(x^2 + 1)$. Show that $\mathbb{R}_{x^2+1}[x]$ is a field. A field obtained by arithmetic modulo a prime is called an *extension field* of the original set.

5 Consider the set $\mathbb{Z}_{2\,x^2+x+1}[x]$, of polynomials with mod 2 integer coefficients, and with addition and multiplication modulo the prime polynomial $x^2 + x + 1$. The set contains the entries $\{0, 1, x, x+1\}$, the remainders produced on division by $x^2 + x + 1$. Prove that $x^2 + x + 1$ is prime. Write the addition and multiplication tables for the set, and show that it satisfies the field axioms. This set is often named $GF(2^2)$ (where GF stands for Galois Field).

6 Show that the set of diagonal 3×3 matrices with elements in $\mathbb{R}$, with ordinary matrix addition and scalar multiplication, is a field.

7 Let $\mathbf{A} \in F^{m \times n}$ and $\mathbf{B} \in F^{n \times r}$ be written

$$\mathbf{A} = [\,\mathbf{A}_1, \cdots \mathbf{A}_n\,], \quad \mathbf{B} = \begin{bmatrix} \mathbf{B}_1 \\ \vdots \\ \mathbf{B}_n \end{bmatrix},$$

respectively, where $\mathbf{A}_j$ is column j of $\mathbf{A}$, and $\mathbf{B}_j$ is row j of $\mathbf{B}$. Show that the product $\mathbf{AB}$ can be computed as $\sum_{j=1}^{n} \mathbf{A}_j \mathbf{B}_j$.

8 Show that if $\mathbf{S}, \mathbf{Q} \in \mathbb{R}^{n \times n}$ and $\mathbf{Q}$ is symmetric, then the product $\mathbf{S}^T \mathbf{Q} \mathbf{S}$ is symmetric.

9   Using elementary row transformations, find a left inverse of the real matrix

$$A = \begin{bmatrix} 0 & 2 \\ -2 & 3 \\ -1 & -1 \end{bmatrix}.$$

10   Using elementary column transformations, find a right inverse of the real matrix

$$A = \begin{bmatrix} 0 & 2 & 3 \\ -2 & 3 & -1 \end{bmatrix}.$$

11   Compute the determinant of the real matrix

$$A = \begin{bmatrix} 1 & 4 & -2 \\ -1 & 0 & 1 \\ 2 & 1 & -2 \end{bmatrix}$$

by (a) Laplace expansion along row 1, (b) by Laplace expansion along column 3, and (c) by the permutation method, and verify that each of these methods produces the same result for this matrix.

12   Compute the determinant of A in Example 16 by Laplace expansion along rows 2 and 3.

13   Using the definition of orthogonal matrices and the properties of determinants, show that if A is an orthogonal matrix, its determinant is ± 1.

14   Find the rank of each of the matrices in Problems 9 and 10.

15   Find nonsingular matrices P, Q, such that PAQ is the normal form under equivalence of the matrix $A = \begin{bmatrix} 0 & 1 & -1 & 1 \\ -2 & 3 & 0 & 4 \\ -2 & 4 & -1 & 5 \end{bmatrix}.$

16   Using a computer program, obtain answers to Problems 9, 10, and 15 by first computing the SVD of A, and then the required matrices in each case.

17   Over the integers mod 2, find nonsingular matrices P, Q, such that PAQ is the normal form under equivalence of the matrix $A = \begin{bmatrix} 0 & 1 & 1 & 1 \\ 0 & 1 & 0 & 0 \\ 0 & 0 & 1 & 1 \end{bmatrix}.$

18   For Example 48, find another $\mathbf{P}$, $\mathbf{Q}$ that satisfy (5.78).

19   Show that if, in (5.10), $\mathbf{AQ}_1$ is in reduced column echelon form, then a suitable $\mathbf{P}$ can be written by inspection of $\mathbf{AQ}_1$.

20   Write the characteristic polynomial for the matrix of Problem 11.

21   Writing the characteristic polynomial $\phi(\lambda)$ of $\mathbf{A} \in F^{n \times n}$ as $\phi(\lambda) = \lambda^n + a_1 \lambda^{n-1} + \cdots a_n = (\lambda - \lambda_1)(\lambda - \lambda_2) \cdots (\lambda - \lambda_n)$, show that in $\phi(\lambda)$, the coefficient a_1 is the negative of the sum of the roots $\lambda_1, \cdots \lambda_n$, and the coefficient a_n is $(-1)^n \times$ their product. Then by inspection of

$$\phi(\lambda) = \det(\lambda \mathbf{I} - \mathbf{A}) = \det \begin{bmatrix} \lambda - a_{11} & -a_{12} & -a_{13} & \cdots \\ -a_{21} & \lambda - a_{22} & -a_{23} & \cdots \\ \cdots & & & \\ & \cdots & & \lambda - a_{nn} \end{bmatrix},$$

show that $a_n = \det(-\mathbf{A}) = (-1)^n \det(\mathbf{A})$, and for $n = 3$, show that $a_1 = -\text{trace}(\mathbf{A}) = -(\sum_{i=1}^{n} a_{ii})$, which is a general result for arbitrary n.

22   For $\mathbf{A} = \begin{bmatrix} 1 & -2 \\ 2 & 5 \end{bmatrix}$, verify by computation that $\phi(\mathbf{A}) = 0$.

23   For the weighting function $h(t) = 0\delta(t) + 3te^{-2t}$, calculate sufficient terms of the associated Markov sequence, and use the Ho algorithm to find matrices $\mathbf{A}$, $\mathbf{B}$, $\mathbf{C}$, $\mathbf{D}$ that have weighting function $h(t)$.

24   Use the Ho algorithm to find a minimal system with weighting sequence $\{\mathbf{H}_k\}_0^\infty = \{1, 2, 3, \cdots\}$.

25   Use the Ho algorithm to find a minimal system with weighting sequence $\{\mathbf{H}_k\}_0^\infty = \{0, 1, 1+\beta, 2\beta+\beta^2, \cdots k\beta^{k-1}+\beta^k, \cdots\}$.

26   Use the Ho algorithm to find a minimal system with periodic mod-2 weighting sequence $\{\mathbf{H}_k\}_0^\infty = \{[0, 0], [0, 1], [1, 1], [0, 0], [0, 1], [1, 1], \cdots\}$.

27 Where possible, find the general solution of the equations $\mathbf{AX} = \mathbf{B}$ for the following pairs of real matrices:

(a) $\mathbf{A} = \begin{bmatrix} 1 & 1 \\ 1 & -1 \end{bmatrix}$, $\mathbf{B} = \begin{bmatrix} 2 \\ 0 \end{bmatrix}$, (b) $\mathbf{A} = \begin{bmatrix} 0 & 1 & 1 \\ 2 & 1 & -1 \end{bmatrix}$, $\mathbf{B} = \begin{bmatrix} 2 \\ 0 \end{bmatrix}$,

(c) $\mathbf{A} = \begin{bmatrix} 0 & 1 & 1 \\ 2 & 1 & -1 \\ 2 & 2 & 0 \end{bmatrix}$, $\mathbf{B} = \begin{bmatrix} 2 \\ 2 \\ 4 \end{bmatrix}$, (d) $\mathbf{A}$ as in (c), $\mathbf{B} = \begin{bmatrix} 2 \\ 2 \\ -4 \end{bmatrix}$.

28 Solve Example 48 by the abbreviated method.

29 When performed using floating-point computation, the Gaussian elimination algorithm of Section 3.3 is often modified to use *partial pivoting,* in which the element with largest absolute value in the leftmost column is brought to the top by row interchanges, rather than choosing any nonzero in the leftmost column of the current submatrix. In so doing, the entries in the resulting echelon form are kept small, thereby avoiding magnification of errors. Use partial pivoting to compute the upper-right row echelon form of the matrix $\mathbf{A}$ of Problem 11.

30 In the presence of inexact data or computation, in order to guarantee that the computed matrix is exact for a matrix near the original, the Gaussian elimination algorithm for row compression is replaced by premultiplication (or postmultiplication, for column compression) by a matrix $\mathbf{H}$ that is orthogonal; that is, $\mathbf{H}^{-1} = \mathbf{H}^T$. The computational requirement is that $\mathbf{HX} = \pm \mathbf{e}_1 \|\mathbf{X}\|$, where $\mathbf{X}$ is a column of the matrix (or submatrix) being compressed; $\mathbf{e}_1$ is a zero vector except for the topmost entry, which is 1; and $\|\mathbf{X}\|$ is the Euclidean length of $\mathbf{X}$. Show that (a) the Householder matrix $\mathbf{H} = \mathbf{I} - (2/\mathbf{V}^T\mathbf{V})\mathbf{V}\mathbf{V}^T$ is orthogonal, where $\mathbf{V}$ is a vector, and that (b) if $\mathbf{V} = \mathbf{X} \mp \|\mathbf{X}\|\mathbf{e}_1$, then $\mathbf{HX}$ is $\pm\|\mathbf{X}\|\mathbf{e}_1$, as required.

Vector spaces

The basic definitions and operations applying to vector spaces will be given. Vector calculations require the matrix operations of Chapter 5 when carried out with respect to a given set of coordinates for a vector space of finite dimension. Since several important results can be expressed in a more abstract, coordinate-free notation, there is often freedom of choice of coordinates, which can be of great convenience when applied to state-space models.

1 Vector-space axioms

A vector space is a set $\mathcal{X}$ in which addition is defined, as well as multiplication of elements of $\mathcal{X}$ by elements of a field F. The axioms that must be satisfied are given below, with $\mathbf{X}$, $\mathbf{Y}$, $\mathbf{Z}$ arbitrary elements of $\mathcal{X}$, and α, β arbitrary elements of F.

A_1 *Uniqueness of addition:* $\mathbf{X} + \mathbf{Y}$ is an element of $\mathcal{X}$, uniquely determined by the pair $(\mathbf{X}, \mathbf{Y})$.

A_2 *Commutative law of addition:* $\mathbf{X} + \mathbf{Y} = \mathbf{Y} + \mathbf{X}$.

A_3 *Associative law of addition:* $\mathbf{X} + (\mathbf{Y} + \mathbf{Z}) = (\mathbf{X} + \mathbf{Y}) + \mathbf{Z}$.

A_4 *Additive identity:* For every element $\mathbf{X}$ in $\mathcal{X}$ there exists an element 0 in $\mathcal{X}$ such that $\mathbf{X} + 0 = 0 + \mathbf{X} = \mathbf{X}$.

A_5 *Additive inverse:* For each element $\mathbf{X}$ in $\mathcal{X}$ there exists a unique element $-\mathbf{X}$ in $\mathcal{X}$ such that $\mathbf{X} + (-\mathbf{X}) = 0$.

SM_1 *Scalar multiplication:* For each scalar $\alpha \in F$ and every vector $\mathbf{X} \in \mathcal{X}$ the operation $\alpha\mathbf{X} = \mathbf{X}\alpha$ produces a vector in $\mathcal{X}$.

SM_2 *Associative law of scalar multiplication:* $(\alpha\beta)\mathbf{X} = \alpha(\beta\mathbf{X})$.

SM_3 *Distributive law:* $\alpha(\mathbf{X} + \mathbf{Y}) = (\alpha\mathbf{X}) + (\alpha\mathbf{Y})$.

SM_4 *Distributive law:* $(\alpha + \beta)\mathbf{X} = (\alpha\mathbf{X}) + (\beta\mathbf{X})$.

SM_5 *Unit scalar multiplication:* $1\mathbf{X} = \mathbf{X}$, with 1 the unit in F.

Example 1
Real n-vectors

The vectors in $\mathbb{R}^n$ used in previous chapters satisfy the above axioms when ordinary vector addition and scalar multiplication is used. That is, let $\mathbf{X}, \mathbf{Y} \in \mathbb{R}^n$ and $\alpha \in \mathbb{R}$, and define vector addition and scalar multiplation to be respectively

$$\mathbf{X} + \mathbf{Y} = \begin{bmatrix} x_1 \\ \vdots \\ x_n \end{bmatrix} + \begin{bmatrix} y_1 \\ \vdots \\ y_n \end{bmatrix} = \begin{bmatrix} x_1 + y_1 \\ \vdots \\ x_n + y_n \end{bmatrix}, \quad \alpha\mathbf{X} = \mathbf{X}\alpha = \begin{bmatrix} \alpha x_1 \\ \vdots \\ \alpha x_n \end{bmatrix}.$$

Example 2
Real matrices

The set of $m \times n$ matrices with elements in field F is a vector space, with element-wise addition and scalar multiplication, as follows. Let $\mathbf{A}, \mathbf{B} \in F^{m \times n}$ and $\alpha \in F$. Define vector addition and scalar multiplation to be respectively

$$\mathbf{A} + \mathbf{B} = \begin{bmatrix} a_{11} & \cdots & a_{1n} \\ \cdots & & \\ a_{m1} & \cdots & a_{mn} \end{bmatrix} + \begin{bmatrix} b_{11} & \cdots & b_{1n} \\ \cdots & & \\ b_{m1} & \cdots & b_{mn} \end{bmatrix}$$

$$= \begin{bmatrix} a_{11}+b_{11} & \cdots & a_{1n}+b_{1n} \\ \cdots & & \\ a_{m1}+b_{m1} & \cdots & a_{mn}+b_{mn} \end{bmatrix},$$

$$\alpha\mathbf{A} = \mathbf{A}\alpha = \begin{bmatrix} \alpha a_{11} & \cdots & \alpha a_{1n} \\ \cdots & & \\ \alpha a_{m1} & \cdots & \alpha a_{mn} \end{bmatrix}.$$

2   Subspaces

A set $\mathcal{V}$ is a subspace of a vector space $\mathcal{X}$ if every element of $\mathcal{V}$ is in $\mathcal{X}$, and if $\mathcal{V}$ is a vector space.

Assume that every member of the set $\mathcal{V}$ is a member of vector space $\mathcal{X}$, and that vector addition and scalar multiplication as defined for $\mathcal{X}$ also apply to $\mathcal{V}$. Then to show that $\mathcal{V}$ is a vector space it is only necessary to verify axioms A_1 and SM_1 for $\mathcal{V}$, since the other axioms follow from the properties of arithmetic in $\mathcal{X}$ over F.

Example 3
Points on a line

Consider the vector space of real column vectors $\begin{bmatrix} x_1 \\ x_2 \end{bmatrix}$ in two dimensions, denoted $\mathbb{R}^2$, with ordinary vector addition and scalar multiplication. Let $\mathcal{V} = \{\begin{bmatrix} x_1 \\ x_2 \end{bmatrix} : x_2 = mx_1 + b\}$ with m and b fixed. That is, $\mathcal{V}$ is the set of points on a straight line with slope m and intercept b. Then if $b = 0$, $\mathcal{V}$ is a vector subspace of $\mathbb{R}^2$, since $\mathcal{V}$ is a vector space. However, if $b \neq 0$, the elements of $\mathcal{V}$ do not satisfy the vector-space axioms.

Example 4
Linear
combination of
two vectors

In $\mathbb{R}^3$, let $\mathbf{X}_1$ and $\mathbf{X}_2$ be constant vectors, and consider the set $\mathcal{V} = \{\mathbf{Y} : \mathbf{Y} = \mathbf{X}_1\alpha + \mathbf{X}_2\beta\}$, for all α, $\beta \in \mathbb{R}$. Then $\mathcal{V}$ is a subset of $\mathbb{R}^3$, since each member of $\mathcal{V}$ is in $\mathbb{R}^3$. For any $\mathbf{Y}_1 = \mathbf{X}_1\alpha_1 + \mathbf{X}_2\beta_1$ and $\mathbf{Y}_2 = \mathbf{X}_1\alpha_2 + \mathbf{X}_2\beta_2$ in $\mathcal{V}$ and any $k \in \mathbb{R}$, vector addition and scalar multiplication give respectively

$$\mathbf{Y}_1 + \mathbf{Y}_2 = (\mathbf{X}_1\alpha_1 + \mathbf{X}_2\beta_1) + (\mathbf{X}_1\alpha_2 + \mathbf{X}_2\beta_2)$$
$$= \mathbf{X}_1(\alpha_1 + \alpha_2) + \mathbf{X}_2(\beta_1 + \beta_2) \in \mathcal{V},$$
$$k\mathbf{Y}_1 = k(\mathbf{X}_1\alpha_1 + \mathbf{X}_2\beta_1) = \mathbf{X}_1(k\alpha_1) + \mathbf{X}_2(k\beta_1) \in \mathcal{V},$$

and the vector-space axioms are easily verified for $\mathcal{V}$, showing that $\mathcal{V}$ is a vector subspace of $\mathbb{R}^3$.

3   Linear dependence of vectors

The definition of linear dependence in Section 3 of Chapter 5 applies to vectors as follows. The vectors $\mathbf{A}_1$, $\mathbf{A}_2$, $\cdots \mathbf{A}_n$ over field F are linearly dependent if there exists a set $\{k_1, \cdots k_n\}$ of elements of F, not all zero, such that

$$(6.1) \quad \mathbf{A}_1 k_1 + \mathbf{A}_2 k_2 + \cdots \mathbf{A}_n k_n = [\mathbf{A}_1, \mathbf{A}_2, \cdots \mathbf{A}_n] \begin{bmatrix} k_1 \\ k_2 \\ \vdots \\ k_n \end{bmatrix} = 0.$$

If no such set of scalars k_i exists, the vectors are linearly independent. Consequently, the vectors are linearly independent if and only if the matrix $\mathbf{A} = [\mathbf{A}_1, \mathbf{A}_2, \cdots \mathbf{A}_n]$ has full column rank, since from Chapter 5, Section 7.3, $k_i = 0$, $i = 1, \cdots n$, is the unique solution of (6.1) if $\text{rank}(\mathbf{A}) = n$; and if $\text{rank}(\mathbf{A}) < n$, then the general solution for the vector $[k_i] = \mathbf{K}$ of (6.1) is of the form $\mathbf{Q}_2\mathbf{Y}$ for nonzero $\mathbf{Q}_2$ and arbitrary $\mathbf{Y}$.

Example 5
Fewer rows
than columns

The columns of any matrix $\mathbf{A} \in F^{m \times n}$ cannot be linearly independent if $m < n$, since $\text{rank}(\mathbf{A}) \leq m < n$, in which case nonzero solutions of (6.1) exist.

4   Range, basis, dimension, and null space

Range

Given a matrix $\mathbf{A} = [\mathbf{A}_1, \mathbf{A}_2 \cdots \mathbf{A}_n] \in F^{m \times n}$, or equivalently, given the column vectors $\mathbf{A}_1$, $\mathbf{A}_2 \cdots \mathbf{A}_n \in F^m$, the *range,* or *image,* written $\mathcal{R}(\mathbf{A})$, is the set of all linear combinations of these vectors:

$$(6.2) \quad \mathcal{R}(\mathbf{A}) = \left\{ \mathbf{X} : \mathbf{X} = [\mathbf{A}_1, \ \mathbf{A}_2 \cdots \mathbf{A}_n] \begin{bmatrix} k_1 \\ k_2 \\ \vdots \\ k_n \end{bmatrix} = \mathbf{AK}, \text{ where } \mathbf{K} \in F^n \right\}.$$

Each vector $\mathbf{A}_i$ above is in $\mathcal{R}(\mathbf{A})$, since, using the notation $\mathbf{e}_i = (i\text{-th column of } \mathbf{I})$, a possible choice of $\mathbf{K}$ is $\mathbf{K} = \mathbf{e}_i$, so that $\mathbf{AK} = \mathbf{Ae}_i = \mathbf{A}_i$.

The columns of $\mathbf{A}$ are said to *span* $\mathcal{R}(\mathbf{A})$.

It is simple to verify that with ordinary matrix addition and scalar multiplication, $\mathcal{R}(\mathbf{A})$ satisfies the vector-space axioms, and is therefore a subspace of F^m.

Basis   If the columns of $\mathbf{A}$ above are linearly independent, they are called a *basis* for $\mathcal{R}(\mathbf{A})$, and since $\mathbf{A}$ then has full column rank n, any vector $\mathbf{X} \in \mathcal{R}(\mathbf{A})$ is a unique linear combination of the basis vectors in $\mathbf{A}$; that is, given $\mathbf{X}$, the equation $\mathbf{X} = \mathbf{AK}$ has a unique solution for $\mathbf{K}$. The entries of $\mathbf{K}$ are called the *coordinates* of $\mathbf{X}$ with respect to the basis $\mathbf{A}$.

Dimension   The *dimension* of a vector space is the number of vectors in a basis for the space.

Standard basis   The set of vectors $[\mathbf{e}_1, \mathbf{e}_2, \cdots \mathbf{e}_n] = \mathbf{I}_n$ is called the *standard basis* for F^n. Then since $\mathbf{X} = \mathbf{I}_n \mathbf{X}$, the entries of vector $\mathbf{X}$ are its coordinates with respect to the standard basis. Other notation is sometimes used, as when $n = 3$ and the substitutions $\mathbf{e}_1 = \mathbf{i}, \mathbf{e}_2 = \mathbf{j}, \mathbf{e}_3 = \mathbf{k}$ are made for the standard basis vectors, and these vectors are associated with unit vectors along axes labeled as illustrated in Figure 6.1.

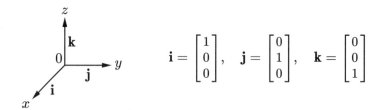

Fig. 6.1   Notation for a standard basis used in three dimensions.

From the previous definition of $\mathcal{R}(\mathbf{A})$, the equation $\mathbf{AX} = \mathbf{B}$ has a solution if and only if the columns of $\mathbf{B}$ (or $\mathbf{B}$ itself, if it contains only one column) are in the range of $\mathbf{A}$. Slightly more generally, $\mathcal{R}(\mathbf{B}) \subset \mathcal{R}(\mathbf{A})$ if and only if $\mathbf{AX} = \mathbf{B}$ has a solution $\mathbf{X}$, and this condition can be tested by constructing the matrix

[**A**, **B**] and computing an upper-row compression, as in Section 7 of Chapter 5,

$$(6.3) \quad [\mathbf{A}, \mathbf{B}] \overset{\text{row}}{\sim} \begin{bmatrix} \mathbf{P}_1\mathbf{A} & \mathbf{P}_1\mathbf{B} \\ 0 & \mathbf{P}_2\mathbf{B} \end{bmatrix},$$

so that $\mathcal{R}(\mathbf{B}) \subset \mathcal{R}(\mathbf{A})$ if and only if $\mathbf{P}_2\mathbf{B} = 0$.

Example 6
Span

The vectors $\mathbf{A}_1 = \begin{bmatrix} 2 \\ 0 \end{bmatrix}$, $\mathbf{A}_2 = \begin{bmatrix} 4 \\ 1 \end{bmatrix}$, $\mathbf{A}_3 = \begin{bmatrix} -2 \\ -1 \end{bmatrix}$ are all in the plane $\mathbb{R}^2$, and span $\mathbb{R}^2$ since the equation

$$[\mathbf{A}_1, \mathbf{A}_2, \mathbf{A}_3]\, \mathbf{K} = \begin{bmatrix} 2 & 0 & -2 \\ 4 & 1 & -1 \end{bmatrix} \begin{bmatrix} k_1 \\ k_2 \\ k_3 \end{bmatrix} = \mathbf{Y} = \begin{bmatrix} y_1 \\ y_2 \end{bmatrix}$$

has a solution $\mathbf{K}$ for any vector $\mathbf{Y} \in \mathbb{R}^2$.

Example 7
Basis

The vectors $\mathbf{A}_1$, $\mathbf{A}_2$ of the previous example are a basis for $\mathbb{R}^2$, as follows. First, since the equation

$$[\mathbf{A}_1, \mathbf{A}_2]\, \mathbf{K}' = \mathbf{Y}$$

has a solution $\mathbf{K}'$ for any $\mathbf{Y} \in \mathbb{R}^2$, these vectors span $\mathbb{R}^2$. Second, since the rank of rank $[\mathbf{A}_1, \mathbf{A}_2]$ is 2, this matrix contains a minimal set of spanning vectors, as follows: suppose a single vector $\mathbf{B}$ were a basis, then $\mathbf{A}_1 = \mathbf{B}k_1$ and $\mathbf{A}_2 = \mathbf{B}k_2$, but rank $[\mathbf{A}_1, \mathbf{A}_2] = \text{rank}(\mathbf{B}[k_1, k_2]) < 2$, a contradiction.

Null space

The *null space*, or *kernel*, of a matrix $\mathbf{A} \in F^{m \times n}$ over a field F is the set

$$(6.4) \quad \mathcal{N}(\mathbf{A}) = \{\mathbf{X} : \mathbf{A}\mathbf{X} = 0\}.$$

With multiplication defined in F, the set $\mathcal{N}(\mathbf{A})$ satisfies the vector-space axioms, and is therefore a subspace of F^n. The dimension of $\mathcal{N}(\mathbf{A})$ is called the *nullity* of $\mathbf{A}$.

From the above definition, the null space of $\mathbf{A}$ is the set of solutions $\mathbf{X}$ of the homogeneous equation $\mathbf{A}\mathbf{X} = 0$, and therefore contains nonzero vectors $\mathbf{X}$ only if rank$(\mathbf{A}) < n$.

If the vector $\mathbf{X}$ is a solution of the equation $\mathbf{A}\mathbf{X} = \mathbf{B}$, then another solution is $\mathbf{X}' = \mathbf{X} + \mathbf{H}$ where $\mathbf{H}$ is any vector in $\mathcal{N}(\mathbf{A})$, for then $\mathbf{A}(\mathbf{X} + \mathbf{H}) = \mathbf{A}\mathbf{X} + 0 = \mathbf{B}$. Hence $\mathbf{X}$ is a unique solution only if $\mathbf{N}(\mathbf{A}) = 0$, which implies that the solution is unique only if $\mathbf{A}$ has full column rank.

4.1	**Bases for the range and null space**

The construction of a basis for the range, or the null space, or both, of a given matrix $\mathbf{A} \in F^{m \times n}$ is a very common computational requirement. Applications to state-space systems are discussed in Chapter 9, where a basis for the range is required, and in Chapters 7 and 9, where bases for the null space are required.

Construct the array $\begin{bmatrix} \mathbf{A} \\ \mathbf{I}_n \end{bmatrix}$, and perform column operations on it, or equivalently postmultiply by a nonsingular matrix $\mathbf{Q} = [\mathbf{Q}_1, \mathbf{Q}_2]$ as described in Section 3 of Chapter 5 to produce

(6.5)
$$\begin{bmatrix} \mathbf{A} \\ \mathbf{I} \end{bmatrix} \mathbf{Q} = \begin{bmatrix} \mathbf{A}\mathbf{Q}_1 & 0 \\ \mathbf{Q}_1 & \mathbf{Q}_2 \end{bmatrix},$$

where $\mathrm{rank}(\mathbf{A}) = \mathrm{rank}(\mathbf{A}\mathbf{Q}_1)$. (When computed with floating-point arithmetic, it is often of advantage for numerical reasons to require $\mathbf{Q}$ to be orthogonal.)

Basis of $\mathcal{R}(\mathbf{A})$ The $m \times r$ matrix $\mathbf{A}\mathbf{Q}_1$ is a basis for $\mathcal{R}(\mathbf{A})$.

Proof: The above statement can be proved as follows. First, $\mathcal{R}(\mathbf{A}\mathbf{Q}_1) \subset \mathcal{R}(\mathbf{A})$, since every vector $\mathbf{X} \in \mathcal{R}(\mathbf{A}\mathbf{Q}_1)$ can be written, for some column $\mathbf{L}_1$, as $\mathbf{X} = (\mathbf{A}\mathbf{Q}_1)\mathbf{L}_1 = \mathbf{A}(\mathbf{Q}_1\mathbf{L}_1)$, which exhibits $\mathbf{X}$ as a vector in both $\mathcal{R}(\mathbf{A}\mathbf{Q}_1)$ and $\mathcal{R}(\mathbf{A})$. Second, $\mathcal{R}(\mathbf{A}) \subset \mathcal{R}(\mathbf{A}\mathbf{Q}_1)$, since every vector $\mathbf{X} \in \mathcal{R}(\mathbf{A})$ can be written, for some $\mathbf{K}$, as

(6.6) $\mathbf{X} = \mathbf{A}\mathbf{K} = (\mathbf{A}\mathbf{Q})(\mathbf{Q}^{-1}\mathbf{K}) = [\mathbf{A}\mathbf{Q}_1, 0] \begin{bmatrix} \mathbf{L}_1 \\ \mathbf{L}_2 \end{bmatrix} = (\mathbf{A}\mathbf{Q}_1)\mathbf{L}_1,$

where $\mathbf{Q}^{-1}\mathbf{K}$ has been rewritten as $\begin{bmatrix} \mathbf{L}_1 \\ \mathbf{L}_2 \end{bmatrix}$. Finally, by construction, $(\mathbf{A}\mathbf{Q}_1)$ contains r independent columns, and is therefore a basis for its range.   □

Example 8
Basis for the range

To compute a basis for $\mathcal{R}(\mathbf{A})$ of Example 23 of Chapter 5, column operations may be performed to compute the lower-left column echelon form, as shown:

$$\mathbf{A} = \begin{bmatrix} -3 & 9 & 1 & -5 \\ -2 & 6 & 0 & -8 \\ -4 & 12 & 1 & -9 \end{bmatrix} \overset{\mathrm{col}}{\sim} \begin{bmatrix} -3 & 0 & 0 & 0 \\ -2 & -2/3 & 0 & 0 \\ -4 & -1/3 & 0 & 0 \end{bmatrix} = \mathbf{A}[\mathbf{Q}_1, \mathbf{Q}_2] = [\mathbf{A}\mathbf{Q}_1, 0].$$

Then $\mathbf{A}\mathbf{Q}_1$ is a basis for $\mathcal{R}(\mathbf{A})$.

Example 9
Selecting independent columns

The columns of a matrix $\mathbf{A}$ of Example 8 are in $\mathcal{R}(\mathbf{A})$, and constitute a basis for their range if they are independent. Therefore, in simple cases such as the previous example, if the rank r of a matrix $\mathbf{A}$ is known in advance, it is often possible to obtain a basis for the range simply by selecting r linearly independent columns of the matrix, rather than by performing column operations.

By inspection of the columns of $\mathbf{A}$ in Example 8, columns 1 and 3 are linearly independent, and since the matrix is known to have rank 2, a basis for $\mathcal{R}(\mathbf{A})$ is the matrix

$$\begin{bmatrix} -3 & 1 \\ -2 & 0 \\ -4 & 1 \end{bmatrix}.$$

Basis of $\mathcal{N}(\mathbf{A})$ The $n\times(n-r)$ matrix $\mathbf{Q}_2$ in (6.5) is a basis for $\mathcal{N}(\mathbf{A})$.

Proof: First, $\mathcal{R}(\mathbf{Q}_2) \subset \mathcal{N}(\mathbf{A})$ since by construction $\mathbf{A}\mathbf{Q}_2 = 0$, so that for every vector $\mathbf{X} \in \mathcal{R}(\mathbf{Q}_2)$, written as $\mathbf{Q}_2\mathbf{K}$, $\mathbf{A}\mathbf{X} = \mathbf{A}(\mathbf{Q}_2\mathbf{K}) = 0\mathbf{K} = 0$. Further, $\mathcal{N}(\mathbf{A}) \subset \mathcal{R}(\mathbf{Q}_2)$, which is proved by supposing, to the contrary, that there is a nonzero vector $\mathbf{X} \in \mathcal{N}(\mathbf{A})$ that is not a linear combination of the columns of $\mathbf{Q}_2$; that is,

$$(6.7) \quad \mathbf{X} = \mathbf{Q} \begin{bmatrix} \mathbf{L}_1 \\ \mathbf{L}_2 \end{bmatrix} = \mathbf{Q}_1\mathbf{L}_1 + \mathbf{Q}_2\mathbf{L}_2$$

with $\mathbf{L}_1 \neq 0$ by assumption. Then

$$(6.8) \quad \mathbf{A}\mathbf{X} = \mathbf{A}\mathbf{Q}_1\mathbf{L}_1 + (\mathbf{A}\mathbf{Q}_2)\mathbf{L}_2 = (\mathbf{A}\mathbf{Q}_1)\mathbf{L}_1 = 0,$$

which implies that the columns of $\mathbf{A}\mathbf{Q}_1$ are linearly dependent, a contradiction by the construction of $\mathbf{Q}$. Finally, $\mathbf{Q}_2$ is a basis for its range, since $\mathbf{Q}$ is nonsingular, and therefore contains independent columns; hence $\mathbf{Q}_2$ contains independent columns.   □

From the above results, the dimension of $\mathcal{R}(\mathbf{A})$ is $r = \text{rank}(\mathbf{A})$, and the dimension of $\mathcal{N}(\mathbf{A})$ is $n-r = \text{nullity}(\mathbf{A})$, so the sum of these two dimensions must equal n, the number of columns of $\mathbf{A}$.

Example 10
Basis for null space

A matrix $\mathbf{Q}$ corresponding to the column operations in Example 8 was computed previously in Example 50, Chapter 5, as shown:

$$[\mathbf{Q}_1, \mathbf{Q}_2] = \begin{bmatrix} 0 & 1 & 3 & -4 \\ 0 & 0 & 1 & 0 \\ 1 & 3 & 0 & -7 \\ 0 & 0 & 0 & 1 \end{bmatrix},$$

so that a basis for $\mathcal{N}(\mathbf{A})$ is the above $\mathbf{Q}_2$.

<table>
<tr><td>

Example 11
Bases from the
SVD

</td><td>

Let $\mathbf{A} \in \mathbb{R}^{m \times n}$ have rank r and SVD factorization $\mathbf{A} = \mathbf{U}_1 \Sigma_r \mathbf{V}_1^T$ from Equation (5.12), where $\mathbf{U}_1$ and $\mathbf{V}_1$ have r columns and full column rank r. Then $\mathbf{U}_1$ is a basis for $\mathcal{R}(\mathbf{A})$ since any vector $\mathbf{AK} \in \mathcal{R}(\mathbf{A})$ can be written $\mathbf{AK} = \mathbf{U}_1(\Sigma_r \mathbf{V}_1^T \mathbf{K})$. Similarly, since $\mathbf{AV} = \mathbf{A}\,[\,\mathbf{V}_1, \mathbf{V}_2\,] = [\,\mathbf{U}_1 \Sigma_r, 0\,]$, a basis for $\mathcal{N}(\mathbf{A})$ is $\mathbf{V}_2$.

</td></tr>
</table>

4.2   Orthogonal bases

In $\mathbb{R}^2$ or $\mathbb{R}^3$, the angle θ between two nonzero vectors $\mathbf{X}$, $\mathbf{Y}$ satisfies

(6.9) $\mathbf{X}^T \mathbf{Y} = \|\mathbf{X}\|_E \|\mathbf{Y}\|_E \cos \theta,$

where, in some contexts, $\mathbf{X}^T \mathbf{Y}$ is called the *inner product* or, when written $\mathbf{X} \cdot \mathbf{Y}$, the *dot product*. Then since the Euclidean norm in $\mathbb{R}^n$ satisfies the Schwarz inequality

(6.10) $|\mathbf{X}^T \mathbf{Y}| \leq \|\mathbf{X}\|_E \|\mathbf{Y}\|_E,$

the angle between two vectors in $\mathbb{R}^n$ is defined via (6.9), and two vectors $\mathbf{X}$, $\mathbf{Y}$ that satisfy $\mathbf{X}^T \mathbf{Y} = 0$ are said to be orthogonal. Let $\mathbf{S} = [\,\mathbf{S}_1, \mathbf{S}_2, \cdots \mathbf{S}_n\,]$ be a set of orthonormal basis vectors for $\mathbb{R}^n$, that is, a set of n vectors for which

(6.11) $\mathbf{S}_i^T \mathbf{S}_j = 0, \quad i \neq j,$

but in addition, for which

(6.12) $\mathbf{S}_i^T \mathbf{S}_i = 1.$

Then entry ij of $\mathbf{S}^T \mathbf{S}$ is $\mathbf{S}_i^T \mathbf{S}_j$, and $\mathbf{S}^T \mathbf{S} = \mathbf{I}_n$. Therefore $\mathbf{S}^T = \mathbf{S}^{-1}$, and from Section 1.2 of Chapter 5, $\mathbf{S}$ is an orthogonal matrix.

<table>
<tr><td>

Example 12
Standard basis

</td><td>

The standard basis vectors $\mathbf{e}_1, \mathbf{e}_2, \cdots \mathbf{e}_n$ are mutually orthogonal since their inner product is $\mathbf{e}_i^T \mathbf{e}_j = 0$, for $i \neq j$. Furthermore $\mathbf{e}_i^T \mathbf{e}_i = 1$.

</td></tr>
</table>

<table>
<tr><td>

Example 13
Two
orthonormal
basis vectors

</td><td>

The vectors $\mathbf{X}_1 = \begin{bmatrix} \cos\theta \\ \sin\theta \end{bmatrix}$, $\mathbf{X}_2 = \begin{bmatrix} -\sin\theta \\ \cos\theta \end{bmatrix}$ both have Euclidean length $\|\mathbf{X}_i\|_E = 1$, and their inner product is $\mathbf{X}_1^T \mathbf{X}_2 = 0$, so these vectors are an orthonormal basis for $\mathbb{R}^2$.

</td></tr>
</table>

Example 14
Gram-Schmidt
procedure

Given a set of vectors $\mathbf{X}_1, \cdots \mathbf{X}_m$ that spans a subspace of $\mathbb{R}^n$, it is sometimes required to find an orthogonal basis for the subspace. The Gram-Schmidt procedure for computing such a basis is as follows. Interchange as necessary so that $\mathbf{X}_1$ is the vector of largest length. For each of the remaining vectors $\mathbf{X}_i$, calculate the orthogonal projection $(\mathbf{X}_i^T \mathbf{X}_1)/(\mathbf{X}_1^T \mathbf{X}_1)\,\mathbf{X}_1$ of $\mathbf{X}_i$ on $\mathbf{X}_1$, and subtract this vector from $\mathbf{X}_i$, as illustrated in Figure 6.2. Then each of the resulting

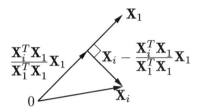

Fig. 6.2   The Gram-Schmidt procedure: finding a vector orthogonal to $\mathbf{X}_1$.

vectors is orthogonal to $\mathbf{X}_1$, as follows:

$$\mathbf{X}_1^T \left(\mathbf{X}_i - \frac{\mathbf{X}_i^T \mathbf{X}_1}{\mathbf{X}_1^T \mathbf{X}_1} \mathbf{X}_1 \right) = \mathbf{X}_1^T \mathbf{X}_i - \frac{\mathbf{X}_i^T \mathbf{X}_1}{\mathbf{X}_1^T \mathbf{X}_1} \mathbf{X}_1^T \mathbf{X}_1 = \mathbf{X}_1^T \mathbf{X}_i - \mathbf{X}_i^T \mathbf{X}_1 = 0.$$

Furthermore since the current $\mathbf{X}_i$ have been obtained by elementary operations on the original set of vectors, they are members of the original subspace. Excluding $\mathbf{X}_1$, let $\mathbf{X}_2$ be the largest of the current vectors, and repeat the above procedure with respect to $\mathbf{X}_2$, then excluding $\mathbf{X}_1, \mathbf{X}_2$, and so on, until the set of nonexcluded vectors contains no nonzero vectors. The resulting nonzero vectors $\mathbf{X}_1, \mathbf{X}_2, \cdots$ are an orthogonal basis for the subspace. Replacing each nonzero vector $\mathbf{X}_i$ by its unit vector $(1/\|\mathbf{X}_i\|_E)\,\mathbf{X}_i$ produces an orthonormal basis.

5 Change of basis

The entries of any $\mathbf{X} \in F^n$ are the coordinates of $\mathbf{X}$ with respect to the standard basis. Let $\mathbf{S} \in F^{n \times n}$ be a nonsingular matrix, so that each of its columns $\mathbf{S}_i$, $i = 1, \cdots n$ is a vector in F^n, and the entries of $\mathbf{S}_i$ are the coordinates of this vector with respect to the standard basis.

Because $\mathbf{S}$ is nonsingular, the equation

(6.13) $\mathbf{X} = \mathbf{S}\mathbf{X_S}$

has a solution $\mathbf{X_S} = (\mathbf{S}^{-1}\mathbf{X}) \in F^n$. The entries of $\mathbf{X_S}$ are the coordinates of $\mathbf{X}$ with respect to the basis $\mathbf{S}$. Conversely, a change of basis from $\mathbf{S}$ to the standard basis is achieved by premultiplying the coordinate vector $\mathbf{X_S}$ by $\mathbf{S}$.

Similarly, for nonsingular matrix $\mathbf{T}$,

(6.14) $\mathbf{X} = \mathbf{TX_T}$,

and by combining (6.14) and (6.13),

(6.15) $\mathbf{X_T} = \mathbf{T}^{-1}\mathbf{X} = (\mathbf{T}^{-1}\mathbf{S})\,\mathbf{X_S}$.

Therefore a change from any coordinate system with basis matrix $\mathbf{S}$ to another with basis matrix $\mathbf{T}$ corresponds to premultiplying the coordinate vector by a nonsingular matrix, given by $\mathbf{T}^{-1}\mathbf{S}$ in (6.15).

Now consider the linear transformation from F^n to F^n represented by the matrix $\mathbf{A}$ in

(6.16) $\mathbf{Y} = \mathbf{AX}$,

and suppose a new basis $\mathbf{S}$ is to be used, so that $\mathbf{X} = \mathbf{SX_S}$ and $\mathbf{Y} = \mathbf{SY_S}$. Then the above equation becomes

(6.17) $\mathbf{Y_S} = (\mathbf{S}^{-1}\mathbf{AS})\,\mathbf{X_S}$,

showing that the transformation matrix $\mathbf{A}$ becomes $\mathbf{S}^{-1}\mathbf{AS}$ when the new basis is used. Construction of the product $\mathbf{S}^{-1}\mathbf{AS}$ is called a *similarity* transformation on $\mathbf{A}$.

Example 15
Coordinates
using different
bases

In Figure 6.3, the coordinates of vector $\mathbf{X}$ with respect to the basis $[\mathbf{S}_1,\, \mathbf{S}_2]$ are the entries of $\mathbf{X_S} = \begin{bmatrix} -1/3 \\ 1 \end{bmatrix}$. With respect to the standard basis $[\mathbf{e}_1,\, \mathbf{e}_2]$, the

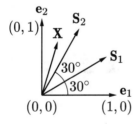

Fig. 6.3   Coordinate systems $\mathbf{e}_1$, $\mathbf{e}_2$, and $\mathbf{S}_1$, $\mathbf{S}_2$.

coordinates of $\mathbf{X}$ are

$$\mathbf{X} = \mathbf{SX_S} = \begin{bmatrix} \cos 30° & \cos 60° \\ \sin 30° & \sin 60° \end{bmatrix} \begin{bmatrix} -1/3 \\ 1 \end{bmatrix} = \begin{bmatrix} \sqrt{3}/2 & 1/2 \\ 1/2 & \sqrt{3}/2 \end{bmatrix} \begin{bmatrix} -1/3 \\ 1 \end{bmatrix}$$

$$= \begin{bmatrix} (3-\sqrt{3})/6 \\ (3\sqrt{3}-1)/6 \end{bmatrix}.$$

Example 16
Transformation
matrix in the
new basis

A transformation matrix $\mathbf{A} = \begin{bmatrix} -1 & 2 \\ -2 & -1 \end{bmatrix}$ in (6.16), written with respect to the standard basis $[\,\mathbf{e}_1, \mathbf{e}_2\,]$ in Figure 6.3, is to be rewritten for vectors expressed with respect to the new basis $\mathbf{S}$ in Figure 6.3. Then (6.17) results in

$$\mathbf{S}^{-1}\mathbf{A}\mathbf{S} = \begin{bmatrix} \sqrt{3} & -1 \\ -1 & \sqrt{3} \end{bmatrix} \begin{bmatrix} -1 & 2 \\ 2 & -1 \end{bmatrix} \begin{bmatrix} \sqrt{3}/2 & 1/2 \\ 1/2 & \sqrt{3}/2 \end{bmatrix} = \begin{bmatrix} -1+2\sqrt{3} & 4 \\ -4 & -1-2\sqrt{3} \end{bmatrix}.$$

Example 17
Rotations in $\mathbb{R}^3$

Changes of coordinate systems in $\mathbb{R}^3$ are of particular importance in the study of kinematics, where a set of three orthogonal unit vectors oriented according to a right-hand convention as shown in Figure 6.4 is called a coordinate *frame*. The figure shows a coordinate frame $0, x_0, y_0, z_0$ with orthonormal basis vectors

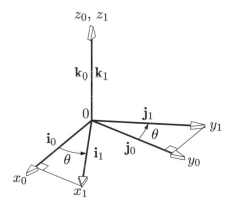

Fig. 6.4   Coordinate frame $0, x_0, y_0, z_0$ and rotated frame $0, x_1, y_1, z_1$.

$\mathbf{i}_0, \mathbf{j}_0, \mathbf{k}_0$, and a second frame $0, x_1, y_1, z_1$ with vectors $\mathbf{i}_1, \mathbf{j}_1, \mathbf{k}_1$, which is rotated by a right-handed angle θ about the z_0-axis. The $0, x_0, y_0, z_0$ coordinates of $\mathbf{i}_1, \mathbf{j}_1, \mathbf{k}_1$, are respectively

$$\mathbf{i}_1 = \begin{bmatrix} \cos\theta \\ \sin\theta \\ 0 \end{bmatrix}, \quad \mathbf{j}_1 = \begin{bmatrix} -\sin\theta \\ \cos\theta \\ 0 \end{bmatrix}, \quad \mathbf{k}_1 = \begin{bmatrix} 0 \\ 0 \\ 1 \end{bmatrix},$$

so given the $0, x_1, y_1, z_1$ coordinates $\mathbf{X}_1$ of a point, its $0, x_0, y_0, z_0$ coordinates $\mathbf{X}$ are

$$\mathbf{X} = \begin{bmatrix} \cos\theta & -\sin\theta & 0 \\ \sin\theta & \cos\theta & 0 \\ 0 & 0 & 1 \end{bmatrix} \mathbf{X}_1 = \mathbf{R}_{z,\theta}\mathbf{X}_1.$$

The rotation matrix $\mathbf{R}_{z,\theta}$ defined in this expression is orthogonal, so that $\mathbf{R}_{z,\theta}^{-1} = \mathbf{R}_{z,\theta}^{T}$, and furthermore $\mathbf{R}_{z,\theta}^{-1} = \mathbf{R}_{z,-\theta}$. Similarly, right-handed rotation about the y_0 axis and the x_0 axis corresponds to the respective orthogonal matrices

$$\mathbf{R}_{y,\theta} = \begin{bmatrix} \cos\theta & 0 & \sin\theta \\ 0 & 1 & 0 \\ -\sin\theta & 0 & \cos\theta \end{bmatrix}, \quad \mathbf{R}_{x,\theta} = \begin{bmatrix} 1 & 0 & 0 \\ 0 & \cos\theta & -\sin\theta \\ 0 & \sin\theta & \cos\theta \end{bmatrix},$$

with inverses $\mathbf{R}_{y,\theta}^{-1} = \mathbf{R}_{y,-\theta}$ and $\mathbf{R}_{x,\theta}^{-1} = \mathbf{R}_{x,-\theta}$.

Example 18
Euler angles

Suppose that in Figure 6.4 the coordinate system is first rotated by ϕ with respect to the original z-axis z_0, then by θ with respect to the current y-axis, then by ψ with respect to the current z-axis. The resulting change of basis is defined by (θ, ϕ, ψ), which are known as the Euler angles. By convention, $\cos\theta$ can be abbreviated as c_θ, and $\sin\theta$ as s_θ. Then the transformation $\mathbf{R}_0^1$ from the rotated coordinates to the original coordinates is given by the product

$$\mathbf{R}_0^1 = \mathbf{R}_{z,\phi}\mathbf{R}_{y,\theta}\mathbf{R}_{z,\psi} = \begin{bmatrix} c_\phi & -s_\phi & 0 \\ s_\phi & c_\phi & 0 \\ 0 & 0 & 1 \end{bmatrix} \begin{bmatrix} c_\theta & 0 & s_\theta \\ 0 & 1 & 0 \\ -s_\theta & 0 & c_\theta \end{bmatrix} \begin{bmatrix} c_\psi & -s_\psi & 0 \\ s_\psi & c_\psi & 0 \\ 0 & 0 & 1 \end{bmatrix}$$

$$= \begin{bmatrix} c_\phi c_\theta c_\psi - s_\phi s_\psi & -c_\phi c_\theta s_\psi - s_\phi c_\psi & c_\phi s_\theta \\ s_\phi c_\theta c_\psi + c_\phi s_\psi & -s_\phi c_\theta s_\psi + c_\phi c_\psi & s_\phi s_\theta \\ -s_\theta c_\psi & s_\theta s_\psi & c_\theta \end{bmatrix}.$$

This matrix must be orthogonal since it is the product of orthogonal matrices.

Example 19
Rigid motion

Coordinate changes that are not pure rotations are used extensively in robotics and computer graphics. Consider the coordinate systems illustrated in Figure 6.5, which illustrates the displacement of a coordinate frame by a vector $\mathbf{d}_0^1$, so that a point with coordinates $\mathbf{X}_1$ in the new coordinate frame has coordinates $\mathbf{X} = \mathbf{X}_1 + \mathbf{d}_0^1$ in the original frame. A rotation in the new frame defined by $\mathbf{R}_0^1$, together with such a translation, gives $\mathbf{X} = \mathbf{R}_0^1\mathbf{X}_1 + \mathbf{d}_0^1$, which defines the set of *rigid motions* of a point $\mathbf{X}_1$ with respect to the original coordinates.

In order to preserve the simplicity of matrix multiplication to describe coordinate changes, augment all vectors in $\mathbb{R}^3$ by a fourth entry equal to 1, so that the above rotation and translation becomes

$$\mathbf{X} = \begin{bmatrix} x \\ y \\ z \\ 1 \end{bmatrix} = \begin{bmatrix} \mathbf{R}_0^1 & \mathbf{d}_0^1 \\ 0 & 1 \end{bmatrix} \begin{bmatrix} x_1 \\ y_1 \\ z_1 \\ 1 \end{bmatrix} = \mathbf{H}_0^1\mathbf{X}_1,$$

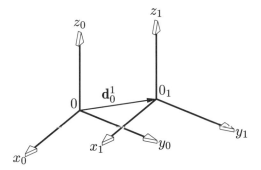

Fig. 6.5   Coordinate frame $0_1, x_1, y_1, z_1$ translated by vector $\mathbf{d}_0^1$.

with inverse

$$\mathbf{X}_1 = (\mathbf{H}_0^1)^{-1}\mathbf{X} = \begin{bmatrix} (\mathbf{R}_0^1)^T & -(\mathbf{R}_0^1)^T\mathbf{d}_0^1 \\ 0 & 1 \end{bmatrix} \mathbf{X}.$$

Vectors described and transformed by using such a convention are said to be in *homogeneous coordinates*. In addition to the rigid motions, scale changes can be accommodated by allowing the bottom-right entry of $\mathbf{H}_0^1$ to be different from 1, and changes of perspective are included by allowing the first three entries of the bottom row of $\mathbf{H}_0^1$ to be nonzero.

6   Further study

More examples of the vector-space results can be found in Ayres [4].

The results for vectors in F^n summarized here can be put into a much more general context, called functional analysis, which is often of use in optimization contexts; see Luenberger [32].

Changes of basis by similarity transformation and other tricks assume major importance in the formulation of the dynamic equations of robot arms and similar mechanical systems; see Spong and Vidyasagar [48].

7   Problems

1   Show that the last sentence of Example 3 is correct.

2   Determine whether each of the following is a vector space:

(a) the set of vectors $\begin{bmatrix} x_1 \\ x_2 \\ x_3 \end{bmatrix}$ with real entries that satisfy $x_1 + x_2 + x_3 = 0$, with ordinary vector addition and scalar multiplication;

(b) the set of all periodic functions with fixed period T, with pointwise addition such that $(f+g)(t) = f(t) + g(t)$ and scalar multiplication such that $(\alpha f)(t) = \alpha(f(t))$;

(c) the set of all solutions of the homogeneous differential equation

$$d^2x/dt^2 + 4dx/dt + 3x = 0.$$

3   Consider the set $P[a, b]$, of functions that are defined over the interval $[a, b] \in \mathbb{R}$, that are continuous except at a finite number of points, and that have both left-hand and right-hand limits at the points of discontinuity. The members of

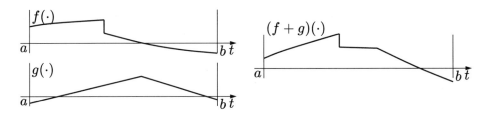

Fig. P6.3   Pointwise addition of piecewise-continuous functions.

P are said to be *piecewise-continuous* functions. For any $f(t)$, $g(t) \in P$, define $f + g$ to be the function with value $f(t) + g(t)$ at all points $t \in [a, b]$ as in Figure P6.3. Likewise for any $\alpha \in \mathbb{R}$, define αf to be the function with value $\alpha f(t)$ at all points $t \in [a, b]$. These operations are called *pointwise* addition and multiplication. Determine whether $P[a, b]$ is a vector space.

4   For the matrices $\mathbf{A}$ shown, find (a) a basis for the range $\mathcal{R}(\mathbf{A})$, and the dimension of $\mathcal{R}(\mathbf{A})$, (b) a basis for the null space $\mathcal{N}(\mathbf{A})$, and the dimension of $\mathcal{N}(\mathbf{A})$:

(i) $\mathbf{A} = \begin{bmatrix} 1 & -1 \\ -3 & -6 \\ 2 & 4 \end{bmatrix}$,   (ii) $\mathbf{A} = \begin{bmatrix} 1 & 2 \\ -3 & -6 \\ 2 & 4 \end{bmatrix}$,   (iii) $\mathbf{A} = \begin{bmatrix} 1 & 2 & -1 \\ -3 & -6 & -6 \\ 2 & 4 & 4 \end{bmatrix}$.

5   For each of the matrices $\mathbf{A}$ in Problem 27 of Chapter 5, find a basis for the range of $\mathbf{A}$, and a basis for the null space of $\mathbf{A}$.

6   Let $\mathbf{A} \in \mathbb{R}^{m \times n}$, and rank$(\mathbf{A}) = r < m, n$. Then from Section 3.6 of Chapter 5, the SVD is $\mathbf{A} = [\mathbf{U}_1, \mathbf{U}_2] \begin{bmatrix} \Sigma_r & 0 \\ 0 & 0 \end{bmatrix} \begin{bmatrix} \mathbf{V}_1^T \\ \mathbf{V}_2^T \end{bmatrix}$, where Σ_r is $r \times r$. Show that a basis for $\mathcal{R}(\mathbf{A})$ is $\mathbf{U}_1 \Sigma_r$, and that a basis for $\mathcal{N}(\mathbf{A})$ is $\mathbf{V}_2$.

7   Apply the Gram-Schmidt method to find an orthogonal basis for the subspaces of $\mathbb{R}^3$ spanned by the following matrices:

(a) $\begin{bmatrix} 0 & 1 & 2 \\ -3 & 2 & 0 \\ 0 & 2 & 4 \end{bmatrix}$,   (b) $\begin{bmatrix} 1 & 0 & 2 \\ 2 & -3 & 0 \\ 0 & 1 & 2 \end{bmatrix}$.

8   Let

$$\mathbf{Y} = \mathbf{A}\mathbf{X} = \begin{bmatrix} 1 & 1 & 3 \\ 1 & 2 & 1 \\ 1 & 3 & 0 \end{bmatrix} \mathbf{X}$$

with respect to the standard basis in $\mathbb{R}^3$, and let

$$[\mathbf{Z}_1, \mathbf{Z}_2, \mathbf{Z}_3] = \begin{bmatrix} 0 & 1 & 0 \\ 1 & 2 & 1 \\ 0 & 1 & 2 \end{bmatrix}$$

be a new basis.

(a) Given the vector $\mathbf{X} = \begin{bmatrix} 3 \\ 0 \\ -2 \end{bmatrix}$, determine its coordinates with respect to the new basis.

(b) Find the linear transformation matrix with respect to the new basis, corresponding to $\mathbf{A}$ above.

9   In two dimensions, show that multiplying a vector by the matrix

$$\mathbf{H} = \begin{bmatrix} \cos\theta & -\sin\theta \\ \sin\theta & \cos\theta \end{bmatrix}$$

rotates the vector counter-clockwise by angle θ.

10   Near a certain point on the earth, positions are recorded by their distances east and north of the given point. Suppose a new pair of coordinate vectors is to be used: magnetic north, which is at an angle θ from true north, and magnetic east, which is 90° from magnetic north. Find a transformation matrix $\mathbf{A}$ such that if $\mathbf{X}$ is a position in geographic coordinates, then $\mathbf{Y} = \mathbf{A}\mathbf{X}$ is the same position in magnetic coordinates.

11   A three-dimensional object can be represented in two dimensions by projecting the points of the object onto a plane. Let the points of the object be given as a set $\{\mathbf{X}\}$ of vectors in a coordinate frame $0, x_0, y_0, z_0$. The object is viewed by sighting inward along the x_1-axis of a second coordinate frame, so that the y_1, z_1 coordinates of any point $\mathbf{X}$ correspond to projecting $\mathbf{X}$ onto the y_1, z_1 plane. The second coordinate frame is obtained from the first by rotating by θ

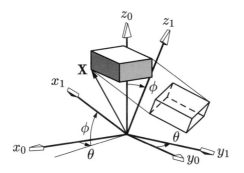

Fig. P6.11   Projection onto the y_1, z_1-plane.

around the z-axis, and then rotating by $-\phi$ around the current y-axis, as shown in Figure P6.11. Then the projection of any point $\mathbf{X}$ is $\mathbf{X}' = \mathbf{PX}$, where $\mathbf{P} \in \mathbb{R}^{2\times 3}$ and where the entries of $\mathbf{X}' \in \mathbb{R}^2$ are the y_1, z_1 coordinates of $\mathbf{X}$. Find $\mathbf{P}$.

12   Let $\mathcal{V}$ be the set of polynomials $a_0 + a_1 t + \cdots a_n t^n$ of degree n with real coefficients and indeterminate t, together with ordinary polynomial addition and ordinary multiplication by scalars in $\mathbb{R}$. Show that $\mathcal{V}$ is a vector space, and find a basis for the space.

7
Similarity transformations

Given the LTI state equations (1.16) for a continuous-time model with matrices $\mathbf{A}, \mathbf{B}, \mathbf{C}, \mathbf{D}$ and state vector $\mathbf{X}$, as shown,

(1.16a) $\quad \dfrac{d}{dt}\mathbf{X}(t) = \mathbf{A}\mathbf{X}(t) + \mathbf{B}\mathbf{U}(t)$

(1.16b) $\quad\quad \mathbf{Y}(t) = \mathbf{C}\mathbf{X}(t) + \mathbf{D}\mathbf{U}(t),$

the change of coordinates

(7.1) $\quad \mathbf{X} = \mathbf{S}\mathbf{X}'$

produces the modified pair of equations

(7.2a) $\quad \dfrac{d}{dt}\left(\mathbf{S}\mathbf{X}'(t)\right) = \mathbf{A}\mathbf{S}\mathbf{X}'(t) + \mathbf{B}\mathbf{U}(t)$

(7.2b) $\quad\quad\quad \mathbf{Y}(t) = \mathbf{C}\mathbf{S}\mathbf{X}'(t) + \mathbf{D}\mathbf{U}(t).$

When $\mathbf{S}$ is constant, premultiplying by $\mathbf{S}^{-1}$ produces the new state-space equations shown:

(7.3a) $\quad \dfrac{d}{dt}\mathbf{X}'(t) = \left(\mathbf{S}^{-1}\mathbf{A}\mathbf{S}\right)\mathbf{X}'(t) + \left(\mathbf{S}^{-1}\mathbf{B}\right)\mathbf{U}(t)$

(7.3b) $\quad\quad \mathbf{Y}(t) = \left(\mathbf{C}\mathbf{S}\right)\mathbf{X}'(t) + \mathbf{D}\mathbf{U}(t).$

Discrete-time systems are transformed similarly, with the left-hand side of (1.16a) replaced by $\mathbf{X}(t+1)$, and the left-hand side of the resulting (7.3a) replaced by $\mathbf{X}'(t+1)$.

For both continuous-time or discrete-time systems, the change of basis results in a new state-space model with matrices $\mathbf{A}' = \mathbf{S}^{-1}\mathbf{A}\mathbf{S}$, $\mathbf{B}' = \mathbf{S}^{-1}\mathbf{B}$, $\mathbf{C}' = \mathbf{C}\mathbf{S}$, $\mathbf{D}' = \mathbf{D}$ and with new state vector $\mathbf{X}'$. The matrices $\mathbf{A}$ and $\mathbf{A}'$ are said to be *similar*, that is, related by a similarity transformation.

Similarity transformations are used to produce new models from old, since the properties of the new models may be useful for analysis or design. It will also be seen that the matrix exponential $e^{t\mathbf{A}}$, the power $\mathbf{A}^k$, and more general functions of $\mathbf{A}$ can be analyzed in detail by using similarity transformations.

In this chapter, methods of computing a diagonal or near-diagonal system will be considered. Normally these results apply only to models with real or complex coefficients.

1 Invariance of the external behavior

One of the great conveniences of similarity transformations is that the response of the transformed system has a direct relationship to the response of the original system.

First consider the solution of (1.16), given in Chapter 2, repeated here with $t_0 = 0$:

$$(7.4) \quad \mathbf{Y}(t) = \mathbf{C}\left(e^{t\mathbf{A}}\mathbf{X}(0) + \int_0^t e^{(t-\tau)\mathbf{A}}\mathbf{B}\,\mathbf{U}(\tau)\,d\tau\right) + \mathbf{D}\mathbf{U}(t).$$

Therefore the output $\mathbf{Y}(t)$ in (7.3) is

$$(7.5) \quad \mathbf{Y}(t) = \mathbf{CS}\left(e^{t\mathbf{S}^{-1}\mathbf{AS}}\mathbf{X}'(0) + \int_0^t e^{(t-\tau)\mathbf{S}^{-1}\mathbf{AS}}\mathbf{S}^{-1}\mathbf{B}\,\mathbf{U}(\tau)\,d\tau\right) + \mathbf{D}\mathbf{U}(t).$$

Replacing $\mathbf{X}'(0)$ by $\mathbf{S}^{-1}\mathbf{X}(0)$ and noting from the definition (2.17) of the exponential that

$$(7.6) \quad e^{t\mathbf{S}^{-1}\mathbf{AS}} = \mathbf{I} + \frac{t}{1!}\mathbf{S}^{-1}\mathbf{AS} + \frac{t^2}{2!}\mathbf{S}^{-1}\mathbf{A}^2\mathbf{S} + \cdots = \mathbf{S}^{-1}\left(\mathbf{I} + \frac{t}{1!}\mathbf{A} + \frac{t^2}{2!}\mathbf{A}^2 + \cdots\right)\mathbf{S}$$
$$= \mathbf{S}^{-1}e^{t\mathbf{A}}\mathbf{S},$$

we see that (7.5) becomes (7.4) identically on cancelation of the matrices $\mathbf{S}$ and $\mathbf{S}^{-1}$.

Identical response The response of a continuous-time LTI system is unaffected by a change of basis $\mathbf{X} = \mathbf{SX}'$ for the space of vectors $\mathbf{X}$, provided that the initial vectors satisfy $\mathbf{X}(0) = \mathbf{SX}'(0)$. In particular, the forced responses, for which $\mathbf{X}(0) = 0$ and $\mathbf{X}'(0) = 0$, are identical. An entirely analogous conclusion results for discrete-time systems.

In addition to the above, it is of interest to see the effect of a similarity transformation on the sequence $\{\mathbf{H}(k)\}_0^\infty$, which characterizes the forced response of both discrete-time and continuous-time systems.

The sequence $\{\mathbf{H}'(k)\}_0^\infty$ for the transformed system has k-th term

$$(7.7) \quad \mathbf{H}'(k) = \mathbf{CS}(\mathbf{S}^{-1}\mathbf{AS})^{k-1}\mathbf{S}^{-1}\mathbf{B} = \mathbf{CSS}^{-1}\mathbf{A}^{k-1}\mathbf{SS}^{-1}\mathbf{B} = \mathbf{CA}^{k-1}\mathbf{B} = \mathbf{H}(k),$$

for $k > 0$, whereas the 0-th term is $\mathbf{D}$, which is unaffected by the transformation. Consequently, as previously shown, the forced output response of the transformed system must be identical to that of the original system for identical inputs.

Example 1
Simplicity via
diagonalization

It may be that calculation of the response $\mathbf{Y}(t)$ is simpler in the form (7.5) than (7.4), for example,

$$\mathbf{A} = \begin{bmatrix} 0 & -2 \\ 1 & -3 \end{bmatrix}, \quad \mathbf{S} = \begin{bmatrix} 2 & 1 \\ 1 & 1 \end{bmatrix}, \quad \mathbf{S}^{-1}\mathbf{A}\mathbf{S} = \begin{bmatrix} -1 & 0 \\ 0 & -2 \end{bmatrix},$$

so that $e^{t\mathbf{S}^{-1}\mathbf{A}\mathbf{S}} = \begin{bmatrix} e^{-t} & 0 \\ 0 & e^{-2t} \end{bmatrix}$, whereas $e^{t\mathbf{A}}$ is a more complicated formula, which will be investigated in Section 4. This is an example of diagonalization as described in the next section.

2   Eigenvalues, eigenvectors, and diagonalization

An important simplification results if a matrix $\mathbf{S}$ can be found such that $\mathbf{A}'$ is diagonal, or as nearly diagonal as possible.

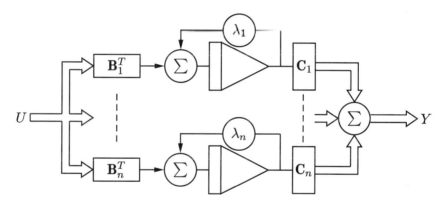

Fig. 7.1   Diagonalized system.

Figure 7.1 is an operational diagram of a system with diagonal $\mathbf{A} = \mathrm{diag}[\lambda_i]$, with the rows of $\mathbf{B}$ written as $\mathbf{B}_i^T$ and the columns of $\mathbf{C}$ written as $\mathbf{C}_i$, and with $\mathbf{D} = 0$ for simplicity. Then the transfer matrix is

$$(7.8) \quad \hat{\mathbf{H}}(s) = \sum_{i=1}^{n} \mathbf{C}_i(s - \lambda_i)^{-1}\mathbf{B}_i^T,$$

and the impulse-response matrix is

$$(7.9) \quad \mathbf{H}(t) = \sum_{i=1}^{n} \mathbf{C}_i e^{t\lambda_i}\mathbf{B}_i^T,$$

showing that, for such a system, $\mathbf{H}(t)$ is the sum of n waveforms, or dynamical modes, which can be calculated independently.

Eigenvalues Let the characteristic polynomial of $\mathbf{A}$ be written

(7.10) $\phi(\lambda) = \det(\lambda \mathbf{I} - \mathbf{A}) = \lambda^n + a_1 \lambda^{n-1} + \cdots a_n$

and denote its roots, called the *eigenvalues* of $\mathbf{A}$, as $\lambda_1, \lambda_2, \cdots \lambda_n$, assuming $\mathbf{A}$ to be real or complex, so these roots always exist. The eigenvalues of a matrix may not be distinct, in which case it is convenient to factor the characteristic polynomial as shown, when there are t distinct values:

(7.11) $\phi(\lambda) = (\lambda - \lambda_1)^{n_1} (\lambda - \lambda_2)^{n_2} \cdots (\lambda - \lambda_t)^{n_t}.$

Thus each eigenvalue λ_i has multiplicity $n_i \geq 1$, and $\sum_{i=1}^{t} n_i = n$.

Similarity transformations The eigenvalues of a matrix are unchanged by a similarity transformation. Let $\mathbf{A}' = \mathbf{S}^{-1}\mathbf{A}\mathbf{S}$, and let $\det \mathbf{S} = \alpha$, which is nonzero since $\mathbf{S}$ is nonsingular. The characteristic polynomial of $\mathbf{A}'$ is

$$\det(\lambda \mathbf{I} - \mathbf{A}') = \det(\lambda \mathbf{I} - \mathbf{S}^{-1}\mathbf{A}\mathbf{S}) = \det(\mathbf{S}^{-1}(\lambda \mathbf{I} - \mathbf{A})\mathbf{S})$$

(7.12) $$= \frac{1}{\alpha} \det(\lambda \mathbf{I} - \mathbf{A})\, \alpha = \det(\lambda \mathbf{I} - \mathbf{A}),$$

which is the characteristic polynomial of $\mathbf{A}$.

Eigenvectors The *eigenvectors* of $\mathbf{A}$ are defined to be nonzero vectors satisfying the equations

(7.13) $(\mathbf{A} - \lambda_i \mathbf{I})\, \mathbf{X}_i = 0, \quad i = 1, \cdots n.$

Diagonalization We shall investigate whether $\mathbf{A}$ can be diagonalized by a similarity transformation, using the following results.

Theorem 7.1 *Let $\mathbf{A} \in \mathbb{R}^{n \times n}$. Then if $\mathbf{A}$ is similar to a diagonal matrix, it possesses n linearly independent eigenvectors.*

Proof: Assume that

(7.14) $\mathbf{A}' = \mathbf{S}^{-1}\mathbf{A}\mathbf{S} = \begin{bmatrix} \lambda_1 & & \\ & \ddots & \\ & & \lambda_n \end{bmatrix}$

for some constants $\lambda_1, \lambda_2, \cdots \lambda_n$, and write $\mathbf{S}$ in columns as $\mathbf{S} = [\,\mathbf{S}_1, \cdots \mathbf{S}_n\,]$. Premultiplying by $\mathbf{S}$ yields

(7.15) $\mathbf{A}\,[\,\mathbf{S}_1 \cdots \mathbf{S}_n\,] = [\,\mathbf{S}_1 \cdots \mathbf{S}_n\,] \begin{bmatrix} \lambda_1 & & \\ & \ddots & \\ & & \lambda_n \end{bmatrix}.$

Then comparing columns on the left and right sides, for $i = 1, \cdots n$, yields

(7.16) $\mathbf{AS}_i = \mathbf{S}_i \lambda_i;$

that is, each vector $\mathbf{S}_i$ satisfies an equation of the form

(7.17) $(\mathbf{A} - \lambda_i \mathbf{I}) \mathbf{S}_i = 0.$

The vector $\mathbf{S}_i$ cannot be zero since $\mathbf{S}$ is nonsingular, so from the above equation, $(\mathbf{A} - \lambda_i \mathbf{I})$ must have less than full rank, and therefore

(7.18) $\det(\mathbf{A} - \lambda_i \mathbf{I}) = (-1)^n \phi(\lambda_i) = 0,$

which shows that λ_i is an eigenvalue of $\mathbf{A}$, and $\mathbf{S}_i$ is an eigenvector. Because $\mathbf{S}$ is nonsingular by assumption, the n eigenvectors of $\mathbf{A}$ are linearly independent. $\square$

Thus the columns of $\mathbf{S}$ are eigenvectors.

Example 2
Checking
eigenvectors
in S

To check that the columns of $\mathbf{S} = [\mathbf{S}_1, \mathbf{S}_2]$ in Example 1 are eigenvectors, first note that $\mathbf{A}$ is similar to a diagonal matrix:

$$\mathbf{A}' = \mathbf{S}^{-1}\mathbf{AS} = \begin{bmatrix} -1 & 0 \\ 0 & -2 \end{bmatrix} = \begin{bmatrix} \lambda_1 & 0 \\ 0 & \lambda_2 \end{bmatrix}.$$

Then, for the first column $\mathbf{S}_1$,

$$(\mathbf{A} - \lambda_1 \mathbf{I}) \mathbf{S}_1 = \left(\begin{bmatrix} 0 & -2 \\ 1 & -3 \end{bmatrix} - (-1) \begin{bmatrix} 1 & 0 \\ 0 & 1 \end{bmatrix} \right) \begin{bmatrix} 2 \\ 1 \end{bmatrix} = \begin{bmatrix} 1 & -2 \\ 1 & -2 \end{bmatrix} \begin{bmatrix} 2 \\ 1 \end{bmatrix} = \begin{bmatrix} 0 \\ 0 \end{bmatrix},$$

and similarly, for the second column,

$$(\mathbf{A} - \lambda_2 \mathbf{I}) \mathbf{S}_2 = \begin{bmatrix} 2 & -2 \\ 1 & -1 \end{bmatrix} \begin{bmatrix} 1 \\ 1 \end{bmatrix} = \begin{bmatrix} 0 \\ 0 \end{bmatrix},$$

so $\mathbf{S}_1$ and $\mathbf{S}_2$ are eigenvectors corresponding to λ_1 and λ_2 respectively.

Theorem 7.2   *If $\mathbf{A}$ has linearly independent eigenvectors, then $\mathbf{A}$ is diagonalizable via a similarity transformation.*

Proof: Let n linearly independent eigenvectors $\mathbf{X}_i$, $i = 1, \cdots n$ be found, and let

(7.19) $\mathbf{S} = [\mathbf{X}_1, \cdots \mathbf{X}_n].$

Then from (7.13), (7.15) is true, and hence $\mathbf{A}'$ has the form shown in (7.14). $\square$

Example 3
Eigenvectors
imply
diagonalization

In the previous example, $\mathbf{S}_1$ and $\mathbf{S}_2$ are eigenvectors, and $\det(\mathbf{S}) \neq 0$, showing that these eigenvectors are linearly independent. Therefore $\mathbf{S}$ is a diagonalizing similarity transformation matrix.

Theorem 7.3   *Let $\mathbf{A} \in \mathbb{R}^{n \times n}$ have distinct eigenvalues $\lambda_i, i = 1, \cdots n$. For each $i = 1, \cdots n$, let $\mathbf{X}_i$ be a nonzero vector satisfying (7.13). Then the vectors $\mathbf{X}_1, \cdots \mathbf{X}_n$ are linearly independent.*

Proof: Since the $\mathbf{X}_i$ are nonzero, the single vector $\mathbf{X}_1$ is independent. Assume that the vectors $\mathbf{X}_1, \cdots \mathbf{X}_k$ are linearly independent, but that the vectors $\mathbf{X}_1, \cdots \mathbf{X}_{k+1}$ are not. Then there exist scalars $\alpha_1 \cdots \alpha_{k+1}$, not all zero, such that

$$(7.20) \quad [\,\mathbf{X}_1, \cdots \mathbf{X}_{k+1}\,] \begin{bmatrix} \alpha_1 \\ \vdots \\ \alpha_{k+1} \end{bmatrix} = 0.$$

Premultiplying (7.20) by $\mathbf{A}$ gives

$$(7.21) \quad \mathbf{X}_1 \alpha_1 \lambda_1 + \cdots \mathbf{X}_k \alpha_k \lambda_k + \mathbf{X}_{k+1} \alpha_{k+1} \lambda_{k+1} = 0$$

whereas multiplying (7.20) by λ_{k+1} gives

$$(7.22) \quad \mathbf{X}_1 \alpha_1 \lambda_{k+1} + \cdots \mathbf{X}_k \alpha_k \lambda_{k+1} + \mathbf{X}_{k+1} \alpha_{k+1} \lambda_{k+1} = 0.$$

Subtracting to remove the last term yields

$$(7.23) \quad \mathbf{X}_1 \alpha_1 (\lambda_{k+1} - \lambda_1) + \cdots \mathbf{X}_k \alpha_k (\lambda_{k+1} - \lambda_k) = 0.$$

The vectors $\mathbf{X}_1, \cdots \mathbf{X}_k$ are linearly independent. Therefore all of the coefficients $\alpha_i (\lambda_{k+1} - \lambda_i)$ above must be zero, which implies that every $\alpha_i = 0$ in (7.23) since the eigenvalues are distinct, and (7.20) reduces to

$$(7.24) \quad \mathbf{X}_{k+1} \alpha_{k+1} = 0,$$

which implies $\alpha_{k+1} = 0$ since $\mathbf{X}_{k+1}$ is nonzero. This means that every co-efficient $\alpha_1, \ldots \alpha_{k+1}$ in (7.20) must be zero, a contradiction. The theorem is therefore true by induction.   □

Example 4
Diagonal result
without further
calculation

For the matrix $\mathbf{A}$ of Example 1, the characteristic polynomial is

$$\det(\lambda \mathbf{I} - \mathbf{A}) = \lambda^2 + 3\lambda + 2 = (\lambda + 1)(\lambda + 2),$$

so that $\lambda_1 = -1$, and $\lambda_2 = -2$. Because these values are distinct, it is known *without calculating* $\mathbf{S}$ that $\mathbf{A}$ is similar to $\text{diag}\,[\,\lambda_i\,] = \begin{bmatrix} -1 & 0 \\ 0 & -2 \end{bmatrix}$.

Example 5
Calculating **S**

The steps required to calculate a diagonalizing matrix **S** for **A** of Example 1 are, first, calculate the λ_i, $i = 1, \cdots n$, as in the above example, and then, for each λ_i, calculate an eigenvector $\mathbf{S}_i$ by solving $(\mathbf{A} - \lambda_i \mathbf{I})\,\mathbf{S}_i = 0$, for example, as follows:

$$\left[\frac{\mathbf{A} - \lambda_1\mathbf{I}}{\mathbf{I}}\right] = \left[\begin{array}{cc} 1 & -2 \\ 1 & -2 \\ \hline 1 & 0 \\ 0 & 1 \end{array}\right] \overset{\text{col}}{\sim} \left[\begin{array}{c|c} 1 & 0 \\ 1 & 0 \\ \hline 1 & 2 \\ 0 & 1 \end{array}\right], \quad \left[\frac{\mathbf{A} - \lambda_2\mathbf{I}}{\mathbf{I}}\right] = \left[\begin{array}{cc} 2 & -2 \\ 1 & -1 \\ \hline 1 & 0 \\ 0 & 1 \end{array}\right] \overset{\text{col}}{\sim} \left[\begin{array}{c|c} 2 & 0 \\ 1 & 0 \\ \hline 1 & 1 \\ 0 & 1 \end{array}\right],$$

from which $\mathbf{S}_1 = \begin{bmatrix} 2 \\ 1 \end{bmatrix}\alpha$ and $\mathbf{S}_2 = \begin{bmatrix} 1 \\ 1 \end{bmatrix}\beta$ are eigenvectors corresponding to λ_1 and λ_2 respectively, where α and β are arbitrary nonzero values.

Example 6
Complex
eigenvectors

Because the roots of a polynomial with real coefficients are complex in general, the matrix $\mathbf{A} - \lambda_i\mathbf{I}$ may contain complex entries, and Equation (7.17) may have complex solutions $\mathbf{S}_i$. For example, the matrix and characteristic polynomial

$$\mathbf{A} = \begin{bmatrix} 0 & -3.25 \\ 1 & -3 \end{bmatrix}, \quad \det(\lambda\mathbf{I} - \mathbf{A}) - \lambda^2 + 3\lambda + 13/4,$$

produce eigenvalues $\lambda = -3/2 \pm j1$. The eigenvectors can be calculated as follows:

$$\left[\begin{array}{cc} 3/2-j & -13/4 \\ 1 & -3/2-j \\ \hline 1 & 0 \\ 0 & 1 \end{array}\right] \overset{\text{col}}{\sim} \left[\begin{array}{c|c} 3/2-j & 0 \\ 1 & 0 \\ \hline 1 & 3/2+j \\ 0 & 1 \end{array}\right],$$

$$\left[\begin{array}{cc} 3/2+j & -13/4 \\ 1 & -3/2+j \\ \hline 1 & 0 \\ 0 & 1 \end{array}\right] \overset{\text{col}}{\sim} \left[\begin{array}{c|c} 3/2+j & 0 \\ 1 & 0 \\ \hline 1 & 3/2-j \\ 0 & 1 \end{array}\right],$$

from which $\mathbf{S}_1 = \begin{bmatrix} 3/2+j \\ 1 \end{bmatrix}$ and $\mathbf{S}_2 = \begin{bmatrix} 3/2-j \\ 1 \end{bmatrix}$ are eigenvectors.

Example 7
Eigenvalues of
symmetric
matrices

Symmetric real matrices have real eigenvalues, a property that simplifies optimization problems and the modeling of mechanical systems. To prove this property, let $\mathbf{A} = \mathbf{A}^T \in \mathbb{R}^{n \times n}$; let λ be an eigenvalue, assumed to be complex; and let $\mathbf{X}$ be the associated eigenvector, so that $\mathbf{A}\mathbf{X} = \lambda\mathbf{X}$. Let the i-th entry of $\mathbf{X}$ be $x_i = \alpha_i + j\omega_i$, write $\bar{x}_i$ for its complex conjugate, and write $\mathbf{X}^H = [\,\bar{x}_1, \cdots \bar{x}_n\,]$.

First, since $\mathbf{X}$ is an eigenvector corresponding to λ,

$$\mathbf{X}^H\mathbf{A}\mathbf{X} = \mathbf{X}^H\lambda\mathbf{X} = \lambda\mathbf{X}^H\mathbf{X},$$

where the complex inner product $\mathbf{X}^H\mathbf{X} = \sum_{i=1}^{n} \bar{x}_i x_i = \sum_{i=1}^{n}(\alpha_i^2 + \omega_i^2)$ is real.

Next we shall show that the term $\mathbf{X}^H\mathbf{A}\mathbf{X}$ is real, so that from the above equation, λ must be real. The detailed expansion of this term is

$$\mathbf{X}^H\mathbf{A}\mathbf{X} = \sum_{i=1}^{n}\sum_{k=1}^{n} \bar{x}_i a_{ik} x_k = \sum_{i=1}^{n} a_{ii}\bar{x}_i x_i + \sum_{i=2}^{n}\sum_{k=1}^{i-1} a_{ik}(\bar{x}_i x_k + \bar{x}_k x_i),$$

which has been separated into a sum containing diagonal entries of $\mathbf{A}$ and one containing the off-diagonal entries, taking note that $a_{ik} = a_{ki}$. The quantities $\bar{x}_i x_i$ and $\bar{x}_i x_k + \bar{x}_k x_i = 2(\alpha_i\alpha_k + \omega_i\omega_k)$ are real, so $\mathbf{X}^H\mathbf{A}\mathbf{X}$ is real as required.

Example 8
Real block
diagonal form

There is a trick that allows a diagonalizable matrix to be transformed to block diagonal form without using complex arithmetic, producing diagonal blocks that are either 1×1 corresponding to real eigenvalues, or 2×2 for complex pairs, with the real and imaginary parts available by inspection.

Suppose that $\lambda_1 = \alpha + j\omega$ and $\lambda_2 = \alpha - j\omega$ are a complex pair. Substitution in Equation (7.13) shows that if $\mathbf{S}_1 = \mathbf{U} + j\mathbf{V}$ is an eigenvector corresponding to λ_1, then its complex conjugate $\mathbf{U} - j\mathbf{V}$ is an eigenvector corresponding to λ_2, as illustrated in Example 6.

Let $\mathbf{S}^{-1}\mathbf{A}\mathbf{S} = \mathbf{J}$, where $\mathbf{S} = [\mathbf{S}_1, \mathbf{S}_2, \cdots]$, producing complex conjugate eigenvalues λ_1, λ_2 as the first two diagonal entries in $\mathbf{J}$. This similarity transformation will be modified by choosing the the transformation matrix $\mathbf{SQ}$ instead of $\mathbf{S}$, producing

$$(\mathbf{SQ})^{-1}\mathbf{A}(\mathbf{SQ}) = \mathbf{Q}^{-1}\mathbf{S}^{-1}\mathbf{A}\mathbf{S}\mathbf{Q} = \mathbf{Q}^{-1}\mathbf{J}\mathbf{Q}.$$

The matrix $\mathbf{Q}$ is given by

$$\mathbf{Q} = \begin{bmatrix} \mathbf{M} & 0 \\ 0 & \mathbf{I}_{n-2} \end{bmatrix}, \quad \text{where } \mathbf{M} = \frac{1}{2}\begin{bmatrix} 1 & j \\ 1 & -j \end{bmatrix},$$

and the modified transformation matrix is

$$\mathbf{SQ} = [\mathbf{U}+j\mathbf{V}, \ \mathbf{U}-j\mathbf{V}, \ \cdots]\begin{bmatrix} \mathbf{M} & 0 \\ 0 & \mathbf{I}_{n-2} \end{bmatrix} = [\mathbf{U}, \ -\mathbf{V}, \cdots],$$

which is $\mathbf{S}$ with $\mathbf{S}_1$ replaced by its real part and with $\mathbf{S}_2$ replaced by the negative of the imaginary part of $\mathbf{S}_1$.

The transformed matrix $\mathbf{Q}^{-1}\mathbf{J}\mathbf{Q}$ is identical to $\mathbf{J}$ except for the top 2×2 block, which is

$$\mathbf{M}^{-1}\begin{bmatrix} \alpha+j\omega & 0 \\ 0 & \alpha-j\omega \end{bmatrix}\mathbf{M} = \begin{bmatrix} 1 & 1 \\ -j & j \end{bmatrix}\left(\frac{1}{2}\begin{bmatrix} \alpha+j\omega & j\alpha-\omega \\ \alpha-j\omega & -j\alpha-\omega \end{bmatrix}\right)$$

$$= \begin{bmatrix} \alpha & -\omega \\ \omega & \alpha \end{bmatrix}.$$

A similar result is obtained for each complex conjugate eigenvalue pair λ_i, λ_{i+1}, with $\mathbf{S}_i$ replaced by its real part, and with $\mathbf{S}_{i+1}$ replaced by the negative of the imaginary part of $\mathbf{S}_i$ in $\mathbf{S}$. Then the result of the similarity transformation is block-diagonal with real 2×2 blocks of the above form in place of the complex eigenvalues.

Example 6 contains one complex pair, and instead of the previously calculated transformation matrix $\mathbf{S} = [\,\mathbf{S}_1, \mathbf{S}_2\,]$, let $\mathbf{S} = [\,\mathrm{Re}(\mathbf{S}_1), -\mathrm{Im}(\mathbf{S}_1)\,] = \begin{bmatrix} 3/2 & -1 \\ 1 & 0 \end{bmatrix}$, so that $\mathbf{S}^{-1}\mathbf{AS} = \begin{bmatrix} -3/2 & -1 \\ 1 & -3/2 \end{bmatrix}$.

Example 9
Insufficient
independent
eigenvectors

For the matrix and characteristic polynomial shown,

$$\mathbf{A} = \begin{bmatrix} 0 & 1 \\ -4 & -4 \end{bmatrix}, \quad \det(\lambda\mathbf{I} - \mathbf{A}) = \lambda^2 + 4\lambda + 4 = (\lambda + 2)^2,$$

the eigenvalues are not distinct, but are both equal to -2. Calculating the eigenvectors as in Example 5 gives

$$\begin{bmatrix} \mathbf{A} - \lambda_1\mathbf{I} \\ \mathbf{I} \end{bmatrix} = \begin{bmatrix} 2 & 1 \\ -4 & -2 \\ 1 & 0 \\ 0 & 1 \end{bmatrix} \overset{\mathrm{col}}{\sim} \begin{bmatrix} 2 & 0 \\ -4 & 0 \\ 1 & -1/2 \\ 0 & 1 \end{bmatrix},$$

so that any vector $\begin{bmatrix} -1/2 \\ 1 \end{bmatrix} \alpha$ for arbitrary $\alpha \neq 0$ is an eigenvector corresponding to both λ_1 and λ_2. The dimension of the null space of $\mathbf{A} - \lambda_1\mathbf{I}$ is 1 as calculated above; consequently there is no more than 1 linearly independent eigenvector, so that $\mathbf{S} = [\,\mathbf{S}_1, \mathbf{S}_2\,]$ for any pair of eigenvectors is a singular matrix, and hence from Theorem 7.1 the matrix $\mathbf{A}$ cannot be diagonalized.

The following theorem shows that there is precisely one independent eigenvector for each eigenvalue of multiplicity 1.

Theorem 7.4   *If λ_i has multiplicity 1, then $rank(\mathbf{A}-\lambda_i\mathbf{I}) = n-1$.*

Proof: Since by assumption $n_i = 1$ in (7.11),

$$(7.25) \quad \frac{d}{d\lambda}\phi(\lambda)\Big|_{\lambda=\lambda_i} = (\lambda_i - \lambda_1)^{n_1}(\lambda_i - \lambda_2)^{n_2} \cdots$$

$$(\lambda_i - \lambda_{i-1})^{n_{i-1}}(\lambda_i - \lambda_{i+1})^{n_{i+1}} \cdots (\lambda_i - \lambda_t)^{n_t},$$

which cannot be zero since the λ_i are distinct. Now, write

$$(7.26) \quad \det(\mathbf{A}-\lambda\mathbf{I}) = \det[\,b_{ij}\,] = (-1)^n\phi(\lambda), \quad \text{where } b_{ij} = \begin{cases} a_{ii} - \lambda, & i = j \\ a_{ij}, & i \neq j \end{cases}.$$

By definition, as in Section 2 of Chapter 5, the determinant can be written symbolically as a sum of products of entries from the rows of $[b_{ij}]$:

(7.27) $\det [b_{ij}] = \sum \pm (b_{1i} b_{2j} \cdots b_{nk}),$

where each sign is determined by the permutation $i, j, \cdots k$ of the integers $1, 2, \cdots n$. Then

(7.28) $(-1)^n \dfrac{d}{d\lambda} \phi(\lambda) = \dfrac{d}{d\lambda} \sum \pm (b_{1i} b_{2j} \cdots b_{nk})$

$= \sum \pm \left(\left(\dfrac{d\,b_{1i}}{d\lambda} b_{2j} \cdots b_{nk} \right) + \cdots \left(b_{1i} b_{2j} \cdots \dfrac{d\,b_{nk}}{d\lambda} \right) \right)$

$= \sum \pm \left(\dfrac{d\,b_{1i}}{d\lambda} b_{2j} \cdots b_{nk} \right) + \cdots \sum \pm \left(b_{1i} b_{2j} \cdots \dfrac{d\,b_{nk}}{d\lambda} \right)$

$= \det \begin{bmatrix} -1 & 0 & 0 & \cdots \\ a_{21} & a_{22}-\lambda & a_{23} & \cdots \\ & \cdots & & \\ & & \cdots & a_{nn}-\lambda \end{bmatrix}$

$+ \det \begin{bmatrix} a_{11}-\lambda & a_{12} & a_{13} & \cdots \\ 0 & -1 & 0 & \cdots \\ a_{31} & a_{32} & a_{33}-\lambda & \cdots \\ & \cdots & & \\ & & \cdots & a_{nn}-\lambda \end{bmatrix}$

$\cdots + \det \begin{bmatrix} a_{11}-\lambda & a_{12} & a_{13} & \cdots \\ a_{21} & a_{22}-\lambda & a_{23} & \cdots \\ & \cdots & & \\ \cdots & 0 & 0 & -1 \end{bmatrix},$

and since, as for (7.25), this expression cannot be zero for $\lambda = \lambda_i$, at least one $n-1$-square minor of $\mathbf{A} - \lambda_i \mathbf{I}$ must be nonzero.   □

Example 10
Rank of $\mathbf{A} - \lambda \mathbf{I}$

The multiplicity of both eigenvalues of Example 1 is 1, and

$\mathrm{rank}(\mathbf{A} - \lambda_1 \mathbf{I}) = \mathrm{rank} \begin{bmatrix} 1 & -2 \\ 1 & -2 \end{bmatrix} = 1, \quad \mathrm{rank}(\mathbf{A} - \lambda_2 \mathbf{I}) = \mathrm{rank} \begin{bmatrix} 2 & -2 \\ 1 & -1 \end{bmatrix} = 1.$

In summary, the columns of a diagonalizing similarity transformation matrix $\mathbf{S} = [\mathbf{S}_1, \cdots \mathbf{S}_n]$ must be linearly independent eigenvectors, any set of eigenvectors $\mathbf{S}_1, \cdots \mathbf{S}_n$ corresponding to distinct eigenvalues $\lambda_1, \cdots \lambda_n$ respectively is linearly independent, and a distinct eigenvalue λ_i produces a null space of $(\mathbf{A} - \lambda_i \mathbf{I})$ of dimension 1. Example 9 shows that a matrix may not be diagonalizable by a similarity transformation if there are fewer than n_i linearly independent eigenvectors for an eigenvalue λ_i of multiplicity $n_i > 1$. Hence the following conclusion.

Theorem 7.5   *The matrix $\mathbf{A} \in \mathbb{R}^n$ is diagonalizable by a similarity transformation $\mathbf{S}^{-1}\mathbf{A}\mathbf{S}$ if and only if, for each eigenvalue λ_i of multiplicity n_i, the null space of $(\mathbf{A} - \lambda_i\mathbf{I})$ has dimension n_i.*

Example 11
Diagonalizable
matrix with
multiple
eigenvalues

The matrix shown

$$\mathbf{A} = \begin{bmatrix} 1 & 2 & -4 \\ -3 & -4 & 4 \\ -3 & -2 & 2 \end{bmatrix}$$

has eigenvalues equal to $3, -2, -2$. Consequently there is exactly one linearly independent eigenvalue corresponding to $\lambda = 3$, since this eigenvalue is distinct, and at least one corresponding to $\lambda = -2$. In this case,

$$\mathbf{A} - (-2)\mathbf{I} = \begin{bmatrix} 3 & 2 & -4 \\ -3 & -2 & 4 \\ -3 & -2 & 4 \end{bmatrix} \overset{\text{col}}{\sim} \begin{bmatrix} 3 & 0 & 0 \\ -3 & 0 & 0 \\ -3 & 0 & 0 \end{bmatrix},$$

so $\text{rank}(\mathbf{A} - (-2)\mathbf{I}) = 1$ and the null space has dimension 2, and *without further calculation* it can be concluded that a diagonalizing similarity transformation matrix $\mathbf{S}$ exists, since two additional linearly independent eigenvectors exist.

Example 12
Calculating
diagonalizing $\mathbf{S}$

To find a diagonalizing $\mathbf{S} = [\,\mathbf{S}_1, \mathbf{S}_2, \mathbf{S}_3\,]$ for the previous example, a basis for the null space of each distinct $\mathbf{A} - \lambda_i\mathbf{I}$ is calculated, as in Chapter 6, Section 4.1. Thus, for $\lambda = 3$,

$$\begin{bmatrix} \mathbf{A} - (3)\mathbf{I} \\ \mathbf{I} \end{bmatrix} = \begin{bmatrix} -2 & 2 & -4 \\ -3 & -7 & 4 \\ -3 & -2 & -1 \\ 1 & 0 & 0 \\ 0 & 1 & 0 \\ 0 & 0 & 1 \end{bmatrix} \overset{\text{col}}{\sim} \begin{bmatrix} -2 & 0 & 0 \\ -3 & -10 & 0 \\ -3 & -5 & 0 \\ 1 & 1 & -1 \\ 0 & 1 & 1 \\ 0 & 0 & 1 \end{bmatrix},$$

and for $\lambda = -2$,

$$\begin{bmatrix} \mathbf{A} - (-2)\mathbf{I} \\ \mathbf{I} \end{bmatrix} = \begin{bmatrix} 3 & 2 & -4 \\ -3 & -2 & 4 \\ -3 & -2 & 4 \\ 1 & 0 & 0 \\ 0 & 1 & 0 \\ 0 & 0 & 1 \end{bmatrix} \overset{\text{col}}{\sim} \begin{bmatrix} 3 & 0 & 0 \\ -3 & 0 & 0 \\ -3 & 0 & 0 \\ 1 & -2/3 & 4/3 \\ 0 & 1 & 0 \\ 0 & 0 & 1 \end{bmatrix},$$

so that

$$\mathbf{S} = \begin{bmatrix} -1 & -2/3 & -4/3 \\ 1 & 1 & 0 \\ 1 & 0 & 1 \end{bmatrix}, \quad \text{and } \mathbf{S}^{-1}\mathbf{A}\mathbf{S} = \begin{bmatrix} 3 & 0 & 0 \\ 0 & -2 & 0 \\ 0 & 0 & -2 \end{bmatrix}.$$

3 **Near-diagonalization: the Jordan canonical form**

If there are n independent eigenvectors then Theorem 7.2 holds and $\mathbf{A}$ is diagonalizable.

If λ_i is not distinct but has multiplicity $n_i > 1$ then a direct extension of Theorem 7.4 can be used to show

Theorem 7.6 $rank(\mathbf{A} - \lambda\mathbf{I})^{n_i} = n - n_i$.

From this theorem, each factor $(\lambda_i - \lambda_i)^{n_i}$ in (7.25) is associated with an n_i-dimensional subspace, the null space of $(\mathbf{A} - \lambda_i\mathbf{I})^{n_i}$. Any basis for the null space contains n_i linearly independent vectors. Thus a direct extension of Theorem 7.3 can be used to show that a total of n linearly independent vectors can be found, but these vectors are not all eigenvectors unless $rank(\mathbf{A} - \lambda_i\mathbf{I}) = n - n_i$, as may happen. With considerably more detail it can be shown how to choose the n_i vectors associated with each λ_i such that the following is true.

Theorem 7.7   *For any matrix $\mathbf{A} \in \mathbb{R}^{n \times n}$ there exists a nonsingular matrix $\mathbf{S} \in \mathbb{C}^{n \times n}$ such that*

$$(7.29) \quad \mathbf{S}^{-1}\mathbf{A}\mathbf{S} = \mathbf{J} = \begin{bmatrix} \mathbf{J}_1 & & \\ & \ddots & \\ & & \mathbf{J}_t \end{bmatrix} = diag\,[\,\mathbf{J}_i\,],$$

where each $\mathbf{J}_i, i = 1, \cdots t$ is n_i-square and block diagonal, containing δ_k-square diagonal blocks, as shown:

$$(7.30) \quad \mathbf{J}_i = diag\,[\,\mathbf{J}_{\delta_k}(\lambda_i)\,], \quad where\ \mathbf{J}_{\delta_k}(\lambda_i) = \begin{bmatrix} \lambda_i & 1 & & \\ & \ddots & \ddots & \\ & & \lambda_i & 1 \\ & & & \lambda_i \end{bmatrix}_{\delta_k \times \delta_k}$$

and where $\sum \delta_k = n_i$.

The matrix $\mathbf{J}$ in the above theorem is called the *Jordan canonical form* of $\mathbf{A}$, and for many applications it is convenient simply to write

$$(7.31) \quad \mathbf{S}^{-1}\mathbf{A}\mathbf{S} = \mathbf{J} = diag\,[\,\mathbf{J}_{\delta_i}\,].$$

Although the Jordan form is an essential theoretical tool, the details of the calculation are omitted because infinitesimally small changes in $\mathbf{A}$ can result in finite changes in the resulting $\mathbf{J}$, which make the computation illposed and rarely practical for floating point computation, except when the Jordan form is a diagonal matrix.

Example 13
2×2 Jordan
block

The Jordan form will be found for the matrix of Example 9, in which it was shown that the eigenvalues are $-2, -2$, and that the given $\mathbf{A}$ cannot be diagonalized since there is only one independent eigenvector. Consequently the Jordan form must be

$$\mathbf{S}^{-1}\mathbf{A}\mathbf{S} = \mathbf{J} = \begin{bmatrix} \lambda_1 & 1 \\ 0 & \lambda_1 \end{bmatrix}.$$

Premultiplying this equation by $\mathbf{S} = [\,\mathbf{S}_2, \mathbf{S}_2\,]$ gives

$$\mathbf{A}\,[\,\mathbf{S}_1, \mathbf{S}_2\,] = [\,\mathbf{S}_1, \mathbf{S}_2\,]\begin{bmatrix} \lambda_1 & 1 \\ 0 & \lambda_1 \end{bmatrix} = [\,\mathbf{S}_1\lambda_1, \mathbf{S}_1 + \mathbf{S}_2\lambda_1\,].$$

Equating columns on the left and right shows that $\mathbf{S}_1$ satisfies $(\mathbf{A} - \lambda_1\mathbf{I})\,\mathbf{S}_1 = 0$ and therefore $\mathbf{S}_1$ is an eigenvector, whereas $\mathbf{S}_2$ satisfies $(\mathbf{A} - \lambda_1\mathbf{I})\,\mathbf{S}_2 = \mathbf{S}_1$. Using the eigenvector $\mathbf{S}_1$ from Example 9 yields

$$(\mathbf{A} - \lambda_1\mathbf{I})\,\mathbf{S}_2 = \begin{bmatrix} 2 & 1 \\ -4 & -2 \end{bmatrix}\mathbf{S}_2 - \begin{bmatrix} -1/2 \\ 1 \end{bmatrix}, \quad \text{so that, e.g., } \mathbf{S}_2 - \begin{bmatrix} -1/4 \\ 0 \end{bmatrix},$$

and $\mathbf{S}_2$ satisfies $(\mathbf{A} - \lambda_1\mathbf{I})^2\mathbf{S}_2 = 0$. When more than one nonscalar Jordan block corresponds to an eigenvalue, the computation contains more detail.

Example 14
Jordan form of
symmetric
matrices

The Jordan form of a symmetric matrix is diagonal, a simplification which can be shown by assuming the contrary. For example assume λ_i has multiplicity 2 and that $(\mathbf{A} - \lambda_i\mathbf{I})\,\mathbf{X}_i \neq 0$, so that $\mathbf{X}_i$ is not an eigenvector, but that $(\mathbf{A} - \lambda_i\mathbf{I})^2\mathbf{X}_i = 0$ from Theorem 7.6. Since λ_i is real by symmetry (Example 7), $(\mathbf{A} - \lambda_i\mathbf{I})^2$ is real and $\mathbf{X}_i$ can be assumed to be real. Then

$$0 = \mathbf{X}_i^T(\mathbf{A} - \lambda_i\mathbf{I})^2\mathbf{X}_i = ((\mathbf{A} - \lambda_i\mathbf{I})\,\mathbf{X}_i)^T(\mathbf{A} - \lambda_i\mathbf{I})\,\mathbf{X}_i,$$

which is the sum of the squares of the entries of $(\mathbf{A} - \lambda_i\mathbf{I})\,\mathbf{X}_i$, which must be zero, a contradiction and the essential argument used in a detailed proof.

4   **Functions of square matrices via the Jordan form**

Let the power series

(7.32) $f(z) = a_0 + a_1 z + a_2 z^2 + \cdots$

converge in a neighborhood of each of the eigenvalues $\lambda_1, \lambda_2, \cdots \lambda_n$, of matrix $\mathbf{A} \in \mathbb{R}^n$, and let the Jordan form $\mathbf{J} = \mathbf{S}^{-1}\mathbf{A}\mathbf{S}$ be known. Then since $\mathbf{A} = \mathbf{S}\mathbf{J}\mathbf{S}^{-1}$, the function $f(\mathbf{A})$ can be found as

(7.33) $f(\mathbf{A}) = a_0\mathbf{A}^0 + a_1\mathbf{A}^1 + a_2\mathbf{A}^2 + \cdots$

$= a_0(\mathbf{S}\mathbf{J}\mathbf{S}^{-1})^0 + a_1(\mathbf{S}\mathbf{J}\mathbf{S}^{-1})^1 + a_2(\mathbf{S}\mathbf{J}\mathbf{S}^{-1})^2 + \cdots$

$= \mathbf{S}(a_0\mathbf{J}^0 + a_1\mathbf{J}^1 + a_2\mathbf{J}^2 + \cdots)\mathbf{S}^{-1}$

$= \mathbf{S}f(\mathbf{J})\mathbf{S}^{-1},$

and finding $f(\mathbf{A})$ reduces to finding $f(\mathbf{J})$. The Jordan matrix is

(7.34) $\mathbf{J} = \begin{bmatrix} \mathbf{J}_{\delta_1} & & \\ & \ddots & \\ & & \mathbf{J}_{\delta_r} \end{bmatrix} = \text{diag}\,[\,\mathbf{J}_{\delta_i}\,]$

and the powers of $\mathbf{J}$ in (7.33) are $\mathbf{J}^k = \text{diag}\,[\,\mathbf{J}_{\delta_i}^k\,]$, so

(7.35) $f(\mathbf{J}) = \text{diag}\,[\,f(\mathbf{J}_{\delta_i})\,],$

and finding $f(\mathbf{J})$ reduces to finding the diagonal blocks $f(\mathbf{J}_{\delta_i})$, where each block $\mathbf{J}_{\delta_i}$ is δ_i-square with repeated eigenvalue λ_i and is of the form

(7.36) $\mathbf{J}_{\delta_i} = \begin{bmatrix} \lambda_i & 1 & & \\ & \ddots & \ddots & \\ & & \lambda_i & 1 \\ & & & \lambda_i \end{bmatrix} = \lambda_i\mathbf{I} + \mathbf{N}, \text{ where } \mathbf{N} = \begin{bmatrix} 0 & 1 & & \\ & \ddots & \ddots & \\ & & 0 & 1 \\ & & & 0 \end{bmatrix}.$

The binomial expansion of $\mathbf{J}_{\delta_i}^k = (\lambda_i\mathbf{I} + \mathbf{N})^k$ is

(7.37) $(\lambda_i\mathbf{I} + \mathbf{N})^k =$

$\lambda_i^k\mathbf{I} + \binom{k}{1}\lambda_i^{k-1}\mathbf{N} + \binom{k}{2}\lambda_i^{k-2}\mathbf{N}^2 + \binom{k}{3}\lambda_i^{k-3}\mathbf{N}^3 + \cdots$

$= \lambda_i^k\mathbf{I} + \frac{k}{1!}\lambda_i^{k-1}\mathbf{N} + \frac{k(k-1)}{2!}\lambda_i^{k-2}\mathbf{N}^2 + \frac{k(k-1)(k-2)}{3!}\lambda_i^{k-3}\mathbf{N}^3 + \cdots$

$= \lambda_i^k\mathbf{I} + \frac{1}{1!}\left(\frac{d}{d\lambda_i}\lambda_i^k\right)\mathbf{N} + \frac{1}{2!}\left(\frac{d^2}{d\lambda_i^2}\lambda_i^k\right)\mathbf{N}^2 + \frac{1}{3!}\left(\frac{d^3}{d\lambda_i^3}\lambda_i^k\right)\mathbf{N}^3 + \cdots,$

of which the first term is diagonal; and from the form of $\mathbf{N}$ in (7.36), the second is nonzero only on the first super-diagonal, the third is nonzero only on the second super-diagonal, and so on. Thus, from (7.32),

(7.38) $f(\mathbf{J}_{\delta_i}) = \sum_{k=0}^{\infty} a_k\mathbf{J}_{\delta_i}^k = \sum_{k=0}^{\infty} a_k(\lambda_i\mathbf{I} + \mathbf{N})^k$

$$= \left(\sum_{k=0}^{\infty} a_k \lambda_i^k \right) \mathbf{I} + \frac{1}{1!} \left(\frac{d}{d\lambda_i} \sum_{k=0}^{\infty} a_k \lambda_i^k \right) \mathbf{N} + \frac{1}{2!} \left(\frac{d^2}{d\lambda_i^2} \sum_{k=0}^{\infty} a_k \lambda_i^k \right) \mathbf{N}^2$$

$$+ \frac{1}{3!} \left(\frac{d^3}{d\lambda_i^3} \sum_{k=0}^{\infty} a_k \lambda_i^k \right) \mathbf{N}^3 + \cdots$$

$$= f(\lambda_i) \mathbf{I} + \frac{1}{1!} \frac{d}{d\lambda_i} f(\lambda_i) \mathbf{N} + \frac{1}{2!} \frac{d^2}{d\lambda_i^2} f(\lambda_i) \mathbf{N}^2 + \frac{1}{3!} \frac{d^3}{d\lambda_i^3} f(\lambda_i) \mathbf{N}^3 + \cdots$$

$$= \begin{bmatrix} f(\lambda_i) & \frac{1}{1!} \frac{d}{d\lambda_i} f(\lambda_i) & \frac{1}{2!} \frac{d^2}{d\lambda_i^2} f(\lambda_i) & \cdots & \frac{1}{(\delta_i-1)!} \frac{d^{(\delta_i-1)}}{d\lambda_i^{(\delta_i-1)}} f(\lambda_i) \\ & f(\lambda_i) & \frac{1}{1!} \frac{d}{d\lambda_i} f(\lambda_i) & \ddots & \\ & & \ddots & & \ddots \end{bmatrix},$$

and using this to compute the blocks in (7.35) gives $f(\mathbf{A})$ via (7.35) and (7.33). The assumption that $f(\cdot)$ has a Taylor series at the eigenvalues of $\mathbf{A}$ guarantees that the derivatives required in (7.38) exist.

The importance of (7.38) is not for numerical computation, but so that the form of the functions in $f(\mathbf{A})$ is available by inspection of (7.38), since the entries of $f(\mathbf{A})$ are linear combinations of the entries of (7.38).

Example 15
Diagonal **J**

If **A** is diagonalizable, an important case, then the $\mathbf{J}_{\delta_i} = \lambda_i$ are scalars, and

(7.39) $\quad f(\mathbf{A}) = \mathbf{S} \ \text{diag} \left[f(\lambda_i) \right] \mathbf{S}^{-1}.$

Example 16
Derivatives in
$f(\mathbf{A})$

If **A** is not diagonalizable then linear combinations of derivatives of $f(z)$, evaluated at the eigenvalues λ_i, appear in $f(\mathbf{A})$. For each distinct eigenvalue λ_i, the highest derivative appearing in the solution is $\frac{d^{(\delta_i-1)}}{d\lambda_i^{(\delta_i-1)}} f(\lambda_i)$, where δ_i is the size of the largest Jordan block containing eigenvalue λ_i.

Example 17
Function $e^{t\mathbf{A}}$

When $f(z) = e^{tz}$, then the coefficients in (7.32) are $a_k = t^k/k!$, giving

(7.40) $\quad e^{t\mathbf{A}} = \mathbf{S} \ \text{diag} \begin{bmatrix} e^{t\lambda_i} & \frac{1}{1!} \frac{d}{d\lambda_i} e^{t\lambda_i} & \frac{1}{2!} \frac{d^2}{d\lambda_i^2} e^{t\lambda_i} & \cdots & \frac{1}{(\delta_i-1)!} \frac{d^{(\delta_i-1)}}{d\lambda_i^{(\delta_i-1)}} e^{t\lambda_i} \\ & e^{t\lambda_i} & \frac{1}{1!} \frac{d}{d\lambda_i} e^{t\lambda_i} & \ddots & \\ & & \ddots & & \ddots \end{bmatrix} \mathbf{S}^{-1}$

$$= \mathbf{S} \ \text{diag} \begin{bmatrix} e^{t\lambda_i} & \frac{t}{1!} e^{t\lambda_i} & \frac{t^2}{2!} e^{t\lambda_i} & \cdots & \frac{t^{\delta_i-1}}{(\delta_i-1)!} e^{t\lambda_i} \\ & e^{t\lambda_i} & \frac{t}{1!} e^{t\lambda_i} & \ddots & \\ & & \ddots & & \ddots \end{bmatrix} \mathbf{S}^{-1},$$

which is convenient for symbolic computation but seldom for floating-point computation because of the sensitivity of the block sizes δ_i with respect to the entries of $\mathbf{A}$. When $\mathbf{A}$ is diagonalizable then all $\delta_i = 1$ and this method often can be used reliably, but no single method is reliable for all possible $t\mathbf{A}$.

Example 18
Function $\mathbf{A}^k$

When $f(z) = z^k$, then $f(\mathbf{A})$ is

$$(7.41) \quad \mathbf{A}^k = \mathbf{S} \, \text{diag} \begin{bmatrix} \lambda_i^k & \binom{k}{1}\lambda_i^{k-1} & \binom{k}{2}\lambda_i^{k-2} & \cdots & \binom{k}{\delta_i-1}\lambda_i^{k-\delta_i+1} \\ & \lambda_i^k & \binom{k}{1}\lambda_i^{k-1} & \ddots \\ & & \ddots & & \ddots \end{bmatrix} \mathbf{S}^{-1}.$$

Example 19
Nonconvergence of series

Requiring $f(\mathbf{A})$ to be written as a converging series can cause difficulties, illustrated by the function $f(z) = 1/(1-z)$, and its series $1 + z + z^2 + \cdots$, which converges for $|z| < 1$. The series diverges when z is replaced by the matrix $\mathbf{A} = \begin{bmatrix} 1/2 & 0 \\ 0 & 2 \end{bmatrix}$.

5   General functions of square matrices

Section 4 requires the function $f(\cdot)$ to have a series representation at the eigenvalues of $\mathbf{A}$. Here a derived property, which results in a more general definition of functions of matrices, will be obtained.

Let the series

$$(7.42) \quad f(z) = \sum_{i=0}^{\infty} \alpha_i z^i$$

converge at the eigenvalues of $\mathbf{A}$, and let the characteristic polynomial be

$$(7.43) \quad \phi(z) = \det(z\mathbf{I} - \mathbf{A}) = z^n + a_1 z^{n-1} + \cdots a_n.$$

Dividing each term of (7.42) by $\phi(z)$ will give a quotient $q_i(z)$ and remainder $r_i(z)$ of degree less than n, for $i = 0, \cdots$. Then, by formally summing,

$$(7.44) \quad f(z) = \sum_{i=0}^{\infty}(\phi(z)\, q_i(z) + r_i(z)) = \phi(z)\sum_{i=0}^{\infty} q_i(z) + \sum_{i=0}^{\infty} r_i(z)$$

$$= \phi(z)\sum_{i=0}^{\infty} q_i(z) + r(z),$$

where $r(z)$ is a polynomial of degree at most $n-1$.

Substituting $\mathbf{A}$ for z in the above expression and using the Cayley-Hamilton theorem gives

(7.45) $f(\mathbf{A}) = 0 + r(\mathbf{A})$,

which is a polynomial of degree $n-1$ in $\mathbf{A}$.

Definition   From the above, it becomes convenient to *define* $f(\mathbf{A})$ to be the polynomial $\alpha_0\mathbf{I} + \alpha_1\mathbf{A} + \cdots \alpha_{n-1}\mathbf{A}^{n-1}$ such that the equation $f(\lambda_i) = \alpha_0 + \alpha_1\lambda_i + \cdots \alpha_{n-1}\lambda_i^{n-1}$ is satisfied at each of the eigenvalues λ_i, $i = 1, \cdots n$ of $\mathbf{A}$. If the λ_i are distinct, writing this equation for each λ_i allows the α_i to be solved. However if $\mathbf{A}$ has multiple eigenvalues, these n equations are not independent, and the calculation of the α_i must be modified.

To obtain further independent equations when λ_i has multiplicity $n_i > 1$, say, write (7.43) as $\phi(z) = (z - \lambda_i)^{n_i} \times$ (other factors), so that its derivative evaluated at λ_i is $d\phi(z)/dz|_{z=\lambda_i} = 0$. Then, from (7.44),

(7.46) $$\left.\frac{df(z)}{dz}\right|_{z=\lambda_i} = \left.\left(\frac{d\phi(z)}{dz}\sum_{i=0}^{\infty}q_i(z) + \phi(z)\frac{d}{dz}\sum_{i=0}^{\infty}q_i(z) + \frac{dr(z)}{dz}\right)\right|_{z=\lambda_i}$$

$$= 0 + 0 + \left.\frac{dr(z)}{dz}\right|_{z=\lambda_i},$$

and similarly,

(7.47) $$\left.\frac{d^2f(z)}{dz^2}\right|_{z=\lambda_i} = \left.\frac{d^2r(z)}{dz^2}\right|_{z=\lambda_i},$$

$$\vdots$$

$$\left.\frac{d^{n_i-1}f(z)}{dz^{n_i-1}}\right|_{z=\lambda_i} = \left.\frac{d^{n_i-1}r(z)}{dz^{n_i-1}}\right|_{z=\lambda_i},$$

giving n_i independent equations for each eigenvalue of multiplicity n_i.

Example 20
sin(A)
For $\mathbf{A} = \begin{bmatrix} -3 & 1 \\ 0 & -2 \end{bmatrix}$ with eigenvalues $\lambda_1 = -3$, $\lambda_2 = -2$, find $\sin(\mathbf{A})$.

Since $\mathbf{A}$ is 2×2, the function is a polynomial of degree 1; thus,

(7.48) $\sin(\mathbf{A}) = \alpha_0\mathbf{I} + \alpha_1\mathbf{A}$,

where $\sin(\lambda_i) = \alpha_0 + \alpha_1\lambda_i$, for $i = 1, 2$, and the equations to be solved are

(7.49) $$\begin{bmatrix} 1 & \lambda_1 \\ 1 & \lambda_2 \end{bmatrix}\begin{bmatrix} \alpha_0 \\ \alpha_1 \end{bmatrix} = \begin{bmatrix} \sin(\lambda_1) \\ \sin(\lambda_2) \end{bmatrix}.$$

Solving for α_0 and α_1 and substituting in (7.48) gives

(7.50) $\sin(\mathbf{A}) = (3\sin(-2)-2\sin(-3))\begin{bmatrix} 1 & 0 \\ 0 & 1 \end{bmatrix} + (\sin(-2)-\sin(-3))\begin{bmatrix} -3 & 1 \\ 0 & -2 \end{bmatrix}$

$= \begin{bmatrix} \sin(-3) & \sin(-2)-\sin(-3) \\ 0 & \sin(-2) \end{bmatrix}.$

Example 21
$f(\mathbf{A})$ for
Example 19

Applying the general definition to the function $f(z) = 1/(1-z)$ and matrix $\mathbf{A} = \begin{bmatrix} 1/2 & 0 \\ 0 & 2 \end{bmatrix}$ of Example 19 gives the following equation to be solved for α_0 and α_1:

$\begin{bmatrix} 1 & 1/2 \\ 1 & 2 \end{bmatrix}\begin{bmatrix} \alpha_0 \\ \alpha_1 \end{bmatrix} = \begin{bmatrix} 1/(1-1/2) \\ 1/(1-2) \end{bmatrix} = \begin{bmatrix} 2 \\ -1 \end{bmatrix}.$

Solving, we see that $\alpha_0 = 3$ and $\alpha_1 = -2$, so that

$f(\mathbf{A}) = \begin{bmatrix} 3 & 0 \\ 0 & 3 \end{bmatrix} + \begin{bmatrix} -1 & 0 \\ 0 & -4 \end{bmatrix} = \begin{bmatrix} 2 & 0 \\ 0 & -1 \end{bmatrix},$

which is a diagonal matrix containing the function of the diagonal elements of $\mathbf{A}$, as is to be expected since $\mathbf{A}$ is diagonal.

Example 22
$\mathbf{A}^k$ has degree
$\leq n-1$

One useful function is the power $\mathbf{A}^k$, which, from the definition, is a linear combination of powers from $\mathbf{A}^0 = \mathbf{I}$ to $\mathbf{A}^{n-1}$. That is, if $f(z) = z^k$, then by long division, $f(z) = \phi(z)q(z) + r(z)$, where $r(z)$ is a polynomial of degree $n-1$, and $\mathbf{A}^k = r(\mathbf{A})$, a polynomial of degree $n-1$ in $\mathbf{A}$, which always exists because the functions λ_i^k exist.

Example 23
Multiple
eigenvalues

Find $e^{t\mathbf{A}}$, for

$\mathbf{A} = \begin{bmatrix} 0 & 1 & 0 \\ 0 & 0 & 1 \\ 27 & -27 & 9 \end{bmatrix},$ for which $\phi(\lambda) = \lambda^3 - 9\lambda^2 + 27\lambda - 27 = (\lambda-3)^3.$

Let

$e^{t\mathbf{A}} = \alpha_0\mathbf{I} + \alpha_1\mathbf{A} + \alpha_2\mathbf{A}^2,$

such that

$e^{tz}\big|_{z=3} = e^{3t} = \alpha_0 + \alpha_1(3) + \alpha_2(9),$

$$\left.\frac{de^{tz}}{dz}\right|_{z=3} = te^{3t} = \alpha_1 + 2\alpha_2(3),$$

$$\left.\frac{d^2 e^{tz}}{dz^2}\right|_{z=3} = t^2 e^{3t} = 2\alpha_2,$$

and solve for α_0, α_1, and α_2, resulting in

$$e^{t\mathbf{A}} = \begin{bmatrix} 1 - 3t + 9t^2/2 & t - 3t^2 & t^2/2 \\ 27t^2/2 & 1 - 3t - 9t^2 & t + 3t^2/2 \\ 27(t + 3t^2/2) & -27(t + t^2) & 1 + 6t + 9t^2/2 \end{bmatrix} e^{3t}.$$

Example 24
Complex
eigenvalues

In the derivation of matrix functions used in these sections, for real $\mathbf{A}$ the eigenvalues are either real or occur in complex-conjugate pairs. Thus, to calculate $e^{t\mathbf{A}}$ for $\mathbf{A} = \begin{bmatrix} \alpha & \omega \\ -\omega & \alpha \end{bmatrix}$, with eigenvalues $\{\lambda_i\} = \{\alpha \pm j\omega\}$, let $e^{t\mathbf{A}} = \alpha_0 \mathbf{I} + \alpha_1 \mathbf{A}$, with

$$\begin{bmatrix} 1 & \lambda_1 \\ 1 & \lambda_2 \end{bmatrix} \begin{bmatrix} \alpha_0 \\ \alpha_1 \end{bmatrix} = \begin{bmatrix} e^{t\lambda_1} \\ e^{t\lambda_2} \end{bmatrix},$$

so that

$$\begin{bmatrix} \alpha_0 \\ \alpha_1 \end{bmatrix} = \frac{1}{\lambda_2 - \lambda_1} \begin{bmatrix} \lambda_2 & -\lambda_1 \\ -1 & 1 \end{bmatrix} \begin{bmatrix} e^{t\lambda_1} \\ e^{t\lambda_2} \end{bmatrix}$$

$$= \frac{e^{t\alpha}}{-2j\omega} \begin{bmatrix} (\alpha - j\omega)e^{tj\omega} - (\alpha + j\omega)e^{-tj\omega} \\ e^{-tj\omega} - e^{tj\omega} \end{bmatrix}$$

$$= \frac{e^{t\alpha}}{\omega} \begin{bmatrix} \omega\cos(\omega t) - \alpha\sin(\omega t) \\ \sin(\omega t) \end{bmatrix}.$$

Substituting the values of α_0 and α_1 gives

$$e^{t\mathbf{A}} = \frac{e^{t\alpha}}{\omega}(\omega\cos(\omega t) - \alpha\sin(\omega t))\mathbf{I} + \frac{e^{t\alpha}}{\omega}\sin(\omega t)\,\mathbf{A}$$

$$= \begin{bmatrix} e^{t\alpha}\cos(\omega t) & e^{t\alpha}\sin(\omega t) \\ -e^{t\alpha}\sin(\omega t) & e^{t\alpha}\cos(\omega t) \end{bmatrix},$$

which contains only real quantities, as in Example 19 of Chapter 2.

6 Further study

Similarity transformations can be found in books on linear algebra such as Strang [51] and also in system theory books such as Chen [9], which also discusses matrix functions. The computation of functions of square matrices is discussed in Golub and Van Loan [19].

1   For the discrete-time system with matrices $\mathbf{A} = \begin{bmatrix} 0 & 1 \\ -2 & -3 \end{bmatrix}$, $\mathbf{B} = \begin{bmatrix} 0 \\ 1 \end{bmatrix}$, $\mathbf{C} = [5, 1]$, $\mathbf{D} = 0$, and state $\mathbf{X}$,

   (a) write the formula for the transfer matrix $\mathbf{H}(z)$;

   (b) find the system matrices that result from the change of state variables $\mathbf{X} = \mathbf{SX}'$, where $\mathbf{S} = \begin{bmatrix} -1 & 1 \\ 1 & -2 \end{bmatrix}$;

   (c) calculate the transfer matrix for the new system, and compare it to the one in (a).

2   Determine whether each of the following matrices can be diagonalized by a similarity transformation $\mathbf{A} \to \mathbf{S}^{-1}\mathbf{AS}$, and determine the resulting diagonal matrix (ω is an arbitrary real parameter):

   (a) $\mathbf{A} = \begin{bmatrix} -5 & 1 \\ -3 & -1 \end{bmatrix}$,   (b) $\mathbf{A} = \begin{bmatrix} -1 & -6 & 5 \\ -1 & 4 & -5 \\ -11 & -6 & 15 \end{bmatrix}$,   (c) $\mathbf{A} = \begin{bmatrix} -2 & -\omega \\ \omega & -2 \end{bmatrix}$.

3   Referring to Example 8, show that for any matrix $\mathbf{A} \in \mathbb{R}^{n \times n}$, if $\mathbf{U} + j\mathbf{V}$ is an eigenvector corresponding to eigenvalue $\alpha + j\omega$, then $\mathbf{U} - j\mathbf{V}$ is an eigenvector corresponding to eigenvalue $\alpha - j\omega$.

4   Find a real similarity matrix to transform

$$\mathbf{A} = \begin{bmatrix} 0 & -5 & 1 \\ 1 & -4 & 0 \\ 0 & 0 & -1 \end{bmatrix}$$

to real block-diagonal form as in Example 8.

5   For the circuit of Figure P7.5 which has neither input nor output, write the state-

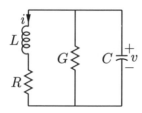

Fig. P7.5   Linear circuit without input or output.

space equations, and for small R and G find a real similarity matrix to transform the system to the form given in Example 8.

6   Calculate the Jordan form of the matrix $\mathbf{A} = \begin{bmatrix} 0 & -9 \\ 1 & 6 \end{bmatrix}$.

7   The eigenvalues λ_i must be computed to diagonalize a real matrix $\mathbf{A}$ and for many other purposes. A standard floating-point method is to find an orthogonal similarity transformation matrix $\mathbf{S}$ such that $\mathbf{S}^T \mathbf{A} \mathbf{S}$ is real and block-triangular, containing blocks on the diagonal that are either 1×1 or 2×2, with zeros below the diagonal blocks, to machine precision. Such a matrix is said to be in *real Schur form,* and its eigenvalues are easily computed from the diagonal blocks. Compute the eigenvalues of the real Schur matrix $\begin{bmatrix} -4 & 1 & 3 \\ 0 & -3 & -2 \\ 0 & 5 & 4 \end{bmatrix}$.

8   Find $\mathbf{A}^{200}$ for $\mathbf{A} = \begin{bmatrix} 1 & 3 \\ 0 & 1 \end{bmatrix}$.

9   Find formulas for $e^{t\mathbf{A}}$ and $\mathbf{A}^k$, for arbitrary t and k, and

$$\mathbf{A} = \begin{bmatrix} 0 & 0 & -8 \\ 1 & 0 & -12 \\ 0 & 1 & -6 \end{bmatrix}.$$

10   Given $\mathbf{A} = \begin{bmatrix} 0 & 1 \\ -\alpha^2 & -2\alpha \end{bmatrix}$ and $\mathbf{B} = \begin{bmatrix} 0 \\ 1 \end{bmatrix}$, find formulas for the exponential $e^{T\mathbf{A}}$ and for the integral $\left(\int_0^T e^{\tau\mathbf{A}} d\tau \right) \mathbf{B}$, and hence verify that with $\alpha = 3$ and $T = 0.1$, the system of Example 6 of Chapter 2 is the discretization of the system of Example 23 in the same chapter.

11   In Section 3 of Chapter 2, a discretization of a time-continuous system with matrices $(\mathbf{A}, \mathbf{B}, \mathbf{C}, \mathbf{D})$ was obtained to have matrices $(\mathbf{F} = e^{T\mathbf{A}}, \mathbf{G} = \int_0^T e^{\tau\mathbf{A}} d\tau, \mathbf{C}, \mathbf{D})$, where T is the sampling interval. Given $\mathbf{F}$, $\mathbf{G}$, and T, show how to find $\mathbf{A}$ and $\mathbf{B}$, and comment on whether they are unique.

12   The free response of the state of a continuous-time system is $\mathbf{X}(t) = e^{t\mathbf{A}}\mathbf{X}(0)$. Suppose that $\mathbf{A}$ can be diagonalized by a similarity transformation, and that vectors $\mathbf{X}_1 = \mathbf{X}(T)$, $\mathbf{X}_0 = \mathbf{X}(0)$ are given and known to satisfy this expression for some $t = T$. How can T be calculated from $\mathbf{X}_0$, $\mathbf{X}_1$, and $\mathbf{A}$?

13   Let λ_i be and eigenvalue of $\mathbf{A}$ and $\mathbf{X}_i$ the corresponding eigenvector; that is, let the equation $\mathbf{A}\mathbf{X}_i = \lambda_i\mathbf{X}_i$ be satisfied. Let $f(\mathbf{A}) = \alpha_0\mathbf{I} + \alpha_1\mathbf{A} + \cdots \alpha_{n-1}\mathbf{A}^{n-1}$ be a known function of $\mathbf{A}$. Calculate $f(\mathbf{A})\mathbf{X}_i$ and thereby show that $f(\lambda_i)$ is an eigenvalue of $f(\mathbf{A})$, with corresponding eigenvector $\mathbf{X}_i$.

Stability

There are many useful definitions of stability for the general nonlinear, time-varying state-space models (1.7) and (1.8), but the situation is much simpler for LTI systems with real or complex matrices. Two subjects are of interest: the algebraic properties of LTI systems that determine their stability, and the energy or energy-like functions associated with the free response of a system, the latter being a fundamental tool for the stability analysis of nonlinear systems as well as linear systems.

First some basic definitions will be given, and the relevant algebraic properties of the eigenvalues and the Jordan form of LTI systems will be derived. Then energy functions will be introduced, with a glimpse of stability theory associated with the name Lyapunov. These functions have several important uses, such as measures of signal energy, as stability-determining functions for nonlinear as well as linear systems, and as cost functions in optimization problems.

1   Basic definitions

A distinction can be made between internal and external stability concepts. External stability corresponds to the idea that the system output eventually should have properties possessed by the input, such as boundedness. LTI systems that are minimal as discussed in Chapter 5 and Chapter 9 and that are internally asymptotically stable (as to be defined) are also externally stable. Therefore internal stability will be considered here.

The internal stability of a system characterizes the size of the response of the state $\mathbf{X}(t)$ with respect to initial state $\mathbf{X}(t_0)$. The Euclidean norm is a suitable measure of size. Consider the state-space equations

$$(8.1) \quad \frac{d}{dt}\mathbf{X} = \mathbf{F}(\mathbf{X}(t)),$$

in continuous time, and

(8.2) $\mathbf{X}(t+1) = \mathbf{F}(\mathbf{X}(t))$,

in discrete time. These systems contain no input $\mathbf{U}(t)$, although the effects of specific input functions can be included by changing $\mathbf{F}$ in the above two equations to $\mathbf{F}(\mathbf{X}(t), t)$, and slightly modifying the definitions and results to be given to account for time-varying systems. Time-invariant systems without input are called *autonomous,* and these will be the focus of attention.

Stability must be defined in general with respect to specific solutions of (8.1) or (8.2). Although time-varying solutions can be included in general, the discussion will be confined to constant solutions $\mathbf{X}(t) = \mathbf{X}_0$, such that

(8.3) $\mathbf{F}(\mathbf{X}_0) = 0$

in the continuous-time equation (8.1), or

(8.4) $\mathbf{X}_0 = \mathbf{F}(\mathbf{X}_0)$

in the discrete-time equation (8.2). It may be that all initial conditions within an arbitrarily small region produce solutions that remain near a single solution, which is then called a stable solution. A nonlinear system may have both unstable and stable solutions.

Only $\mathbf{X}_0 = 0$ need be considered, since the change of variables $\mathbf{X}'(t) = \mathbf{X}(t) - \mathbf{X}_0$ and the substitution $\mathbf{X} = \mathbf{X}' + \mathbf{X}_0$ in (8.1) change this equation to

(8.5) $\dfrac{d}{dt}\mathbf{X}' = \mathbf{F}(\mathbf{X}' + \mathbf{X}_0) = \mathbf{F}'(\mathbf{X}')$,

with a similar change for (8.2).

Stability The origin is *stable* if, for any given value $\epsilon > 0$, there is a value $\delta(\epsilon) > 0$ such that if $\|\mathbf{X}(0)\| < \delta$ then $\|\mathbf{X}(t)\| < \epsilon$ for all $t > 0$, as illustrated in Figure 8.1. Otherwise the origin is *unstable.*

Asymptotic stability The origin is *asymptotically stable* if it is stable, and if there exists $\delta' > 0$ such that whenever $\|\mathbf{X}(0)\| < \delta'$ then $\lim_{t \to \infty} \|\mathbf{X}(t)\| = 0$.

Instability of the origin does not necessarily imply that solutions become indefinitely large, only that arbitrarily small deviations from the origin do not remain arbitrarily small. In a nonlinear system, for example, all solutions starting near the origin may approach stable solutions a finite distance away.

The only constant solution of interest for LTI systems is the origin, so that it is meaningful to speak of the stability of the system, rather than of a solution.

The above definition of stability is sometimes referred to as stability in the sense of Lyapunov, and also as *marginal stability* in the literature, with asymptotic stability simply called stability. Further definitions of greatest applicability to nonlinear and time-varying systems are also used in the literature.

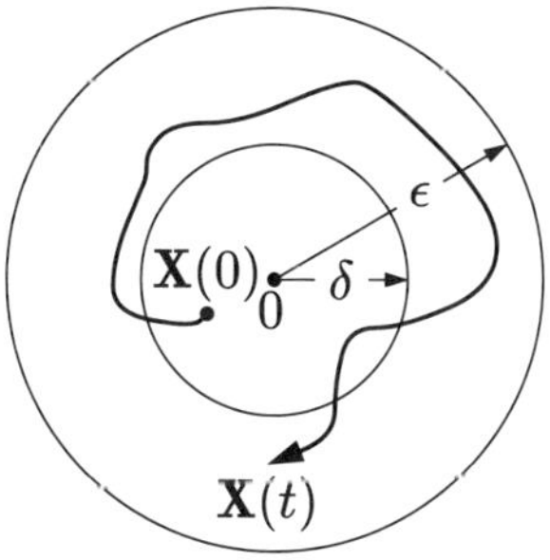

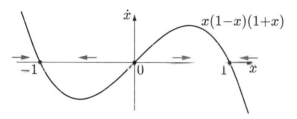

Fig. 8.1   Stability: The initial state $\mathbf{X}(0)$ within distance δ of the origin produces a solution that remains within distance ϵ of the origin.

Example 1
Stable and
unstable
constant
solutions

Consider the system described by $\dot{x} = x(1 - x)(1 + x)$. Solving $0 = x(1 -$

$$\dot{x} \qquad x(1-x)(1+x)$$

Fig. 8.2   A nonlinear system with three constant solutions, of which two are stable and one is unstable. The gray arrows show the direction of motion of x along the horizontal axis.

$x)(1 + x)$ gives three constant solutions at $x_0 = -1,\ 0,\ 1$. For $0 < x < 1$, for example, the quantity $x(1-x)(1+x)$ is positive, and therefore $x(t)$ is increasing, as shown in Figure 8.2. Similarly, the arrows show the direction of motion in the other intervals along the horizontal axis. The solution $x_0 = 0$ is unstable, and solutions $x_0 = \pm 1$ are stable. The solutions $x_0 = \pm 1$ are also asymptotically stable. Although the origin is unstable, solutions do not approach infinity.

2   LTI systems

Consider the linear version of (8.1), which is

(8.6) $\dfrac{d}{dt}\mathbf{X} = \mathbf{A}\mathbf{X},$

and of (8.2), which is

(8.7) $\mathbf{X}(t+1) = \mathbf{A}\mathbf{X}(t).$

The detailed information about the solutions of these equations obtained in Chapter 7 will be used to determine algebraic properties of $\mathbf{A}$ that determine whether the origin is stable, asymptotically stable, or unstable.

2.1   LTI Continuous-time systems

The free response will be discussed using the Jordan form, calculated in Section 4 of Chapter 7. Given $\mathbf{A}$ with eigenvalues λ_i, $i = 1, \ldots n$, and $\mathbf{X}(0)$, there exists a nonsingular matrix $\mathbf{S}$ such that the free response is

$$(8.8) \quad \mathbf{X}(t) = e^{t\mathbf{A}}\mathbf{X}(0) = \mathbf{S}e^{t\mathbf{S}^{-1}\mathbf{A}\mathbf{S}}\mathbf{S}^{-1}\mathbf{X}(0) = \mathbf{S}e^{t\mathbf{J}}\mathbf{S}^{-1}\mathbf{X}(0),$$

where $\mathbf{J} = \operatorname{diag}\left[\mathbf{J}_{\delta_k}\right]$ is the Jordan form of $\mathbf{A}$, containing Jordan blocks $\mathbf{J}_{\delta_k}$ with eigenvalue λ_k. Then from Equation (7.40), which shows the form of of $e^{t\mathbf{J}_{\delta_k}}$, the entries of $\mathbf{X}(t)$ are linear combinations of functions of the form $e^{t\lambda_k}$ for 1×1 Jordan blocks and of the form $t^{\gamma}e^{t\lambda_k}$ for nonscalar Jordan blocks, where the γ are nonnegative integers.

The functions $e^{t\lambda_k}$ and $t^{\gamma}e^{t\lambda_k}$ approach 0 as $t \to \infty$ if λ_k has negative real part $\operatorname{Re}(\lambda_k)$. If $\operatorname{Re}(\lambda_k)$ is 0 then the function is bounded for $\gamma = 0$, increasing if $\gamma > 0$. If $\operatorname{Re}(\lambda_k)$ is positive then the function is increasing.

Summary   From the above discussion, if all $\operatorname{Re}(\lambda_i) < 0$ then the system is asymptotically stable. If $\operatorname{Re}(\lambda_i) \leq 0$ for all eigenvalues of scalar Jordan blocks, and if $\operatorname{Re}(\lambda_i) < 0$ for all eigenvalues of nonscalar Jordan blocks, then the system is stable. If any $\operatorname{Re}(\lambda_i) > 0$ for any root λ_i, or $\operatorname{Re}(\lambda_i) \geq 0$ for any nonscalar Jordan block, then the system is unstable.

Diagonal system   For LTI systems with scalar Jordan blocks, that is, with diagonalizable matrix $\mathbf{A}$, the above conclusions simplify as follows: if any $\operatorname{Re}(\lambda_i) > 0$ then the system is unstable, if all $\operatorname{Re}(\lambda_i) \leq 0$ then the system is stable, and if all $\operatorname{Re}(\lambda_i) < 0$ then the system is asymptotically stable.

Example 2
Multiple
marginally
stable
eigenvalues
For the LTI system $\frac{d}{dt}\mathbf{X} = \mathbf{A}\mathbf{X}$, if $\mathbf{A} = \begin{bmatrix} 0 & 0 \\ 0 & 0 \end{bmatrix}$, the system is stable, but if $\mathbf{A} = \begin{bmatrix} 0 & 1 \\ 0 & 0 \end{bmatrix}$, the system is unstable, although for both of these matrices the eigenvalues λ_k satisfy $\operatorname{Re}(\lambda_k) \leq 0$. In neither case is the system asymptotically stable.

2.2   LTI Discrete-time systems

The discussion in the previous section is easily modified to apply to discrete-time systems. The definitions of stability and asymptotic stability must be interpreted to apply to sequences, rather than to continuous functions of time, and the state-transition matrix is $\mathbf{A}^{t-t_0}$, rather than an exponential; so from Equation (7.41) the free state response for $t_0 = 0$ is

(8.9) $\mathbf{X}(t) = \mathbf{A}^t \mathbf{X}(0) = \mathbf{S} \operatorname{diag} [\mathbf{J}_{\delta_k}^t] \mathbf{S}^{-1} \mathbf{X}(0),$

where the entries of $\mathbf{X}(t)$ are linear combinations of functions which have the form $\binom{t}{\gamma} \lambda_k^{t-\gamma}$ where $\gamma \geq 0$ is an integer. This function approaches 0 as $t \to \infty$ if $|\lambda_k| < 1$. If $|\lambda_k| = 1$ then the function is bounded for $\gamma = 0$, increasing if $\gamma > 0$. If $|\lambda_k| > 1$, the function is increasing.

Summary   In summary, if all $|\lambda_i| < 1$ then the system is asymptotically stable. If $|\lambda_i| \leq 1$ for all eigenvalues of scalar Jordan blocks, and if $|\lambda_i| < 1$ for all eigenvalues of nonscalar Jordan blocks, then the system is stable. If any $|\lambda_i| > 1$, or if $|\lambda_i| \geq 1$ for any nonscalar Jordan block, then the system is unstable.

Diagonal system   For LTI systems with diagonalizable matrix $\mathbf{A}$, the above conclusions simplify as follows: if any $|\lambda_i| > 1$ then the system is unstable, if all $|\lambda_i| \leq 1$ then the system is stable, and if all $|\lambda_i| < 1$ then the system is asymptotically stable.

Example 3
Multiple marginally stable eigenvalues   For the LTI system $\mathbf{X}(t+1) = \mathbf{A}\mathbf{X}(t)$, if $\mathbf{A} = \begin{bmatrix} 1 & 0 \\ 0 & 1 \end{bmatrix}$, the system is stable, but if $\mathbf{A} = \begin{bmatrix} 1 & 1 \\ 0 & 1 \end{bmatrix}$, the system is unstable, although the eigenvalues of both matrices satisfy $|\lambda_k| \leq 1$. In neither case is the system asymptotically stable.

---

3   Energy functions and Lyapunov stability

In the previous section it was shown that knowledge of the eigenvalues of an LTI system are sufficient to determine asymptotic stability. Then given a nonlinear system such as (8.1) or (8.2), the first-order character of the solutions in a small neighborhood of each of the constant operating points can be inferred from the LTI linearization, computed as in Section 6 of Chapter 1, at each of these points. More specifically, it can be shown that if the linearization is asymptotically stable, then the operating point of the nonlinear system is asymptotically stable, meaning that solutions starting near enough to the operating point approach it. Likewise if the linearization is unstable then the operating point of the nonlinear system is unstable. This is known as the first method of Lyapunov for analyzing nonlinear systems. However, nothing is guaranteed about solutions which are merely stable, since higher-order detail is required, and unless the system is linear, nothing is guaranteed about solutions which are a large distance away from the operating point.

The second method of Lyapunov is a powerful qualitative tool for analysing nonlinear systems, but it also has important consequences for LTI systems, as will be shown.

Consider a mechanical system with no energy sources but in which there is friction, which converts mechanical energy to heat. The total mechanical energy of the system decreases over time, and this decreasing energy corresponds to diminishing movement of the mechanical variables. A Lyapunov function is a formalization of this principle, and is illustrated in Figure 8.3.

Suppose one can find a scalar function $v(\mathbf{X})$ and a region Ω for which

1. Ω is a bounded closed region of $\mathbb{R}^n$ defined by $v(\mathbf{X}) \leq c$, where c is a constant;

2. $v(\mathbf{X})$ is continuously differentiable in Ω;

3. at the origin $v(\mathbf{X})$ has a minimum which is unique and global within Ω.

Then one statement of the elements of Lyapunov stability theory is contained in the following.

Theorem 8.1 *Along every solution $\mathbf{X}(t)$ of (8.1) or solution sequence $\{\mathbf{X}(t)\}$ of (8.2) contained in Ω,*

1. *if $v(\mathbf{X}(t))$ is decreasing except at the origin, then the origin is asymptotically stable, and all initial conditions inside Ω produce trajectories that approach the origin;*

2. *if $v(\mathbf{X}(t))$ is increasing except at the origin, then the origin is unstable.*

In continuous time, the change in v along a solution trajectory is obtained from the total derivative

$$(8.10) \quad \frac{dv}{dt} = \sum_{i=1}^{n} \frac{\partial v}{\partial x_i} \frac{dx_i}{dt} = \left[\frac{\partial v}{\partial x_1}, \cdots \frac{\partial v}{\partial x_n} \right] \begin{bmatrix} \frac{dx_1}{dt} \\ \vdots \\ \frac{dx_n}{dt} \end{bmatrix} = \left[\frac{\partial v}{\partial \mathbf{X}} \right]^T \frac{d\mathbf{X}}{dt}$$

$$= \left[\frac{\partial v}{\partial \mathbf{X}} \right]^T \mathbf{F}(\mathbf{X}).$$

In discrete time, the change in v at each step is

$$(8.11) \quad \Delta v = v(\mathbf{X}(t+1)) - v(\mathbf{X}(t)) = v(\mathbf{F}(\mathbf{X})) - v(\mathbf{X}).$$

A function $v(\mathbf{X})$ satisfying the conditions of the theorem is called a Lyapunov function. The interest in the above theorem is that solutions of the dynamic equations (8.1) or (8.2) are not required. It is only necessary to verify the sign of a formula in a region. For nonlinear systems the difficulty is that there is no guarantee of finding a suitable function $v(\cdot)$ that has the required properties in

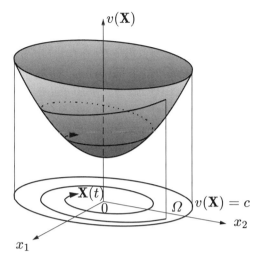

Fig. 8.3   Illustrating a trajectory with $v(\mathbf{X})$ decreasing everywhere in Ω.

a region Ω. In addition, the theorem gives only necessary conditions. If a given function $v(\mathbf{X})$ does not satisfy the conditions of the theorem, there is no information about whether there is such a function. For physical systems, the total energy is the logical choice in the search for a function satisfying the theorem. There are various statements of similar results which account for time-varying systems, and for specialized definitions of stability.

Example 4
Stability of a
dlssipative
circuit
Consider the linear circuit shown in Figure 8.4. If the inductor current i and

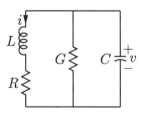

Fig. 8.4   Linear circuit with energy storage and dissipation.

capacitor voltage v are chosen as state variables, then the state equations are

$$\frac{d}{dt}\begin{bmatrix} i \\ v \end{bmatrix} = \begin{bmatrix} -R/L & 1/L \\ -1/C & -G/C \end{bmatrix}\begin{bmatrix} i \\ v \end{bmatrix},$$

and the stored energy is

$$w(i, v) = \frac{Li^2}{2} + \frac{Cv^2}{2},$$

where the notation $w(\cdot)$ is used to avoid confusion with voltage v. The rate of change of w is

$$\dot{w} = \frac{\partial w}{\partial i}\frac{di}{dt} + \frac{\partial w}{\partial v}\frac{dv}{dt} = [\,Li,\; Cv\,]\begin{bmatrix} di/dt \\ dv/dt \end{bmatrix}$$

$$= [\,Li,\; Cv\,]\begin{bmatrix} -R/L & 1/L \\ -1/C & -G/C \end{bmatrix}\begin{bmatrix} i \\ v \end{bmatrix} = -Ri^2 - Gv^2,$$

which is the rate that energy is dissipated as heat. This function is negative everywhere in the state space except the origin, so Ω can be the interior of an arbitrarily large ellipse $w(i, v) = c$, and any initial state in Ω will result in a solution trajectory that asymptotically approaches the origin.

Example 5
Effect of
negative
resistance

Suppose that R is negative in the previous example, so that energy is added to the system through the negative resistance. Then for the function $w(i, v)$ used previously, $\dot{w}$ is no longer negative everywhere away from the origin, since if $i \neq 0$ and $v = 0$ then $\dot{w} > 0$, whereas if $i = 0$ and $v \neq 0$ then $\dot{w} < 0$. Therefore this $w(i, v)$ is no longer a Lyapunov function. One can be constructed as follows. From the state equations, assuming small damping so that the eigenvalues are complex, the eigenvalues are given by

$$\lambda = -\frac{R/L + G/C}{2} \pm \frac{j}{2}\sqrt{4/(LC) - (R/L - G/C)^2} = -\alpha \pm j\omega,$$

where $-\alpha$ is the real part and ω is the imaginary part. Then from Example 8 (and Problem 5) of Chapter 7, a similarity transformation can be found to transform the above system to the form

$$\frac{d}{dt}\mathbf{X}' = \begin{bmatrix} -\alpha & -\omega \\ \omega & -\alpha \end{bmatrix}\mathbf{X}'.$$

Now choose the function $w(\mathbf{X}') = (1/2)\mathbf{X}'^T\mathbf{X}'$, which is positive for all $\mathbf{X}' \neq 0$. The rate of change of this function is

$$\dot{w} = \frac{1}{2}\dot{\mathbf{X}}'^T\mathbf{X}' + \frac{1}{2}\mathbf{X}'^T\dot{\mathbf{X}}' = \mathbf{X}'^T\dot{\mathbf{X}}' = \mathbf{X}'^T\begin{bmatrix} -\alpha & -\omega \\ \omega & -\alpha \end{bmatrix}\mathbf{X}' = -\alpha\,\mathbf{X}'^T\mathbf{X}',$$

which is negative everywhere except the origin, provided α is positive. Therefore it can be concluded that the origin is asymptotically stable if $\alpha > 0$; that is, if $(R/L + G/C)/2 > 0$.

3.1   Energy functions for LTI systems

Energy-like functions have an importance for stability analysis, as has been seen, or for other applications. Suppose the stability of (8.1) or (8.2) is being investigated, with Lyapunov function candidates of the form

(8.12) $v(\mathbf{X}) = \mathbf{X}^T \mathbf{P} \mathbf{X},$

where $\mathbf{P}$ is real and symmetric and has the property that $\mathbf{X}^T \mathbf{P} \mathbf{X} > 0$ for all $\mathbf{X} \neq 0$. Such a function of $\mathbf{X}$ is said to be *positive-definite,* and $\mathbf{P}$ is called a positive-definite matrix. For example if $\mathbf{P} = \text{diag}[\lambda_i]$, $i = 1, \cdots n$ with real positive λ_i then in (8.12),

(8.13) $v(\mathbf{X}) = \lambda_1 x_1^2 + \cdots \lambda_n x_n^2,$

which is positive for all nonzero $\mathbf{X}$, and $\mathbf{P}$ is positive-definite. Given a positive-definite diagonal matrix $\mathbf{P}$, if $\mathbf{Z} = \mathbf{S}\mathbf{X}$, where $\mathbf{S}$ is orthogonal, then $\mathbf{Z}^T \mathbf{P} \mathbf{Z} = \mathbf{X}^T (\mathbf{S}^T \mathbf{P} \mathbf{S}) \mathbf{X}$ is positive-definite, and $\mathbf{S}^T \mathbf{P} \mathbf{S}$ is symmetric and positive-definite but is not diagonal in general. Conversely it is a fact that any symmetric real matrix has real eigenvalues and can be diagonalized by a similarity transformation (Chapter 7, Example 7 and Example 14) that is orthogonal (Chapter 7, Theorem 7.1, and Chapter 6, Example 14).

If $\mathbf{P}$ is a positive-definite matrix, then $v(\mathbf{X})$ of the form of (8.12) is like the square of a weighted norm of $\mathbf{X}$. In fact, a weighted norm can be defined with respect to such a matrix $\mathbf{P}$ as $\|\mathbf{X}\|_\mathbf{P} = (\mathbf{X}^T \mathbf{P} \mathbf{X})^{1/2}$. The quantity $v(\mathbf{X})$ can also be said to be the generalized system energy associated with the state $\mathbf{X}$.

Example 6
Energy in
physical
components

There are numerous physical examples in which the stored energy in linear components has an expression of the form discussed in this section. As in Example 4, the energy in an inductor with current i is $(L/2)\, i^2$; the energy in a capacitor with voltage v is $(C/2)\, v^2$. The energy in a linear spring with constant k and deflection x is $(k/2)\, x^2$. The kinetic energy of a particle of mass m and velocity v is $(m/2)\, v^2$. Similar expressions exist for many other physical laws and elements. Therefore the total stored energy in a physical system containing such components will be a weighted linear combination of the squares of the variables.

3.2   Lyapunov equations for LTI continuous-time systems

The rate of change of $v(\mathbf{X})$ along trajectories of (8.1) will be investigated, as required in stability analysis. Differentiating $v(\mathbf{X})$ gives

(8.14) $\dot{v}(\mathbf{X}) = \dot{\mathbf{X}}^T \mathbf{P} \mathbf{X} + \mathbf{X}^T \mathbf{P} \dot{\mathbf{X}} = \mathbf{X}^T \mathbf{A}^T \mathbf{P} \mathbf{X} + \mathbf{X}^T \mathbf{P} \mathbf{A} \mathbf{X} = \mathbf{X}^T (\mathbf{A}^T \mathbf{P} + \mathbf{P} \mathbf{A})\, \mathbf{X},$

and for this function to be negative everywhere except at the origin, the quantity $\mathbf{A}^T\mathbf{P} + \mathbf{PA} = (\mathbf{PA})^T + \mathbf{PA}$, which is a symmetric matrix, must satisfy

$$(8.15) \quad \mathbf{A}^T\mathbf{P} + \mathbf{PA} = -\mathbf{Q},$$

where $\mathbf{Q}$ is symmetric and positive-definite, and defines the rate at which energy is extracted from the system. An equation of the above form is called a continuous-time Lyapunov equation.

The connection between $\mathbf{Q}$ and $\mathbf{P}$ in (8.15) is the following. Let $\mathbf{Q}$ be any positive-definite matrix and let the rate of change of v be $\dot{v} = -\mathbf{X}(t)^T\mathbf{Q}\mathbf{X}(t)$, where $\mathbf{X}(t) = e^{t\mathbf{A}}\mathbf{X}(0)$. Then the total energy dissipated is

$$(8.16) \quad \int_0^\infty (-\dot{v})\,dt = \int_0^\infty \mathbf{X}(t)^T\mathbf{Q}\,\mathbf{X}(t)\,dt = \int_0^\infty \mathbf{X}(0)^T e^{t\mathbf{A}^T}\mathbf{Q}e^{t\mathbf{A}}\mathbf{X}(0)\,dt$$

$$= \mathbf{X}(0)^T \left(\int_0^\infty e^{t\mathbf{A}^T}\mathbf{Q}e^{t\mathbf{A}}dt \right)\mathbf{X}(0) = \mathbf{X}(0)^T\mathbf{P}\,\mathbf{X}(0),$$

where

$$(8.17) \quad \mathbf{P} = \int_0^\infty e^{t\mathbf{A}^T}\mathbf{Q}e^{t\mathbf{A}}\,dt,$$

provided this integral exists. But since from Section 4 of Chapter 7 the integrand contains linear combinations of products of functions of the form $t^\gamma e^{t\lambda_i}$ for integers $\gamma \geq 0$, the integral exists if and only if all $\mathrm{Re}(\lambda_i) < 0$. Then substituting (8.17) in (8.15) gives

$$(8.18) \quad \mathbf{A}^T\mathbf{P} + \mathbf{PA} = \int_0^\infty (\mathbf{A}^T e^{t\mathbf{A}^T}\mathbf{Q}e^{t\mathbf{A}} + e^{t\mathbf{A}^T}\mathbf{Q}e^{t\mathbf{A}}\mathbf{A})\,dt = \int_0^\infty \frac{d}{dt}\left(e^{t\mathbf{A}^T}\mathbf{Q}e^{t\mathbf{A}}\right)dt$$

$$= \lim_{t\to\infty} e^{t\mathbf{A}^T}\mathbf{Q}e^{t\mathbf{A}} - \mathbf{Q} = -\mathbf{Q},$$

which shows that if $\mathbf{P}$ is defined by (8.17) then $\mathbf{P}$ satisfies (8.15).

Using a detailed version of the above argument we can show that if an arbitrary positive-definite $\mathbf{Q}$ is chosen and (8.15) is solved for $\mathbf{P}$, then if $\mathbf{P}$ is positive-definite the system is asymptotically stable. The details will not be considered because this is usually a much less efficient method for testing asymptotic stability than simply computing the system eigenvalues. Rather, solutions of this equation are important because they define the generalized energy of a system at state $\mathbf{X}$.

Example 7
Signal energy

The energy in a signal $y(t)$ over $[0, \infty)$ may be defined as

$$w = \int_0^\infty y^2(t)\,dt.$$

If the signal is the free response $y(t) = \mathbf{C}e^{t\mathbf{A}}\mathbf{X}(0)$ of an asymptotically stable linear system, then from (8.17) this energy is given by

$$w = \int_0^\infty (\mathbf{C}\mathbf{X}(t))^T (\mathbf{C}\mathbf{X}(t))\, dt = \int_0^\infty \mathbf{X}(0)^T e^{t\mathbf{A}^T} \mathbf{C}^T \mathbf{C} e^{t\mathbf{A}} \mathbf{X}(0)\, dt$$

$$= \mathbf{X}(0)^T \left(\int_0^\infty e^{t\mathbf{A}^T} \mathbf{C}^T \mathbf{C} e^{t\mathbf{A}}\, dt \right) \mathbf{X}(0) = \mathbf{X}(0)^T \mathbf{P}\mathbf{X}(0),$$

where $\mathbf{P}$ is the solution of the Lyapunov equation

$$\mathbf{A}^T \mathbf{P} + \mathbf{P}\mathbf{A} = -\mathbf{C}^T \mathbf{C}.$$

Example 8
Optimal control

A standard method of stabilizing an LTI system with matrices $(\mathbf{A}, \mathbf{B}, \mathbf{C}, \mathbf{D})$ with $\mathbf{D} = 0$ is to find a feedback matrix $\mathbf{K}$ such that the input is $\mathbf{U} = -\mathbf{K}\mathbf{X} + \mathbf{U}'$, so that the resulting system with matrices $(\hat{\mathbf{A}} = \mathbf{A} - \mathbf{B}\mathbf{K}, \mathbf{B}, \mathbf{C}, 0)$ and input $\mathbf{U}'$ is stable. The matrix $\mathbf{K}$ is found by letting $\mathbf{X}(0) = \mathbf{X}_0$ be arbitrary, setting $\mathbf{U}' = 0$, and minimizing the cost function

$$q = \frac{1}{2} \int_0^\infty (\mathbf{Y}^T \mathbf{Q}\mathbf{Y} + \mathbf{U}^T \mathbf{R}\mathbf{U})\, dt,$$

where the cost of control signal $\mathbf{U}$ is defined by $\mathbf{R}$, which is symmetric and positive-definite, and where the cost of deviation of $\mathbf{Y}$ from the origin is defined by $\mathbf{Q}$, which is symmetric and at least positive semidefinite; that is, $\mathbf{Y}^T \mathbf{Q}\mathbf{Y} \geq 0$ for all $\mathbf{Y}$.

Provided the resulting system is asymptotically stable, the resulting cost is given by the solution of a Lyapunov equation, as follows, since $\mathbf{Y} = \mathbf{C}\mathbf{X}$ and $\mathbf{U} = -\mathbf{K}\mathbf{X}$:

$$q = \frac{1}{2} \int_0^\infty (\mathbf{X}^T \mathbf{C}^T \mathbf{Q}\mathbf{C}\mathbf{X} + \mathbf{X}^T \mathbf{K}^T \mathbf{R}\mathbf{K}\mathbf{X})\, dt$$

$$= \frac{1}{2} \int_0^\infty \mathbf{X}^T (\mathbf{C}^T \mathbf{Q}\mathbf{C} + \mathbf{K}^T \mathbf{R}\mathbf{K})\, \mathbf{X}\, dt$$

$$= \frac{1}{2} \mathbf{X}_0^T \left(\int_0^\infty e^{t\hat{\mathbf{A}}^T} (\mathbf{C}^T \mathbf{Q}\mathbf{C} + \mathbf{K}^T \mathbf{R}\mathbf{K})\, e^{t\hat{\mathbf{A}}} dt \right) \mathbf{X}_0 = \frac{1}{2} \mathbf{X}_0^T \mathbf{P}\mathbf{X}_0,$$

where

$$\hat{\mathbf{A}}^T \mathbf{P} + \mathbf{P}\hat{\mathbf{A}} = -(\mathbf{C}^T \mathbf{Q}\mathbf{C} + \mathbf{K}^T \mathbf{R}\mathbf{K}).$$

3.3   **Solving continuous Lyapunov equations**

Suppose that $\mathbf{Q}$ in (8.15) is known. The left-hand side of this equation is a matrix of which the entries are linear combinations of the unknowns in $\mathbf{P}$. Because $\mathbf{P}$ is symmetric, there are $n(n+1)/2$ unknowns, and (8.15) can be rewritten as $n(n+1)/2$ scalar linear equations in these unknowns.

Example 9
2 × 2 continuous
Lyapunov
equation

Writing (8.15) in detail for $\mathbf{A} \in \mathbb{R}^{2 \times 2}$ and symmetric $\mathbf{P}$ and $\mathbf{Q}$ gives

$$\begin{bmatrix} a_{11} & a_{21} \\ a_{12} & a_{22} \end{bmatrix} \begin{bmatrix} p_{11} & p_{12} \\ p_{12} & p_{22} \end{bmatrix} + \begin{bmatrix} p_{11} & p_{12} \\ p_{12} & p_{22} \end{bmatrix} \begin{bmatrix} a_{11} & a_{12} \\ a_{21} & a_{22} \end{bmatrix} = - \begin{bmatrix} q_{11} & q_{12} \\ q_{12} & q_{22} \end{bmatrix}.$$

Then the unknowns are p_{11}, p_{12}, p_{22}, and equating matrix entries on the left and right results in the following three simultaneous equations:

$$\begin{bmatrix} 2a_{11} & 2a_{21} & 0 \\ a_{12} & a_{11}+a_{22} & a_{21} \\ 0 & 2a_{12} & 2a_{22} \end{bmatrix} \begin{bmatrix} p_{11} \\ p_{12} \\ p_{22} \end{bmatrix} = \begin{bmatrix} -q_{11} \\ -q_{12} \\ -q_{22} \end{bmatrix}.$$

Existence

The existence of solutions of (8.15) will be investigated. First, if $\mathbf{A} = \text{diag}[\lambda_i]$, then entry ij of matrix equation (8.15) is

(8.19) $\quad \lambda_i p_{ij} + p_{ij} \lambda_j = -q_{ij},$

which is nonsingular if and only if $\lambda_i + \lambda_j \neq 0$. Next, if $\mathbf{A} = \mathbf{J} = \text{diag}[\mathbf{J}_{\delta_i}]$ is in Jordan form, then (8.15) is of the form

(8.20)
$$\begin{bmatrix} \mathbf{J}_{\delta_1}^T & 0 & \cdots \\ 0 & \mathbf{J}_{\delta_2}^T & \\ \vdots & & \ddots \end{bmatrix} \begin{bmatrix} \mathbf{P}_{11} & \mathbf{P}_{12} & \cdots \\ \mathbf{P}_{21} & \mathbf{P}_{22} & \\ \vdots & & \ddots \end{bmatrix} + \begin{bmatrix} \mathbf{P}_{11} & \mathbf{P}_{12} & \cdots \\ \mathbf{P}_{21} & \mathbf{P}_{22} & \\ \vdots & & \ddots \end{bmatrix} \begin{bmatrix} \mathbf{J}_{\delta_1} & 0 & \cdots \\ 0 & \mathbf{J}_{\delta_2} & \\ \vdots & & \ddots \end{bmatrix}$$

$$= \begin{bmatrix} \mathbf{Q}_{11} & \mathbf{Q}_{12} & \cdots \\ \mathbf{Q}_{21} & \mathbf{Q}_{22} & \\ \vdots & & \ddots \end{bmatrix},$$

where $\mathbf{P} = [\mathbf{P}_{ij}]$ and $\mathbf{Q} = [\mathbf{Q}_{ij}]$ are compatible blockwise decompositions of the symmetric matrices $\mathbf{P}$ and $\mathbf{Q}$, giving a set of equations

(8.21) $\quad \mathbf{J}_{\delta_i}^T \mathbf{P}_{ij} + \mathbf{P}_{ij} \mathbf{J}_{\delta_j} = -\mathbf{Q}_{ij}.$

These equations may be solved individually, and are simplified by noting that $\mathbf{P}_{ij} = \mathbf{P}_{ji}^T$ and $\mathbf{Q}_{ij} = \mathbf{Q}_{ji}^T$. If λ_i is the eigenvalue of $\mathbf{J}_{\delta_i}$ and λ_j is the eigenvalue of $\mathbf{J}_{\delta_j}$ then because these matrices are triangular, detailed inspection of (8.21), illustrated in Example 10, yields the same conclusion as for (8.19); that is, Equation (8.21) is nonsingular if and only if $\lambda_i + \lambda_j \neq 0$. Finally, given any $\mathbf{A}$, the similarity transformation $\mathbf{J} = \mathbf{S}^{-1}\mathbf{A}\mathbf{S}$ with premultiplication of (8.15) by $\mathbf{S}^T$ and postmultiplication by $\mathbf{S}$ changes (8.15) to

(8.22) $\quad \mathbf{S}^T \mathbf{A}^T (\mathbf{S}^{-1})^T \mathbf{S}^T \mathbf{P} \mathbf{S} + \mathbf{S}^T \mathbf{P} \mathbf{S} \mathbf{S}^{-1} \mathbf{A} \mathbf{S} = \mathbf{J}^T (\mathbf{S}^T \mathbf{P} \mathbf{S}) + (\mathbf{S}^T \mathbf{P} \mathbf{S}) \mathbf{J} = -(\mathbf{S}^T \mathbf{Q} \mathbf{S}),$

so solution of (8.15) reduces to a problem for which the known matrix on the left-hand side is in Jordan form. This completes an outline of the proof of the following result.

Solution conditions A solution of (8.15) exists and is unique if and only if $\lambda_i + \lambda_j \neq 0$ for all possible pairs λ_i, λ_j of eigenvalues of $\mathbf{A}$, including the pairs λ_i, λ_i.

In fact, although the Jordan form is very useful in proving the above result, it is not recommended for computation. Instead, a similarity transformation producing a real upper block-triangular form with 1×1 or 2×2 diagonal blocks, called the real Schur form (see Problem 7 of Chapter 7) is typically preferred for reasons of computational stability.

Example 10
Jordan blocks
An example of (8.21) will be investigated. Suppose $\delta_1 = 2$ and $\delta_2 = 3$, giving a block equation of the form

$$\mathbf{J}_{\delta_1}^T \mathbf{P}_{12} + \mathbf{P}_{12} \mathbf{J}_{\delta_2}$$

$$= \begin{bmatrix} \lambda_1 & 0 \\ 1 & \lambda_1 \end{bmatrix} \begin{bmatrix} p_{11} & p_{12} & p_{13} \\ p_{21} & p_{22} & p_{23} \end{bmatrix} + \begin{bmatrix} p_{11} & p_{12} & p_{13} \\ p_{21} & p_{22} & p_{23} \end{bmatrix} \begin{bmatrix} \lambda_2 & 1 & 0 \\ 0 & \lambda_2 & 1 \\ 0 & 0 & \lambda_2 \end{bmatrix}$$

$$= \begin{bmatrix} q_{11} & q_{12} & q_{13} \\ q_{21} & q_{22} & q_{23} \end{bmatrix} = -\mathbf{Q}_{12}.$$

By equating entries on the left and right, the following can be solved in order:

$$(\lambda_1 + \lambda_2)\, p_{11} = -q_{11}$$
$$(\lambda_1 + \lambda_2)\, p_{21} = -q_{21} - p_{11}$$
$$(\lambda_1 + \lambda_2)\, p_{12} = -q_{12} - p_{11}$$
$$(\lambda_1 + \lambda_2)\, p_{22} = -q_{22} - p_{12} - p_{21}$$
$$(\lambda_1 + \lambda_2)\, p_{13} = -q_{13} - p_{12}$$
$$(\lambda_1 + \lambda_2)\, p_{23} = -q_{23} - p_{13} - p_{22}.$$

Unstable systems It should not be concluded that Equation (8.15) has solutions only for asymptotically stable systems. As shown above, the sole requirement for the equations to be nonsingular is that $\lambda_i + \lambda_j \neq 0$ for all pairs of eigenvalues of $\mathbf{A}$. This condition is true if all $\mathrm{Re}(\lambda_i) < 0$, but it may also be true for an unstable matrix $\mathbf{A}$. For example, suppose all eigenvalues of $\mathbf{A}$ have positive real parts. Defining $\mathbf{P}$ as

(8.23) $$\mathbf{P} = -\int_{-\infty}^{0} e^{t\mathbf{A}^T} \mathbf{Q} c^{t\mathbf{A}} dt,$$

then, in a similar way as for (8.18),

(8.24) $$\mathbf{A}^T \mathbf{P} + \mathbf{P}\mathbf{A} = -\int_{-\infty}^{0} \frac{d}{dt}\left(e^{t\mathbf{A}^T}\mathbf{Q}e^{t\mathbf{A}}\right) dt = \lim_{t \to -\infty} e^{t\mathbf{A}^T}\mathbf{Q}e^{t\mathbf{A}} - \mathbf{Q} = -\mathbf{Q},$$

so in this case the solution $\mathbf{P}$ of (8.15) is such that $\mathbf{X}(0)^T\mathbf{P}\mathbf{X}(0)$ is the free-response energy added to the system in the past.

Example 11
Unstable A

As an illustration of an unstable matrix $\mathbf{A}$, in (8.15) let

$$\mathbf{A} = \frac{1}{3}\begin{bmatrix} 13 & -7 \\ 14 & -8 \end{bmatrix}, \quad \mathbf{Q} = \begin{bmatrix} 1 & 2 \\ 2 & 4 \end{bmatrix},$$

where the eigenvalues of $\mathbf{A}$ are $2, -1/3$. The solution matrix is

$$\mathbf{P} = \begin{bmatrix} 64.5 & -60 \\ -60 & 53.25 \end{bmatrix},$$

which is not positive-definite, as can be determined (since $\mathbf{P}$ is real and symmetric) by computing its eigenvalues. These are 119.1 and -1.388, and since one is not positive, from the discussion in Section 3.1, $\mathbf{P}$ is not positive-definite.

3.4   Discrete-time Lyapunov equations

The analysis of discrete-time systems requires changes of detail, but is otherwise analogous. The change in the scalar function $v(\mathbf{X})$ of Equation (8.12) will be found along a solution $\mathbf{X}(t) = \mathbf{A}^t\mathbf{X}(0)$ of Equation (8.7). From Equation (8.11), the change in one step is

$$(8.25) \quad \Delta v = \mathbf{X}^T(t+1)\,\mathbf{P}\,\mathbf{X}(t+1) - \mathbf{X}(t)^T\mathbf{P}\,\mathbf{X}(t) = \mathbf{X}(t)^T(\mathbf{A}^T\mathbf{P}\mathbf{A} - \mathbf{P})\,\mathbf{X}(t).$$

For this quantity to be negative everywhere except at the origin, the symmetric matrix $\mathbf{A}^T\mathbf{P}\mathbf{A} - \mathbf{P}$ must satisfy

$$(8.26) \quad \mathbf{A}^T\mathbf{P}\mathbf{A} - \mathbf{P} = -\mathbf{Q},$$

where $\mathbf{Q}$ is symmetric and positive-definite. This is called a discrete-time Lyapunov equation.

Solution
From an analysis similar to that for continuous-time systems, a solution of (8.26) exists and is unique if and only if $\lambda_i\lambda_j \neq 1$ for all possible pairs λ_i, λ_j of eigenvalues of $\mathbf{A}$, including the pairs λ_i, λ_i.

Analogously to (8.17), if all eigenvalues of $\mathbf{A}$ satisfy $\lambda_i < 1$ and $\mathbf{P}$ is defined as

$$(8.27) \quad \mathbf{P} = \sum_{k=0}^{\infty}(\mathbf{A}^T)^k\mathbf{Q}\,\mathbf{A}^k,$$

then substitution in (8.26) gives

$$(8.28) \quad \mathbf{A}^T\mathbf{P}\mathbf{A} - \mathbf{P} = \sum_{k=1}^{\infty}(\mathbf{A}^T)^k\mathbf{Q}\,\mathbf{A}^k - \sum_{k=0}^{\infty}(\mathbf{A}^T)^k\mathbf{Q}\,\mathbf{A}^k = -\mathbf{Q},$$

showing that $\mathbf{P}$ is the solution of (8.26).

Example 12
Scalar equation

If $\mathbf{A} = a$, $\mathbf{P} = p$, and $\mathbf{Q} = q$ are scalars, (8.26) is

$$apa - p = -q,$$

which has a unique solution if and only if $a^2 \neq 1$.

Example 13
Diagonal
system

If $\mathbf{A} = \text{diag}[\lambda_i]$, then (8.26) is equivalent to a set of scalar equations of the form

$$\lambda_i p_{ij} \lambda_j - p_{ij} = -q_{ij},$$

which has a unique solution if and only if $\lambda_i \lambda_j \neq 1$ for all eigenvalues λ_i, λ_j.

Example 14
2×2 discrete
equation

Writing (8.26) in detail for $\mathbf{A} \in \mathbb{R}^{2 \times 2}$ and symmetric $\mathbf{P}$ and $\mathbf{Q}$,

$$\begin{bmatrix} a_{11} & a_{21} \\ a_{12} & a_{22} \end{bmatrix} \begin{bmatrix} p_{11} & p_{12} \\ p_{12} & p_{22} \end{bmatrix} \begin{bmatrix} a_{11} & a_{12} \\ a_{21} & a_{22} \end{bmatrix} - \begin{bmatrix} p_{11} & p_{12} \\ p_{12} & p_{22} \end{bmatrix} = - \begin{bmatrix} q_{11} & q_{12} \\ q_{12} & q_{22} \end{bmatrix},$$

which is equivalent to three equations in three unknowns as in Example 9, but in this case the equations to be solved are

$$\begin{bmatrix} a_{11}^2 - 1 & 2a_{21}a_{11} & a_{21}^2 \\ a_{12}a_{11} & a_{11}a_{22} + a_{12}a_{21} - 1 & a_{22}a_{21} \\ a_{12}^2 & 2a_{12}a_{22} & a_{22}^2 - 1 \end{bmatrix} \begin{bmatrix} p_{11} \\ p_{12} \\ p_{22} \end{bmatrix} = \begin{bmatrix} -q_{11} \\ -q_{12} \\ -q_{22} \end{bmatrix}.$$

Example 15
Energy of a
sequence

By analogy with Example 7, the energy in a sequence $\{y(t)\}_0^\infty$ is often defined as

$$w = \sum_{t=0}^{\infty} y^2(t)\, dt.$$

If the signal is the free response $y(t) = \mathbf{C}\mathbf{A}^t \mathbf{X}(0)$ of an asymptotically stable discrete-time linear system, then by using (8.27) this energy is given by

$$w = \sum_{t=0}^{\infty} (\mathbf{C}\mathbf{X}(t))^T (\mathbf{C}\mathbf{X}(t)) = \sum_{t=0}^{\infty} \mathbf{X}(0)^T (\mathbf{A}^T)^t \mathbf{C}^T \mathbf{C} \mathbf{A}^t \mathbf{X}(0)$$

$$= \mathbf{X}(0)^T \left(\sum_{t=0}^{\infty} (\mathbf{A}^T)^t \mathbf{C}^T \mathbf{C} \mathbf{A}^t \right) \mathbf{X}(0) = \mathbf{X}(0)^T \mathbf{P} \mathbf{X}(0),$$

where $\mathbf{P}$ is the solution of the Lyapunov equation

$$\mathbf{A}^T \mathbf{P} \mathbf{A} - \mathbf{P} = -\mathbf{C}^T \mathbf{C}.$$

Example 16
Unstable **A**

The matrix **A** of Example 11 is unstable when taken as the **A**-matrix of a discrete-time system. If the matrix **Q** of the same example is used in the discrete-time Lyapunov equation (8.26), the solution is

$$\mathbf{P} = \begin{bmatrix} -19.87 & 4.875 \\ 4.875 & 7.125 \end{bmatrix},$$

which exists as expected, but its eigenvalues are -20.7 and 7.98, showing that **P** is not positive-definite.

---

4 | **Further study**

Among the references where detailed treatments of stability theory for linear systems can be found are Antsaklis and Michel [2] and Kailath [25]. The standard reference for Lyapunov equations is Gantmacher [17]. Nonlinear systems are discussed by Minorsky [38], and a more recent reference is Vidyasagar [54].

---

5 | **Problems**

1 For continuous-time systems with matrix **A** shown, determine the internal stability in each case.

$$\text{(a)} \begin{bmatrix} -9 & -18 & 7 \\ 3 & 5 & -3 \\ -4 & -8 & 3 \end{bmatrix}, \quad \text{(b)} \begin{bmatrix} -25 & -10 & 47 \\ 7 & 3 & -13 \\ -12 & -4 & 23 \end{bmatrix}, \quad \text{(c)} \begin{bmatrix} 6 & 4 & -12 \\ -2 & -1 & 4 \\ 3 & 2 & -6 \end{bmatrix}.$$

---

2 Assuming (a) a continuous-time system, and then (b) a discrete-time system, determine the internal stability of the system with matrices

$$\mathbf{A} = \begin{bmatrix} -4 & -4 & 7 \\ 1 & -1 & 0 \\ -3 & -4 & 6 \end{bmatrix}, \quad \mathbf{B} = \begin{bmatrix} 2 \\ -1 \\ 1 \end{bmatrix}, \quad \mathbf{C} = \begin{bmatrix} 1 & 0 & -1 \end{bmatrix}, \quad \mathbf{D} = 0.$$

---

3 Computer simulations are discrete-time processes that may be unstable independently of the continuous-time state-space equations they are intended to approximate. Consider Euler's forward method, described in Example 11 of Chapter 2, applied to $\frac{d}{dt}\mathbf{X} = \mathbf{A}\mathbf{X}$, where **A** is asymptotically stable, with eigenvalues $\lambda_i = -\alpha_i + j\omega_i$ in the left half-plane. Euler's forward method uses the approximation $\mathbf{X}(k+1) = \mathbf{X}(k) + h\mathbf{X}'(k) = (\mathbf{I} + h\mathbf{A})\mathbf{X}(k)$ for step size h, which must remain stable for the approximate solution to be valid. In Problem 13 of Chapter 7, the eigenvalues of $\mathbf{I} + h\mathbf{A}$ are $1 + h\lambda_i$, which must satisfy $|1 + h\lambda_i| < 1$

for the simulation to be stable. Assume that all eigenvalues of $\mathbf{A}$ are real, and find the maximum allowable step size h. Because h is limited by the speed of the fastest (that is, largest) eigenvalue λ_i, this method is often very inefficient. If h is naïvely increased beyond the allowable maximum for reasons of efficiency, then the solution becomes unstable and therefore meaningless.

4 In the simulation of $\frac{d}{dt}\mathbf{X} = \mathbf{AX}$, show that Euler's *backward* method, given by $\mathbf{X}(k+1) = \mathbf{X}(k) + h\mathbf{X}'(k+1) = \mathbf{X}(k) + h\mathbf{AX}(k+1)$, is asymptotically stable for all positive step sizes h provided $\mathbf{A}$ is asymptotically stable.

5 Solve Problem 4 for the *trapezoidal* method, given by $\mathbf{X}(k+1) = \mathbf{X}(k) + \frac{h}{2}(\mathbf{X}'(k) + \mathbf{X}'(k+1))$.

6 For the mass, spring, dashpot system of Example 5 of Chapter 4 with zero external force, the system model is

$$m\ddot{x} + c\dot{x} + kx = 0.$$

Find a Lyapunov function for the system and thereby show that it is asymptotically stable.

7 If possible, solve the continuous-time Lyapunov equation (8.15) for $\mathbf{P}$ given the following matrix pairs:

(a) $\mathbf{A} = \begin{bmatrix} -1/2 & -1 \\ 1 & -1/2 \end{bmatrix}$, $\mathbf{Q} = \begin{bmatrix} 2 & 1 \\ 1 & 2 \end{bmatrix}$; (b) $\mathbf{A} = \begin{bmatrix} -1 & 1 \\ 0 & 1 \end{bmatrix}$, $\mathbf{Q} = \begin{bmatrix} 1 & 0 \\ 0 & 1 \end{bmatrix}$;

(c) $\mathbf{A} = \begin{bmatrix} -1 & 1 \\ 0 & 2 \end{bmatrix}$, $\mathbf{Q} = \begin{bmatrix} 1 & 0 \\ 0 & 1 \end{bmatrix}$.

8 If possible, solve the discrete-time Lyapunov equation (8.26) for $\mathbf{P}$ for the matrix pairs of Problem 7.

9 Applying the similarity transformation $\mathbf{S}^{-1}\mathbf{AS} = \mathbf{A}'$, the continuous-time Lyapunov equation $\mathbf{A}^T\mathbf{P} + \mathbf{PA} = -\mathbf{Q}$ reduces to

$$(\mathbf{A}')^T(\mathbf{S}^T\mathbf{PS}) + (\mathbf{S}^T\mathbf{PS})\mathbf{A}' = -(\mathbf{S}^T\mathbf{QS})$$

as in Equation (8.22). If $\mathbf{A}'$ is in real Schur form (see Problem 7 of Chapter 7)

$$\mathbf{A}' = \begin{bmatrix} \mathbf{A}_{11} & \mathbf{A}_{12} & \mathbf{A}_{13} \\ 0 & \mathbf{A}_{22} & \mathbf{A}_{23} \\ 0 & 0 & \mathbf{A}_{33} \end{bmatrix},$$

show how to partition $\mathbf{P}' = \mathbf{S}^T \mathbf{P} \mathbf{S}$ and $\mathbf{Q}' = \mathbf{S}^T \mathbf{Q} \mathbf{S}$, and write the block equations of the form

$$\mathbf{A}_{ii}^T \mathbf{P}_{ij} + \mathbf{P}_{ij} \mathbf{A}_{jj} = \text{(known quantities)}$$

that can be solved in order to obtain the solution $\mathbf{P}'$ and hence $\mathbf{P}$.

10   One way of solving the discrete-time Lyapunov equation $\mathbf{A}^T \mathbf{P} \mathbf{A} - \mathbf{P} = -\mathbf{Q}$ is to use a similarity transformation to transform $\mathbf{Q}$ to real Schur form, as for the continuous-time case, and then to solve a sequence of equations obtained from the upper-triangular structure of $\mathbf{A}$. An alternative which has been used for stable matrices $\mathbf{A}$ is to compute the sum given by Equation (8.27); that is, $\mathbf{P} = \sum_{k=0}^{\infty} (\mathbf{A}^T)^k \mathbf{Q} \mathbf{A}^k$. This sum can be computed efficiently by using the recursion

$$\mathbf{P}_0 = \mathbf{Q}, \quad \mathbf{A}_0 = \mathbf{A},$$
$$\mathbf{P}_{k+1} = \mathbf{A}_k^T \mathbf{P}_k \mathbf{A}_k + \mathbf{P}_k, \quad \mathbf{A}_{k+1} = \mathbf{A}_k^2.$$

Then k applications of this recursion produce $\mathbf{P}_k$, which is the sum of the required series up to term $\ell(k)$. Determine $\ell(k)$.

11   Using a Lyapunov equation of appropriate type, compute the energy in the following:

(a) the signal $y(t) = 2te^{-2t}$ over $[0, \infty)$;

(b) the sequence $\{y_k\}_0^{\infty} = \{1/(2^k) + 1/(3^k)\}$.

9

Minimality via similarity transformations

Given a state-space system $(\mathbf{A}, \mathbf{B}, \mathbf{C}, \mathbf{D})$ of order n, that is, for which $\mathbf{A}$ is $n \times n$, it is useful to be able to find a system $(\hat{\mathbf{A}}, \hat{\mathbf{B}}, \hat{\mathbf{C}}, \hat{\mathbf{D}})$ of smallest order $\hat{n}$ such that the two systems are externally equivalent. As in Section 6 of Chapter 5, such a system will be called a *minimal* representation of the external behavior. External equivalence means that the input-output behaviors of the two systems are identical, so that

$$(9.1) \quad \mathbf{C}(\lambda\mathbf{I} - \mathbf{A})^{-1}\mathbf{B} + \mathbf{D} = \hat{\mathbf{C}}(\lambda\mathbf{I} - \hat{\mathbf{A}})^{-1}\hat{\mathbf{B}} + \hat{\mathbf{D}},$$

or equivalently, that $\hat{\mathbf{D}} = \mathbf{D}$, and

$$(9.2) \quad \mathbf{C}\mathbf{A}^{k-1}\mathbf{B} = \hat{\mathbf{C}}\hat{\mathbf{A}}^{k-1}\hat{\mathbf{B}}, \quad k = 1, 2 \cdots.$$

Methods for determining the minimal order $\hat{n}$ and for finding a system of minimal order are important in principle for understanding, and in practice for economy of design. Such methods are also useful following realization algorithms as given in Chapter 4, where state-space models that are not necessarily the smallest possible are obtained.

One method of finding a minimal realization has already been discussed in Chapter 5, Section 6: given $(\mathbf{A}, \mathbf{B}, \mathbf{C}, \mathbf{D})$, calculate $\{\mathbf{H}(k)\}_0^{2n}$, and apply the Ho algorithm.

The strategy to be invoked in this chapter is first to find, when possible, a similarity matrix $\mathbf{S}$ such that $(\mathbf{A}, \mathbf{B}, \mathbf{C}, \mathbf{D})$ is transformed into the system

$$(9.3a) \quad \mathbf{A}' = \mathbf{S}^{-1}\mathbf{A}\mathbf{S} = \begin{bmatrix} \mathbf{A}_{11} & 0 \\ \mathbf{A}_{21} & \mathbf{A}_{22} \end{bmatrix}, \quad \mathbf{B}' = \mathbf{S}^{-1}\mathbf{B} = \begin{bmatrix} 0 \\ \mathbf{B}_2 \end{bmatrix},$$

$$(9.3b) \quad \mathbf{C}' = \mathbf{C}\mathbf{S} = [\, \mathbf{C}_1, \, \mathbf{C}_2 \,], \quad \mathbf{D}' = \mathbf{D},$$

where $\mathbf{A}', \mathbf{B}', \mathbf{C}'$ are shown partitioned conformably. Then by inspection, as will be demonstrated, the system $(\mathbf{A}_{22}, \mathbf{B}_2, \mathbf{C}_2, \mathbf{D})$ is externally equivalent to the original system, and is smaller provided the zero blocks shown in (9.3a) exist. The process is illustrated in Figure 9.1.

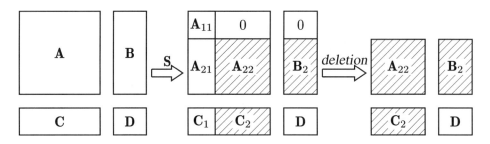

Fig. 9.1 Constructing an equivalent controllable system.

Equations (9.3) provide an easy introduction to the concept of controllability, which is of crucial importance in the design of input functions, or of feedback control schemes.

A second similarity transformation $\mathbf{T}$ will be found to transform the result of the above first step to the form

(9.4a) $\quad \tilde{\mathbf{A}} = \mathbf{T}\mathbf{A}_{22}\mathbf{T}^{-1} = \begin{bmatrix} \tilde{\mathbf{A}}_{11} & \tilde{\mathbf{A}}_{12} \\ 0 & \tilde{\mathbf{A}}_{22} \end{bmatrix}, \quad \tilde{\mathbf{B}} = \mathbf{T}\mathbf{B}_2 = \begin{bmatrix} \tilde{\mathbf{B}}_1 \\ \tilde{\mathbf{B}}_2 \end{bmatrix},$

(9.4b) $\quad \tilde{\mathbf{C}} = \mathbf{C}_2\mathbf{T}^{-1} = [\,0,\ \tilde{\mathbf{C}}_2\,], \quad \tilde{\mathbf{D}} = \mathbf{D},$

from which, by inspection, the system $(\tilde{\mathbf{A}}_{22}, \tilde{\mathbf{B}}_2, \tilde{\mathbf{C}}_2, \tilde{\mathbf{D}})$ is externally equivalent, and is also smaller if the zero blocks shown in (9.4) exist. Figure 9.2 illustrates the computation.

Equations (9.4) introduce the concept of observability, which is an essential property for applications such as optimal estimation and filtering of dynamical quantities.

It will be shown that if the above steps are performed in order, then the result is a system $\hat{\mathbf{A}} = \tilde{\mathbf{A}}_{22}, \hat{\mathbf{B}} = \tilde{\mathbf{B}}_2, \hat{\mathbf{C}} = \tilde{\mathbf{C}}_2, \hat{\mathbf{D}} = \mathbf{D}$ of smallest order $\hat{n}$ among the systems externally equivalent to $(\mathbf{A}, \mathbf{B}, \mathbf{C}, \mathbf{D})$.

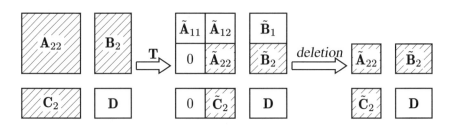

Fig. 9.2 Constructing an equivalent observable system.

1   **Step 1: Controllability**

There is a very practical need as well as a theoretical motivation for the change of variables $\mathbf{X} = \mathbf{S}\mathbf{X}'$ corresponding to the similarity transformation of Equation (9.3). Consider the discrete-time LTI system described by Equation (1.17), repeated here:

(1.17a) $\mathbf{X}(t+1) = \mathbf{A}\mathbf{X}(t) + \mathbf{B}\mathbf{U}(t)$

(1.17b) $\mathbf{Y}(t) = \mathbf{C}\mathbf{X}(t) + \mathbf{D}\mathbf{U}(t).$

We want to find a finite input sequence $\{\mathbf{U}\}_0^{\ell-1}$ that will take the initial state $\mathbf{X}(0)$ to the final state $\mathbf{X}(\ell) = \mathbf{X}_f$, where $\mathbf{X}(0)$ and $\mathbf{X}_f$ are arbitrary vectors in the state space. The input sequence will depend on $\mathbf{X}(0)$ and $\mathbf{X}_f$, but if there is a positive integer ℓ for which such a sequence can be found in every case, the system will be called *controllable* (but see Example 1 for another possibility).

The general formula (2.2) for the complete response of the state can be rewritten as

(9.5) $\mathbf{X}(\ell) - \mathbf{A}^{\ell}\mathbf{X}(0) = [\,\mathbf{B},\ \mathbf{AB},\ \cdots \mathbf{A}^{\ell-1}\mathbf{B}\,] \begin{bmatrix} \mathbf{U}(\ell-1) \\ \mathbf{U}(\ell-2) \\ \vdots \\ \mathbf{U}(0) \end{bmatrix},$

so that the required input sequence exists provided the vector left-hand side of Equation (9.5) is in the column range of the $n \times m\ell$ matrix $\mathscr{C}_{\ell} = [\,\mathbf{B},\ \mathbf{AB},\ \cdots \mathbf{A}^{\ell-1}\mathbf{B}\,]$. A method for finding the smallest ℓ is to start with $\ell = 1$ and see if there is a solution for the input sequence; if not try $\ell = 2$, and so on. A solution will exist for arbitrary left-hand side if and only if there are n independent columns in the coefficient matrix $\mathscr{C}_{\ell}$; that is, if and only if $\mathrm{rank}\,\mathscr{C}_{\ell} = n$.

A crucial observation is that an upper limit for ℓ can be found, using the Cayley-Hamilton theorem. First, define the *controllability matrix*

(9.6) $\mathscr{C} = [\,\mathbf{B},\ \mathbf{AB},\ \cdots \mathbf{A}^{n-1}\mathbf{B}\,].$

Proposition 1   *For any $\ell \geq n$, $rank(\mathscr{C}_{\ell}) = rank(\mathscr{C})$.*

Proof: Let $\phi(\lambda)$ be the characteristic polynomial of $\mathbf{A}$, and consider any block entry $\mathbf{A}^r\mathbf{B}$ in $\mathscr{C}_{\ell} = [\,\mathbf{B},\ \mathbf{AB},\ \cdots \mathbf{A}^{n-1}\mathbf{B},\ \cdots \mathbf{A}^{\ell-1}\mathbf{B}\,]$, with $r \geq n$. From the Cayley-Hamilton theorem (see Section 5 and Example 22 of Chapter 7), $\mathbf{A}^r = \beta_0\mathbf{I} + \beta_1\mathbf{A} + \cdots \beta_{n-1}\mathbf{A}^{n-1}$ for some $\beta_0, \cdots \beta_{n-1}$. Consequently, the columns of $\mathbf{A}^r\mathbf{B}$ are linear combinations of the columns of $\mathscr{C}$, with coefficients $\beta_0, \cdots \beta_{n-1}$. Therefore, $\mathrm{rank}(\mathscr{C}_{\ell}) = \mathrm{rank}(\mathscr{C})$. ☐

From the above analysis, a system is controllable if and only if $\mathrm{rank}(\mathscr{C}) = n$.

The equivalent to (9.5) for controllability of continuous-time systems is

(9.7) $$\mathbf{X}(T) - e^{T\mathbf{A}}\mathbf{X}(0) = \int_0^T e^{(T-\tau)\mathbf{A}}\mathbf{B}\mathbf{U}(\tau)\,d\tau,$$

and since $e^{T\mathbf{A}}$ has inverse $e^{-T\mathbf{A}}$ for any finite T and $\mathbf{A}$, and therefore has full rank, the left-hand side of this equation may be any vector in $\mathbb{R}^n$, and consequently the range of the right-hand side must be all of $\mathbb{R}^n$ for the system to be controllable. By expanding the exponential function under the integral as a series, it is easy to show that if the system is controllable, then $\operatorname{rank}(\mathscr{C}) = n$. Because the goal of this section is primarily to motivate the use of the word *controllability*, the details of necessity and sufficiency are omitted.

Example 1
Reachability

Some authors define controllability to imply $\mathbf{X}(\ell) = 0$ in (9.5), with $\mathbf{X}(\ell)$ arbitrary in (9.5) defining *reachability*. For discrete-time systems, the two properties are not identical. For example, the system $\mathbf{A} = \begin{bmatrix} 0 & 0 \\ 1 & 1/4 \end{bmatrix}$, $\mathbf{B} = \begin{bmatrix} 0 \\ 1 \end{bmatrix}$, does not have a solution of (9.5) for arbitrary $\mathbf{X}(\ell)$, because $\operatorname{rank}(\mathscr{C}) = \operatorname{rank}[\mathbf{B}, \mathbf{A}\mathbf{B}] = \operatorname{rank}\begin{bmatrix} 0 & 0 \\ 1 & 1/4 \end{bmatrix} = 1 < 2 = n$. However, any state $\mathbf{X}(0)$ can be taken to the origin in one step, for which (9.5) becomes $\mathbf{X}(1) - \mathbf{A}\mathbf{X}(0) = \mathbf{B}\mathbf{U}(0)$, with $\mathbf{X}(1) = 0$. Substituting the values of $\mathbf{A}$ and $\mathbf{B}$ in this equation gives

$$\begin{bmatrix} 0 \\ -[1, 1/4]\mathbf{X}(0) \end{bmatrix} = \begin{bmatrix} 0 \\ 1 \end{bmatrix} u(0),$$

which has a scalar solution $\mathbf{U}(0) = u(0)$ for any $\mathbf{X}(0)$. It turns out that for continuous-time systems, there is no distinction to be made between controllability and reachability. Similar differences of terminology can also be found for observability, to be introduced later.

1.1   Construction of the controllability transformation

It will be shown that if the original system is not completely controllable, then a similarity transformation can be constructed to produce an equivalent system from which, by inspection, an equivalent system of smaller order can be obtained.

First, find a basis $\mathbf{U}$ for the (column) range of the controllability matrix $\mathscr{C}$. That is, by a column compression on $\mathscr{C}$, or otherwise, find a nonsingular matrix $[\mathbf{Q}_1, \mathbf{Q}_2]$ such that $\mathbf{U} = \mathscr{C}\mathbf{Q}_1$ and $\operatorname{rank}\mathscr{C} = n' = \operatorname{rank}\mathbf{U}$. If $n' = n$ then the system is completely controllable; otherwise let $\mathbf{V}$ be any matrix such that

(9.8) $$\mathbf{S} = [\mathbf{V}, \mathbf{U}], \quad \text{and } \mathbf{S}^{-1} = \begin{bmatrix} \mathbf{M} \\ \mathbf{N} \end{bmatrix},$$

where $\mathbf{S}$ is nonsingular. Then by partitioning the inverse as shown with n' rows in $\mathbf{N}$, by definition,

$$(9.9) \quad \mathbf{S}^{-1}\mathbf{S} = \begin{bmatrix} \mathbf{M} \\ \mathbf{N} \end{bmatrix} [\mathbf{V}, \mathbf{U}] = \begin{bmatrix} \mathbf{MV} & \mathbf{MU} \\ \mathbf{NV} & \mathbf{NU} \end{bmatrix} = \begin{bmatrix} \mathbf{I}_{n-n'} & 0 \\ 0 & \mathbf{I}_{n'} \end{bmatrix}.$$

The matrix $\mathbf{V}$ can be constructed as a basis for the null space of $\mathbf{U}^T$, as in Example 2.

Second, consider the transformed system

$$(9.10a) \quad \mathbf{A}' = \mathbf{S}^{-1}\mathbf{AS} = \begin{bmatrix} \mathbf{MAV} & \mathbf{MAU} \\ \mathbf{NAV} & \mathbf{NAU} \end{bmatrix} = \begin{bmatrix} \mathbf{A}_{11} & \mathbf{A}_{12} \\ \mathbf{A}_{21} & \mathbf{A}_{22} \end{bmatrix},$$

$$(9.10b) \quad \mathbf{B}' = \mathbf{S}^{-1}\mathbf{B} = \begin{bmatrix} \mathbf{MB} \\ \mathbf{NB} \end{bmatrix} = \begin{bmatrix} \mathbf{B}_1 \\ \mathbf{B}_2 \end{bmatrix},$$

$$(9.10c) \quad \mathbf{C}' = \mathbf{CS} = [\mathbf{CV}, \mathbf{CU}] = [\mathbf{C}_1, \mathbf{C}_2],$$

$$(9.10d) \quad \mathbf{D}' = \mathbf{D}.$$

Proposition 2   *In the above system, $\mathbf{A}_{12} = 0$ and $\mathbf{B}_1 = 0$.*

Proof: Because $\mathbf{U} = \mathscr{C}\mathbf{Q}_1$ is a basis for the range of $\mathscr{C}$,

$$(9.11) \quad \mathbf{AU} = \mathbf{A}\mathscr{C}\mathbf{Q}_1 = [\mathbf{AB}, \mathbf{A}^2\mathbf{B}, \cdots \mathbf{A}^n\mathbf{B}]\mathbf{Q}_1$$

and from the Cayley-Hamilton theorem the columns of $\mathbf{A}^n\mathbf{B}$ are linear combinations of the columns of $\mathscr{C}$. Therefore all columns of the right-hand side of the above equation are in the range of $\mathscr{C}$, and hence of $\mathbf{U}$; that is, for some matrix $\mathbf{K}$,

$$(9.12) \quad \mathbf{AU} = \mathbf{UK}.$$

Therefore from (9.9), $\mathbf{A}_{12} = \mathbf{MAU} = \mathbf{MUK} = 0\,\mathbf{K} = 0$. Similarly, since $\mathbf{B}$ is a submatrix of $\mathscr{C}$ and $\mathbf{U}$ is a basis for the range of $\mathscr{C}$, $\mathbf{B} = \mathbf{UL}$ for some matrix $\mathbf{L}$, and $\mathbf{B}_1 = \mathbf{MB} = \mathbf{MUL} = 0\,\mathbf{L} = 0$. $\square$

The consequence of Proposition 2 is that the transformed system has the form of (9.3), as illustrated in Figure 9.1.

Proposition 3   *The system $(\mathbf{A}_{22}, \mathbf{B}_2, \mathbf{C}_2, \mathbf{D})$ with matrices computed as above is controllable and externally equivalent to $(\mathbf{A}, \mathbf{B}, \mathbf{C}, \mathbf{D})$.*

Proof: The rank n' of the controllability matrix of the original system is invariant under premultiplication by a nonsingular matrix; thus

$$(9.13) \quad \begin{aligned} \text{rank}[\mathbf{B}, \mathbf{AB}, \cdots \mathbf{A}^{n-1}\mathbf{B}] &= \text{rank}(\mathbf{S}^{-1}[\mathbf{B}, \mathbf{AB}, \cdots \mathbf{A}^{n-1}\mathbf{B}]) \\ &= \text{rank}[\mathbf{S}^{-1}\mathbf{B}, \mathbf{S}^{-1}\mathbf{ASS}^{-1}\mathbf{B}, \cdots (\mathbf{S}^{-1}\mathbf{AS})^{n-1}\mathbf{S}^{-1}\mathbf{B}] \\ &= \text{rank}[\mathbf{B}', \mathbf{A}'\mathbf{B}', \cdots (\mathbf{A}')^{n-1}\mathbf{B}']. \end{aligned}$$

From (9.3a) this new controllability matrix is

$$(9.14) \quad [\,\mathbf{B}', \, \mathbf{A}'\mathbf{B}', \cdots (\mathbf{A}')^{n-1}\mathbf{B}'\,] = \begin{bmatrix} 0 & 0 & & 0 \\ \mathbf{B}_2 & \mathbf{A}_{22}\mathbf{B}_2 & \cdots & (\mathbf{A}'_{22})^{n-1}\mathbf{B}_2 \end{bmatrix},$$

the rank n' of which is, by inspection and the Cayley-Hamilton theorem, equal to the rank of $[\,\mathbf{B}_2, \, \mathbf{A}_{22}\mathbf{B}_2, \cdots \mathbf{A}_{22}^{n'-1}\mathbf{B}_2\,]$, the controllability matrix of the system $(\mathbf{A}_{22}, \mathbf{B}_2, \mathbf{C}_2, \mathbf{D})$, which is therefore controllable.

Furthermore for $k \geq 0$, the $(k{+}1)$-th term of the weighting sequence is

$$(9.15) \quad \mathbf{CA}^k\mathbf{B} = \mathbf{CSS}^{-1}\mathbf{A}^k\mathbf{SS}^{-1}\mathbf{B} = \mathbf{CS}(\mathbf{S}^{-1}\mathbf{AS})^k\mathbf{S}^{-1}\mathbf{B} = \mathbf{C}'(\mathbf{A}')^k\mathbf{B}'$$

$$= [\,\mathbf{C}_1, \, \mathbf{C}_2\,]\begin{bmatrix} \mathbf{A}_{11} & 0 \\ \mathbf{A}_{21} & \mathbf{A}_{22} \end{bmatrix}^k \begin{bmatrix} 0 \\ \mathbf{B}_2 \end{bmatrix} = \mathbf{C}_2\mathbf{A}_{22}^k\mathbf{B}_2,$$

so the reduced system is externally equivalent to the original system. □

Careful inspection of the above development shows that if only the reduced controllable system $(\mathbf{A}_{22}, \mathbf{B}_2, \mathbf{C}_2, \mathbf{D})$ is to be computed, the matrix $\mathbf{V}$ need not be computed explicitly. In practice, other algorithms are often preferred.

Example 2
Computing **V**

From Section 4.1 of Chapter 6, a basis $\mathbf{V}$ for the null space of $\mathbf{U}^T$ can be obtained by a column compression:

$$(9.16) \quad \begin{bmatrix} \mathbf{U}^T \\ \mathbf{I} \end{bmatrix} \overset{\text{col}}{\sim} \begin{bmatrix} \mathbf{U}^T\mathbf{V}_1 & 0 \\ \mathbf{V}_1 & \mathbf{V} \end{bmatrix},$$

where $\mathbf{U}^T\mathbf{V}_1$ is square and has full rank $n' = \text{rank } \mathbf{U}$. Because $\mathbf{V}$ is $n \times (n{-}n')$, the matrix $\mathbf{S} = [\,\mathbf{V}, \mathbf{U}\,]$ is square, but it remains to be shown that this matrix is nonsingular. Premultiplying $\mathbf{S}$ by $[\,\mathbf{V}_1, \, \mathbf{V}\,]^T$ gives

$$(9.17) \quad \begin{bmatrix} \mathbf{V}_1^T \\ \mathbf{V}^T \end{bmatrix}\mathbf{S} = \begin{bmatrix} \mathbf{V}_1^T\mathbf{V} & \mathbf{V}_1^T\mathbf{U} \\ \mathbf{V}^T\mathbf{V} & 0 \end{bmatrix},$$

in which, by construction, $\mathbf{V}_1^T\mathbf{U} = (\mathbf{U}^T\mathbf{V}_1)^T$ is square and nonsingular.

The submatrix $\mathbf{V}^T\mathbf{V}$ is also nonsingular since $\mathbf{V}$ has full column rank. To prove this, assume that $\mathbf{V}^T\mathbf{V}$ has less than full rank, so there must exist a nonzero vector $\mathbf{X}$ such that $(\mathbf{V}^T\mathbf{V})\mathbf{X} = 0$ and $\mathbf{X}^T(\mathbf{V}^T\mathbf{V})\mathbf{X} = (\mathbf{VX})^T(\mathbf{VX}) = 0$, which is the sum of the squares of the entries of vector $\mathbf{VX}$. This sum of squares can only be 0 if $\mathbf{VX} = 0$, implying that $\mathbf{V}$ has less than full column rank, a contradiction.

The matrices $\mathbf{V}^T\mathbf{V}$ and $\mathbf{V}_1^T\mathbf{U}$ are square and nonsingular. Therefore the determinant $\pm \det(\mathbf{V}^T\mathbf{V}) \det(\mathbf{V}_1^T\mathbf{U})$ of the right side of (9.17) is nonzero, and since $[\,\mathbf{V}_1, \, \mathbf{V}\,]^T$ is nonsingular on the left of (9.17), $\mathbf{S}$ must also be nonsingular.

Example 3
Controllability transformation

Let the matrices be given as

$$
\mathbf{A} = \begin{bmatrix} 3 & 3 & 3 \\ -8 & -7 & -4 \\ -2 & -1 & -4 \end{bmatrix}, \quad \mathbf{B} = \begin{bmatrix} 2 \\ -2 \\ -1 \end{bmatrix}, \quad \mathbf{C} = [3, 2, 1].
$$

The controllability matrix $\mathscr{C}$ and resulting column echelon form, which is not unique, are

$$
\mathscr{C} = \begin{bmatrix} 2 & -3 & 3 \\ -2 & 2 & 2 \\ -1 & 2 & -4 \end{bmatrix} \overset{\text{col}}{\sim} \left[\begin{array}{cc|c} 2 & 0 & 0 \\ -2 & -1 & 0 \\ -1 & .5 & 0 \end{array} \right] = [\, \mathbf{U}, \mathbf{0} \,],
$$

so that a basis for $\mathcal{R}(\mathscr{C})$ is $\mathbf{U}$, as shown. Because $\mathbf{U}$ has two columns, an externally equivalent system of order $n' = 2$ can be found; had the system been completely controllable, the transformation defined by $\mathbf{S}$ would result in no change of dimension and could be omitted. A basis matrix $\mathbf{V}$ for the null space of $\mathbf{U}^T$ can be constructed as

$$
\begin{bmatrix} \mathbf{U}^T \\ \mathbf{I} \end{bmatrix} = \begin{bmatrix} 2 & -2 & -1 \\ 0 & -1 & .5 \\ 1 & 0 & 0 \\ 0 & 1 & 0 \\ 0 & 0 & 1 \end{bmatrix} \overset{\text{col}}{\sim} \left[\begin{array}{cc|c} 2 & 0 & 0 \\ 0 & -1 & 0 \\ 1 & 1 & 1 \\ 0 & 1 & .5 \\ 0 & 0 & 1 \end{array} \right] = \begin{bmatrix} \mathbf{U}^T\mathbf{V}_1 & \mathbf{0} \\ \mathbf{V}_1 & \mathbf{V} \end{bmatrix},
$$

producing the matrix $\mathbf{V}$ is as illustrated, and a nonsingular matrix $\mathbf{S} = [\, \mathbf{V}, \mathbf{U} \,]$. The transformed system of the form of (9.3) is

$$
\mathbf{A}' = \left[\begin{array}{c|cc} -3 & 0 & 0 \\ \hline 5.25 & -1.5 & -.75 \\ 3.5 & 1 & -3.5 \end{array} \right], \quad \mathbf{B}' = \begin{bmatrix} 0 \\ 1 \\ 0 \end{bmatrix}, \quad \mathbf{C}' = [\, .5 \mid 1 \quad 1.5 \,],
$$

so an externally equivalent, completely controllable system of order 2 is

$$
\mathbf{A}_{22} = \begin{bmatrix} -1.5 & -.75 \\ 1 & -3.5 \end{bmatrix}, \quad \mathbf{B}_2 = \begin{bmatrix} 1 \\ 0 \end{bmatrix}, \quad \mathbf{C}_2 = [1 \quad -1.5].
$$

Example 4
$\hat{\mathbf{H}}(s)$ is independent of $\mathbf{A}_{11}$

From (9.15), the sequence $\{\mathbf{H}(k)\}_0^\infty = \{\mathbf{D}, \mathbf{CB}, \mathbf{CAB}, \cdots\}$ is independent of the uncontrollable part $\mathbf{A}_{11}$ of the block-triangular matrix $\mathbf{A}'$ in (9.3a). The transfer matrix can be expanded as the series $\hat{\mathbf{H}}(s) = \sum_{k=0}^\infty s^{-k}\mathbf{H}(k)$, which is independent of $\mathbf{A}_{11}$. Therefore the poles of $\hat{\mathbf{H}}(s)$ are eigenvalues of $\mathbf{A}_{22}$, and the eigenvalues of $\mathbf{A}_{11}$ do not appear as poles of $\hat{\mathbf{H}}(s)$.

| 2 | **Step 2: Observability** |

Two methods for extracting the observable part of a system will be given, but first, the use of the word *observable* will be explained.

The general formula (2.4) for the output of a discrete-time system can be rewritten, for $t = \{0, 1, \cdots \ell\}$,

$$(9.18) \quad \begin{bmatrix} \mathbf{Y}(0) \\ \mathbf{Y}(1) \\ \mathbf{Y}(2) \\ \vdots \\ \mathbf{Y}(\ell) \end{bmatrix} - \begin{bmatrix} \mathbf{D} & & & \\ \mathbf{CB} & \mathbf{D} & & \\ \mathbf{CAB} & \ddots & \ddots & \\ \vdots & & \ddots & \\ \mathbf{CA}^{\ell-1}\mathbf{B} & \cdots & \cdots & \mathbf{CB} & \mathbf{D} \end{bmatrix} \begin{bmatrix} \mathbf{U}(0) \\ \mathbf{U}(1) \\ \vdots \\ \mathbf{U}(\ell) \end{bmatrix} = \begin{bmatrix} \mathbf{C} \\ \mathbf{CA} \\ \vdots \\ \mathbf{CA}^{\ell} \end{bmatrix} \mathbf{X}(0).$$

If it is possible, for all possible known $\{\mathbf{U}(k)\}_0^{\ell}$ and $\{\mathbf{Y}(k)\}_0^{\ell}$, to determine the value of the initial state $\mathbf{X}(0)$, the system is said to be *observable*. Solving for $\mathbf{X}(0)$ uniquely in (9.18) is possible, given an arbitrary left-hand side, if and only if the matrix on the right side, denoted $\mathscr{O}_{\ell}$, has full column rank n. From the Cayley-Hamilton theorem, any submatrix $\mathbf{CA}^r$ is a linear combination of the submatrices of the *observability* matrix

$$(9.19) \quad \mathscr{O} = \begin{bmatrix} \mathbf{C} \\ \mathbf{CA} \\ \vdots \\ \mathbf{CA}^{n-1} \end{bmatrix},$$

from which follows

Proposition 4   *For any $\ell \geq n-1$, $rank(\mathscr{O}_{\ell}) = rank(\mathscr{O})$.*

Consequently, the system is observable if and only if $rank(\mathscr{O}) = n$.

| 2.1 | **Direct transformation** |

A transformation related to (9.10), sometimes called its dual transformation, will be described. Let $\tilde{\mathbf{U}}^T$ be a basis for the row range of $\mathscr{O}$, constructed for the system $(\mathbf{A}_{22}, \mathbf{B}_2, \mathbf{C}_2, \mathbf{D})$ rather than the original system $(\mathbf{A}, \mathbf{B}, \mathbf{C}, \mathbf{D})$, and let

$$(9.20) \quad \mathbf{T} = \begin{bmatrix} \tilde{\mathbf{V}}^T \\ \tilde{\mathbf{U}}^T \end{bmatrix}$$

be nonsingular. Then using a similarity transformation as illustrated in Figure 9.2, $(\mathbf{A}_{22}, \mathbf{B}_2, \mathbf{C}_2, \mathbf{D})$ is transformed to the form of (9.4), from which the system $(\tilde{\mathbf{A}}_{22}, \tilde{\mathbf{B}}_2, \tilde{\mathbf{C}}_2, \tilde{\mathbf{D}})$ is equivalent and completely observable.

The order of the two steps can be reversed, first constructing an observable equivalent of $(\mathbf{A}, \mathbf{B}, \mathbf{C}, \mathbf{D})$, and then constructing the controllable equivalent of the result.

Example 5
Observability
transformation

In Example 3, the observability matrix and its row echelon form are

$$\begin{bmatrix} \mathbf{C}_2 \\ \mathbf{C}_2\mathbf{A}_{22} \end{bmatrix} = \begin{bmatrix} 1 & -1.5 \\ -3 & 4.5 \end{bmatrix} \overset{\text{row}}{\sim} \begin{bmatrix} 1 & -1.5 \\ 0 & 0 \end{bmatrix},$$

of which the top row is $\tilde{\mathbf{U}}^T$, and finding $\tilde{\mathbf{V}}^T$ in (9.20) by a row compression,

$$[\tilde{\mathbf{U}}, \mathbf{I}] = \begin{bmatrix} 1 & 1 & 0 \\ -1.5 & 0 & 1 \end{bmatrix} \overset{\text{row}}{\sim} \begin{bmatrix} 1 & 1 & 0 \\ 0 & 1.5 & 1 \end{bmatrix} = \begin{bmatrix} 1 & 1 & 0 \\ 0 & \tilde{\mathbf{V}}^T \end{bmatrix},$$

gives the transformed system

$$\tilde{\mathbf{A}} = \mathbf{TAT}^{-1} = \begin{bmatrix} -2 & 1.75 \\ 0 & -3 \end{bmatrix}, \quad \tilde{\mathbf{B}} = \mathbf{TB} = \begin{bmatrix} 1.5 \\ 1 \end{bmatrix}, \quad \tilde{\mathbf{C}} = \mathbf{CT}^{-1} = [0 \mid 1],$$

of which the observable part is

$$\tilde{\mathbf{A}}_{22} = -3, \quad \tilde{\mathbf{B}}_2 = 1, \quad \tilde{\mathbf{C}}_2 = 1.$$

2.2   Observability by constructing the dual system

An equivalent sequence of operations to those above is to first construct the system

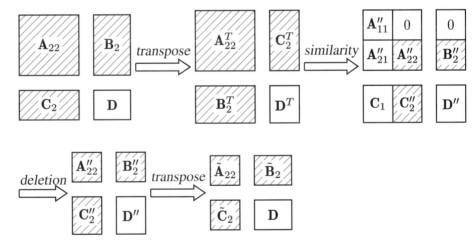

Fig. 9.3   Extracting the observable part by the controllability algorithm.

(9.21) $(\mathbf{A}'', \mathbf{B}'', \mathbf{C}'', \mathbf{D}'') = (\mathbf{A}^T, \mathbf{C}^T, \mathbf{B}^T, \mathbf{D}^T),$

for which

(9.22) $\mathbf{C}''(\mathbf{A}'')^k\mathbf{B}'' = (\mathbf{CA}^k\mathbf{B})^T,$

and compute its controllable part $(\mathbf{A}_{22}'', \mathbf{B}_2'', \mathbf{C}_2'', \mathbf{D}'')$ for which

(9.23) $\quad (\tilde{\mathbf{A}}_{22}, \tilde{\mathbf{B}}_2, \tilde{\mathbf{C}}_2, \tilde{\mathbf{D}}) = ((\mathbf{A}_{22}'')^T, (\mathbf{C}_2'')^T, (\mathbf{B}_2'')^T, (\mathbf{D}'')^T).$

The computation is illustrated in Figure 9.3.

3 Minimality

A useful tool for visualizing the controllable and observable parts referred to in previous sections is to suppose that, as in Section 1.1, a controllability transformation has been used to produce system matrices of the form of Equations (9.3). The corresponding change of basis is $\mathbf{X} = \mathbf{S}\mathbf{X}'$. Then, for a discrete-time system, the equations become

(9.24a) $\quad \mathbf{X}'(t{+}1) = \begin{bmatrix} \mathbf{X}_1(t{+}1) \\ \mathbf{X}_2(t{+}1) \end{bmatrix} = \begin{bmatrix} \mathbf{A}_{11} & 0 \\ \mathbf{A}_{21} & \mathbf{A}_{22} \end{bmatrix} \begin{bmatrix} \mathbf{X}_1(t) \\ \mathbf{X}_2(t) \end{bmatrix} + \begin{bmatrix} 0 \\ \mathbf{B}_2 \end{bmatrix} \mathbf{U}(t)$

(9.24b) $\quad \mathbf{Y}(t) = [\mathbf{C}_1, \mathbf{C}_2] \begin{bmatrix} \mathbf{X}_1(t) \\ \mathbf{X}_2(t) \end{bmatrix} + \mathbf{D}\mathbf{U}(t),$

where $\mathbf{X}'(t)$ has been partitioned conformably with the matrices into two subvectors as shown. Then subvector $\mathbf{X}_1(t)$ is unaffected by the input $\mathbf{U}(t)$, and is called the uncontrollable subvector and associated with a subsystem $\bar{\mathrm{C}}$ described by the top block equation of (9.24a), illustrated in Figure 9.4(a). Subvector $\mathbf{X}_2(t)$ is the state vector of the controllable subsystem C described by the lower block equation of (9.24a). The analogous observability decomposition is shown in Figure 9.4(b).

Minimality Previous sections have shown how similarity transformations can be used to find a system $(\hat{\mathbf{A}}, \hat{\mathbf{B}}, \hat{\mathbf{C}}, \hat{\mathbf{D}})$ that is the observable part of the controllable part of the original system, or vice versa. First it will be checked that the result is a model that is both controllable and observable. Then it will be shown that a controllable, observable system is of minimal order.

Proposition 5 *If* $(\mathbf{A}_{22}, \mathbf{B}_2)$ *in (9.4) is a controllable pair, then* $(\tilde{\mathbf{A}}_{22}, \tilde{\mathbf{B}}_2)$ *is a controllable pair.*

Proof: The controllability matrix for $(\mathbf{A}_{22}, \mathbf{B}_2)$ has the form

$$\mathscr{C} = \begin{bmatrix} \tilde{\mathbf{B}}_1 & \tilde{\mathbf{A}}_{11}\tilde{\mathbf{B}}_1 + \tilde{\mathbf{A}}_{12}\tilde{\mathbf{B}}_2 & \cdots & \\ \tilde{\mathbf{B}}_2 & \tilde{\mathbf{A}}_{22}\tilde{\mathbf{B}}_2 & \tilde{\mathbf{A}}_{22}^2\tilde{\mathbf{B}}_2 & \cdots \end{bmatrix}.$$

Assume that this matrix has full row rank. Then its rows are linearly independent, and the bottom block row contains linearly independent rows. Therefore the bottom block row has full row rank, and $(\tilde{\mathbf{A}}_{22}, \tilde{\mathbf{B}}_2)$ is a controllable pair. $\square$

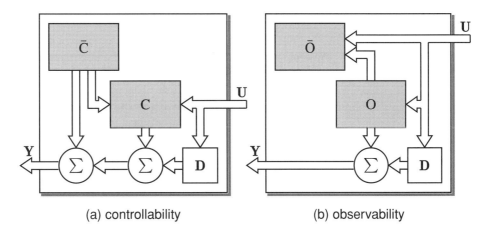

(a) controllability   (b) observability

Fig. 9.4   System decompositions: (a) controllability decomposition; (b) observability decomposition.

From the above result, extracting the observable part from the controllable part results in a controllable, observable system. A similar conclusion can be reached for performing these operations in reverse order.

Proposition 6   *A system that is both controllable and observable is minimal.*

Proof: Assume to the contrary that the system $(\mathbf{A}, \mathbf{B}, \mathbf{C}, \mathbf{D})$ of order n is controllable and observable, and that the system $(\tilde{\mathbf{A}}, \tilde{\mathbf{B}}, \tilde{\mathbf{C}}, \mathbf{D})$ is of order $\tilde{n} < n$ and externally equivalent, so that for $k = 1, 2, \cdots$,

(9.25) $\mathbf{H}_k = \mathbf{C}\mathbf{A}^{k-1}\mathbf{B} = \tilde{\mathbf{C}}\tilde{\mathbf{A}}^{k-1}\tilde{\mathbf{B}}.$

Let the observability matrix $\mathcal{O}$ and the controllability matrix $\mathcal{C}$ of $(\mathbf{A}, \mathbf{B}, \mathbf{C}, \mathbf{D})$ have full rank n, and define

(9.26) $\tilde{\mathcal{O}} = \begin{bmatrix} \tilde{\mathbf{C}} \\ \tilde{\mathbf{C}}\tilde{\mathbf{A}} \\ \vdots \\ \tilde{\mathbf{C}}\tilde{\mathbf{A}}^{n-1} \end{bmatrix}, \quad \tilde{\mathcal{C}} = [\, \tilde{\mathbf{B}}, \ \tilde{\mathbf{A}}\tilde{\mathbf{B}}, \ \cdots \tilde{\mathbf{A}}^{n-1}\tilde{\mathbf{B}} \,],$

which have dimension $np \times n'$ and $n' \times nm$ respectively. Then consider the Hankel matrix $\mathbf{M}$ that satisfies

(9.27) $\mathbf{M} = \mathcal{O}\mathcal{C} = \begin{bmatrix} \mathbf{C} \\ \mathbf{C}\mathbf{A} \\ \vdots \\ \mathbf{C}\mathbf{A}^{n-1} \end{bmatrix} [\, \mathbf{B}, \ \mathbf{A}\mathbf{B}, \ \cdots \mathbf{A}^{n-1}\mathbf{B} \,] = \begin{bmatrix} \mathbf{H}_1 & \mathbf{H}_2 & \cdots & \mathbf{H}_n \\ \mathbf{H}_2 & \mathbf{H}_3 & \cdots & \mathbf{H}_{n+1} \\ \cdots & & & \\ \mathbf{H}_n & \mathbf{H}_{n+1} & \cdots & \mathbf{H}_{2n-1} \end{bmatrix}$

$= \tilde{\mathcal{O}}\tilde{\mathcal{C}}.$

The full rank n of $\mathcal{O}$ and $\mathcal{C}$ in (9.26) implies that there exist nonsingular matrices $\mathbf{P}, \mathbf{Q}$ such that

$$(9.28) \quad \mathbf{P}\mathcal{O} = \begin{bmatrix} \mathbf{I}_n \\ 0 \end{bmatrix}, \quad \mathcal{C}\mathbf{Q} = [\,\mathbf{I}_n, \, 0\,], \quad \text{and} \quad \mathbf{PMQ} = \begin{bmatrix} \mathbf{I}_n & 0 \\ 0 & 0 \end{bmatrix},$$

showing that rank $\mathbf{M} = n$. But by using the properties of the rank of a product, rank $\mathbf{M} = \text{rank}(\tilde{\mathcal{O}}\tilde{\mathcal{C}}) \leq \min(\text{rank } \tilde{\mathcal{O}}, \text{rank } \tilde{\mathcal{C}}) \leq \tilde{n}$; that is, $n \leq \tilde{n}$, but by assumption $\tilde{n} < n$, a contradiction. $\qquad\square$

Example 6
Constructing a
minimal
equivalent
system

The transfer matrix

$$\hat{\mathbf{H}}(s) = \begin{bmatrix} 2/(s+3) & 2/(s+3) \\ 2/(s+3) & 2/(s+3) \end{bmatrix}$$

has a realization that can be written by the method of Section 9 of Chapter 4 as

$$\mathbf{A} = \begin{bmatrix} -3 & & & \\ & -3 & & \\ & & -3 & \\ & & & -3 \end{bmatrix}, \quad \mathbf{B} = \begin{bmatrix} 1 \\ 1 \\ & 1 \\ & 1 \end{bmatrix},$$

$$\mathbf{C} = \begin{bmatrix} 2 & & 2 & \\ & 2 & & 2 \end{bmatrix}, \quad \mathbf{D} = \begin{bmatrix} 0 & 0 \\ 0 & 0 \end{bmatrix}.$$

This realization does not have minimal dimension. The controllability matrix is

$$[\,\mathbf{B}, \, \mathbf{AB}, \, \mathbf{A}^2\mathbf{B}, \, \mathbf{A}^3\mathbf{B}\,] = \begin{bmatrix} 1 & 0 & -3 & 0 & 9 & 0 & -27 & 0 \\ 1 & 0 & -3 & 0 & 9 & 0 & -27 & 0 \\ 0 & 1 & 0 & -3 & 0 & 9 & 0 & -27 \\ 0 & 1 & 0 & -3 & 0 & 9 & 0 & -27 \end{bmatrix},$$

which has rank 2. One possible basis $\mathbf{U}$ for the range of this matrix is the first two columns, which are independent. Finding a basis $\mathbf{V}$ for $\mathcal{N}(\mathbf{U}^T)$ gives

$$\mathbf{S} = [\,\mathbf{V}, \mathbf{U}\,] = \begin{bmatrix} -1 & 0 & 1 & 0 \\ 1 & 0 & 1 & 0 \\ 0 & -1 & 0 & 1 \\ 0 & 1 & 0 & 1 \end{bmatrix}, \quad \mathbf{S}^{-1} = (1/2) \begin{bmatrix} -1 & 1 & 0 & 0 \\ 0 & 0 & -1 & 1 \\ 1 & 1 & 0 & 0 \\ 0 & 0 & 1 & 1 \end{bmatrix},$$

so that

$$\mathbf{S}^{-1}\mathbf{AS} = \mathbf{A}, \quad \mathbf{S}^{-1}\mathbf{B} = \begin{bmatrix} 0 & 0 \\ 0 & 0 \\ 1 & 0 \\ 0 & 1 \end{bmatrix}, \quad \mathbf{CS} = \begin{bmatrix} -2 & -2 & 2 & 2 \\ 2 & 2 & 2 & 2 \end{bmatrix}.$$

The controllable part of the system is

$$\mathbf{A}_{22} = \begin{bmatrix} -3 & \\ & -3 \end{bmatrix}, \quad \mathbf{B}_2 = \begin{bmatrix} 1 & 0 \\ 0 & 1 \end{bmatrix}, \quad \mathbf{C}_2 = \begin{bmatrix} 2 & 2 \\ 2 & 2 \end{bmatrix},$$

and its dual is $\mathbf{A}'' = \mathbf{A}_{22}^T$, $\mathbf{B}'' = \mathbf{C}_2^T$, $\mathbf{C}'' = \mathbf{B}_2^T$, of which the controllability matrix is

$$[\,\mathbf{B},\mathbf{AB}\,] = \begin{bmatrix} 2 & 2 & -6 & -6 \\ 2 & 2 & -6 & -6 \end{bmatrix},$$

which has rank 1. A transformation matrix and resulting transformed system are

$$\mathbf{S} = \begin{bmatrix} 1 & 1 \\ -1 & 1 \end{bmatrix}, \quad \mathbf{S}^{-1}\mathbf{A}''\mathbf{S} = \begin{bmatrix} -3 & \\ & -3 \end{bmatrix}, \quad \mathbf{S}^{-1}\mathbf{B}'' = \begin{bmatrix} 0 & 0 \\ 2 & 2 \end{bmatrix},$$

$$\mathbf{C}''\mathbf{S} = \begin{bmatrix} 1 & 1 \\ -1 & 1 \end{bmatrix}.$$

The controllable part of this system is

$$\mathbf{A}_{22}'' = -3, \quad \mathbf{B}_2'' = [\,2 \quad 2\,], \quad \mathbf{C}_2'' = \begin{bmatrix} 1 \\ 1 \end{bmatrix},$$

and transposing to construct the dual, a minimal system with transfer matrix $\hat{\mathbf{H}}(s)$ is seen to be

$$\tilde{\mathbf{A}}_{22} = -3, \quad \tilde{\mathbf{B}}_2 = [\,1 \quad 1\,], \quad \tilde{\mathbf{C}}_2 = \begin{bmatrix} 2 \\ 2 \end{bmatrix}, \quad \mathbf{D} = \begin{bmatrix} 0 & 0 \\ 0 & 0 \end{bmatrix}.$$

Example 7
Controllability
and
observability
grammians

From Section 1, an LTI system is defined to be controllable if rank $\mathscr{C} = n$. However, for other than textbook systems, the calculation of $\mathscr{C}$ in order to estimate the rank of this matrix is considered naïve for reasons of accuracy, efficiency, and computational stability. One alternative for continuous-time LTI systems is to first solve the Lyapunov equation

(9.29) $\mathbf{A}\mathbf{W}_\mathrm{C} + \mathbf{W}_\mathrm{C}\mathbf{A}^T = -\mathbf{B}\mathbf{B}^T,$

for the matrix $\mathbf{W}_\mathrm{C}$, called the *controllability grammian,* and then to compute its rank. Similarly, to test observability, the *observability grammian* $\mathbf{W}_\mathrm{O}$ is computed from the equation

(9.30) $\mathbf{A}^T\mathbf{W}_\mathrm{O} + \mathbf{W}_\mathrm{O}\mathbf{A} = -\mathbf{C}^T\mathbf{C},$

and its rank is tested. The equivalent discrete-time Lyapunov equations are, respectively,

(9.31a) $\mathbf{A}\mathbf{W}_\mathrm{C}\mathbf{A}^T - \mathbf{W}_\mathrm{C} = -\mathbf{B}\mathbf{B}^T,$

(9.31b) $\mathbf{A}^T\mathbf{W}_\mathrm{O}\mathbf{A} - \mathbf{W}_\mathrm{O} = -\mathbf{C}^T\mathbf{C}.$

We shall show that rank $\mathscr{C} = n$ if and only if rank $\mathbf{W}_\mathrm{C} = n$ as follows, assuming that (9.29) is nonsingular. The other cases are similar.

First, assume rank $\mathscr{C} < n$, so that the system is not completely controllable. With a similarity transformation, (9.29) becomes

(9.32) $(\mathbf{S}^{-1}\mathbf{A}\mathbf{S})(\mathbf{S}^{-1}\mathbf{W}_C\mathbf{S}^{-T}) + (\mathbf{S}^{-1}\mathbf{W}_C\mathbf{S}^{-T})(\mathbf{S}^T\mathbf{A}^T\mathbf{S}^{-T}) = -\mathbf{S}^{-1}\mathbf{B}\mathbf{B}^T\mathbf{S}^{-T},$

where $(\mathbf{S}^{-1})^T$ has been written $\mathbf{S}^{-T}$. Rewriting yields

(9.33) $\tilde{\mathbf{A}}\tilde{\mathbf{W}}_C + \tilde{\mathbf{W}}_C\tilde{\mathbf{A}}^T = -\tilde{\mathbf{B}}\tilde{\mathbf{B}}^T,$

where $\tilde{\mathbf{A}}$, $\tilde{\mathbf{B}}$, and $\tilde{\mathbf{W}}_C$ are defined term-wise from (9.32). Using the development of Section 1, choose $\mathbf{S}$ to produce the form of Equation (9.3), so that (9.33) has the form

$$\begin{bmatrix} \mathbf{A}_{11} & 0 \\ \mathbf{A}_{21} & \mathbf{A}_{22} \end{bmatrix} \begin{bmatrix} \mathbf{W}_{11} & \mathbf{W}_{12} \\ \mathbf{W}_{12}^T & \mathbf{W}_{22} \end{bmatrix} + \begin{bmatrix} \mathbf{W}_{11} & \mathbf{W}_{12} \\ \mathbf{W}_{12}^T & \mathbf{W}_{22} \end{bmatrix} \begin{bmatrix} \mathbf{A}_{11}^T & \mathbf{A}_{21}^T \\ 0 & \mathbf{A}_{22}^T \end{bmatrix} =$$
$$- \begin{bmatrix} 0 & 0 \\ 0 & \mathbf{B}_2\mathbf{B}_2^T \end{bmatrix}.$$

The top-left block equation is

(9.34) $\mathbf{A}_{11}\mathbf{W}_{11} + \mathbf{W}_{11}\mathbf{A}_{11}^T = 0,$

which is nonsingular by assumption; therefore $\mathbf{W}_{11} = 0$. Similarly $\mathbf{W}_{12} = 0$ from the top-right block equation, so that rank $\tilde{\mathbf{W}}_C < n$, and hence rank $\mathbf{W}_C < n$.

Now suppose that rank $\mathbf{W}_C < n$. Since $\mathbf{W}_C$ is symmetric, let $\mathbf{S}^{-1}$ be chosen (Chapter 8, Section 3.1) such that $\mathbf{S}^{-1}\mathbf{W}_C\mathbf{S}^{-T} = \tilde{\mathbf{W}}_C$ is the normal form of $\mathbf{W}_C$ defined in Chapter 5. Then partitioning $\tilde{\mathbf{A}}$ and $\tilde{\mathbf{B}}$ conformably, (9.33) is of the form

(9.35) $\begin{bmatrix} \mathbf{A}_{11} & \mathbf{A}_{12} \\ \mathbf{A}_{21} & \mathbf{A}_{22} \end{bmatrix} \begin{bmatrix} \mathbf{I} & 0 \\ 0 & 0 \end{bmatrix} + \begin{bmatrix} \mathbf{I} & 0 \\ 0 & 0 \end{bmatrix} \begin{bmatrix} \mathbf{A}_{11}^T & \mathbf{A}_{21}^T \\ \mathbf{A}_{12}^T & \mathbf{A}_{22}^T \end{bmatrix} = - \begin{bmatrix} \mathbf{B}_1\mathbf{B}_1^T & \mathbf{B}_1\mathbf{B}_2^T \\ \mathbf{B}_2\mathbf{B}_1^T & \mathbf{B}_2\mathbf{B}_2^T \end{bmatrix}.$

The lower-right block of the above equation is $0 = \mathbf{B}_2\mathbf{B}_2^T$. The diagonal entries of $\mathbf{B}_2\mathbf{B}_2^T$, which are zero, are the squares of the lengths of the rows of $\mathbf{B}_2$, and therefore $\mathbf{B}_2 = 0$. Hence the lower-left block of (9.35) becomes $\mathbf{A}_{21} = -\mathbf{B}_2\mathbf{B}_1^T = 0$, so that $\tilde{\mathbf{A}}$, $\tilde{\mathbf{B}}$, and corresponding controllability matrix $\tilde{\mathscr{C}} = [\tilde{\mathbf{B}}, \cdots \tilde{\mathbf{A}}^{n-1}\tilde{\mathbf{B}}]$ have the form

(9.36) $\tilde{\mathbf{A}} = \begin{bmatrix} \mathbf{A}_{11} & \mathbf{A}_{12} \\ 0 & \mathbf{A}_{22} \end{bmatrix}, \quad \tilde{\mathbf{B}} = \begin{bmatrix} \mathbf{B}_1 \\ 0 \end{bmatrix}, \quad \tilde{\mathscr{C}} = \begin{bmatrix} \mathbf{B}_1 & \mathbf{A}_{11}\mathbf{B}_1 & \mathbf{A}_{11}^2\mathbf{B}_1 & \cdots \\ 0 & 0 & 0 & \cdots \end{bmatrix},$

so that rank $\tilde{\mathscr{C}} < n$ by inspection, and consequently rank $\mathscr{C} < n$.

Example 8
Balanced
systems

An extension of Example 7 leads to a method of approximating a large system by one of smaller dimension.

From Section 1, the controllable part of a system for which (9.36) applies is given by submatrices $\mathbf{A}_{11}$, $\mathbf{B}_1$, and $\mathbf{C}_1 = $ (leftmost r columns of $\mathbf{CS}$), where $\mathbf{A}_{11}$ is $r \times r$. If, however, the system in (9.35) is completely controllable but $\tilde{\mathbf{W}}_{\mathrm{C}} = \begin{bmatrix} \Sigma_1 & 0 \\ 0 & \Sigma_2 \end{bmatrix}$ is diagonal and contains the singular values of $\mathbf{W}_{\mathrm{C}}$ in descending order by size, and if the values in $\Sigma_2 \in \mathbb{R}^{(n-r) \times (n-r)}$ are near zero, then $\mathbf{A}_{11}$, $\mathbf{B}_1$, $\mathbf{C}_1$ is sometimes taken to be the "strongly" controllable part of the system, and deleting the corresponding "weakly" controllable part from the model, as for (9.35), may be considered.

The transformed observability grammian is obtained, given $\mathbf{S}$, by premultiplying (9.30) by $\mathbf{S}^T$ and postmultiplying by $\mathbf{S}$ to produce

(9.37) $(\mathbf{S}^T \mathbf{A}^T \mathbf{S}^{-T})(\mathbf{S}^T \mathbf{W}_{\mathrm{O}} \mathbf{S}) + (\mathbf{S}^T \mathbf{W}_{\mathrm{O}} \mathbf{S})(\mathbf{S}^{-1} \mathbf{A} \mathbf{S}) = -\mathbf{S}^T \mathbf{C}^T \mathbf{C} \mathbf{S},$

or, in abbreviated form, where $\tilde{\mathbf{W}}_{\mathrm{O}}$ and $\tilde{\mathbf{C}}$ are defined term-wise,

(9.38) $\tilde{\mathbf{A}}^T \tilde{\mathbf{W}}_{\mathrm{O}} + \tilde{\mathbf{W}}_{\mathrm{O}} \tilde{\mathbf{A}} = -\tilde{\mathbf{C}}^T \tilde{\mathbf{C}}.$

As in the analysis of controllability, a transformation $\mathbf{S}$ which exhibits strongly and weakly observable parts can be obtained. However, strong observability of a subsystem might compensate for weak controllability or vice versa, so that controllability and observability considered separately may not be sufficient criteria for dividing the system into parts to be retained and deleted.

Fortunately if the required Lyapunov equations have solutions, it is possible to find a similarity transformation for the system such that the resulting controllability and observability grammians $\tilde{\mathbf{W}}_{\mathrm{C}}$ and $\tilde{\mathbf{W}}_{\mathrm{O}}$ are identical and diagonal. Their diagonal entries are singular values of a known matrix, and give a possible measure of the distance of the system from systems of lower order. Such a transformed LTI system is said to be *balanced*.

Several methods can be used to calculate the balancing transformation $\mathbf{S}$. The system will be assumed asymptotically stable, so that the symmetric matrices $\mathbf{W}_{\mathrm{C}}$ and $\mathbf{W}_{\mathrm{O}}$ are nonnegative definite. First solve Equations (9.29) and (9.30) for $\mathbf{W}_{\mathrm{C}}$ and $\mathbf{W}_{\mathrm{O}}$, and then compute the orthogonal eigenvector matrices $\mathbf{V}_{\mathrm{C}}$, $\mathbf{U}$ and diagonal matrices Λ_{C}, Λ respectively, containing eigenvalues ordered by size on the diagonal, corresponding to the decompositions shown:

(9.39a) $\mathbf{W}_{\mathrm{C}} = \mathbf{V}_{\mathrm{C}} \Lambda_{\mathrm{C}} \mathbf{V}_{\mathrm{C}}^T$

(9.39b) $\Lambda_{\mathrm{C}}^{1/2} \mathbf{V}_{\mathrm{C}}^T \mathbf{W}_{\mathrm{O}} \mathbf{V}_{\mathrm{C}} \Lambda_{\mathrm{C}}^{1/2} = \mathbf{U} \Lambda \mathbf{U}^T.$

Let the transformation matrix $\mathbf{S}$ be

(9.40) $\mathbf{S} = \mathbf{V}_{\mathrm{C}} \Lambda_{\mathrm{C}}^{1/2} \mathbf{U} \Lambda^{-1/4}.$

Then, from (9.32) and the orthogonality of $\mathbf{V}_C$ and $\mathbf{U}$,

(9.41) $\quad \tilde{\mathbf{W}}_C = \mathbf{S}^{-1}\mathbf{W}_C\mathbf{S}^{-T} = \Lambda^{1/4}\mathbf{U}^T\Lambda_C^{-1/2}\mathbf{V}_C^T\mathbf{V}_C\Lambda_C\mathbf{V}_C^T\mathbf{V}_C\Lambda_C^{-1/2}\mathbf{U}\Lambda^{1/4} = \Lambda^{1/2},$

and furthermore from (9.37) and (9.39b),

$$\begin{aligned}
\tilde{\mathbf{W}}_O &= \mathbf{S}^T\mathbf{W}_O\mathbf{S} = \Lambda^{-1/4}\mathbf{U}^T\Lambda_C^{1/2}\mathbf{V}_C^T\mathbf{W}_O\mathbf{V}_C\Lambda_C^{1/2}\mathbf{U}\Lambda^{-1/4} \\
&= \Lambda^{-1/4}\mathbf{U}^T\mathbf{U}\Lambda\mathbf{U}^T\mathbf{U}\Lambda^{-1/4} = \Lambda^{1/2}.
\end{aligned}$$

Thus $\tilde{\mathbf{W}}_C$ and $\tilde{\mathbf{W}}_O$ are identical and diagonal. Factor $\mathbf{W}_O$ as for (9.39a) to get

(9.42) $\quad \mathbf{W}_O = \mathbf{V}_O\Lambda_O\mathbf{V}_O^T.$

Then from (9.39b),

(9.43) $\quad \mathbf{U}\Lambda\mathbf{U}^T = \Lambda_C^{1/2}\mathbf{V}_C^T\mathbf{W}_O\mathbf{V}_C\Lambda_C^{1/2} = (\Lambda_C^{1/2}\mathbf{V}_C^T\mathbf{V}_O\Lambda_O^{1/2})(\Lambda_O^{1/2}\mathbf{V}_O^T\mathbf{V}_C\Lambda_C^{1/2})$
$\qquad\qquad = \mathbf{K}^T\mathbf{K},$

where $\mathbf{K} = \Lambda_O^{1/2}\mathbf{V}_O^T\mathbf{V}_C\Lambda_C^{1/2}$. It is a fact that the nonzero singular values of any matrix $\mathbf{K} \in \mathbb{R}^{m \times n}$ are the square roots of the nonzero eigenvalues of $\mathbf{K}^T\mathbf{K}$, and consequently that if $\mathbf{K}^T = \mathbf{K}$, the singular values are the absolute values of the eigenvalues. In the above equation, Λ contains the eigenvalues of $\mathbf{K}^T\mathbf{K}$. Therefore the diagonal entries of $\Lambda^{1/2}$ are the singular values of $\mathbf{K}$, which contains products of the singular values $\Lambda_C^{1/2}$ of $\mathbf{W}_C$ and $\Lambda_O^{1/2}$ of $\mathbf{W}_O$, and of the orthogonal matrices $\mathbf{V}_O^T$ and $\mathbf{V}_C$. These products can be interpreted in terms of signal gains.

The use of the above results for system approximation requires consideration of the algorithmic details, and sometimes adjustment of the steady-state gain of the reduced system to equal that of the approximated system.

Example 9
BPH tests

The following is an alternative to other tests for controllability. A system is completely controllable if and only if rank $[\mathbf{A}-\lambda\mathbf{I},\ \mathbf{B}] = n$ for all values of λ, and a system is completely observable if and only if rank $\begin{bmatrix} \mathbf{C} \\ \mathbf{A}-\lambda\mathbf{I} \end{bmatrix} = n$ for all values of λ. The ranks need only be tested for values of λ equal to eigenvalues of $\mathbf{A}$. This is called the BPH (Popov-Belevitch-Hautus) test for controllability or observability. The controllability result is proven in the following; observability is analogous.

Assume the system is not controllable. Then use a transformation $\mathbf{S}$ as in Equation (9.3), noting that premultiplication or postmultiplication by a nonsingular matrix preserves matrix rank. Thus,

$$\text{rank}\,[\mathbf{A}-\lambda\mathbf{I},\ \mathbf{B}] = \text{rank}\left(\mathbf{S}^{-1}[\mathbf{A}-\lambda\mathbf{I},\ \mathbf{B}]\begin{bmatrix} \mathbf{S} & 0 \\ 0 & \mathbf{I} \end{bmatrix}\right)$$

$$= \text{rank} \left[\mathbf{S}^{-1}\mathbf{A}\mathbf{S} - \lambda \mathbf{I}, \ \mathbf{S}^{-1}\mathbf{B} \right]$$

$$= \text{rank} \begin{bmatrix} \mathbf{A}_{11} - \lambda \mathbf{I} & 0 & 0 \\ \mathbf{A}_{21} & \mathbf{A}_{22} - \lambda \mathbf{I} & \mathbf{B}_2 \end{bmatrix},$$

where, by assumption, $\mathbf{A}_{11}$ has nonzero dimension. Then the upper block rows of the above resulting matrix are linearly dependent when λ is an eigenvalue λ_1 of $\mathbf{A}_{11}$, so that rank $[\mathbf{A} - \lambda_1\mathbf{I}, \ \mathbf{B}] < n$.

Now let the system be controllable, but assume that rank $[\mathbf{A} - \lambda\mathbf{I}, \ \mathbf{B}] < n$ for some $\lambda = \lambda_1$. Then for $\lambda = \lambda_1$ the rows of this matrix are not linearly independent, and there exists a nonzero vector $\mathbf{V}$ such that $\mathbf{V}^T[\mathbf{A} - \lambda_1\mathbf{I}, \ \mathbf{B}] = 0$. Therefore $\mathbf{V}^T\mathbf{B} = 0$ and $\mathbf{V}^T(\mathbf{A} - \lambda_1\mathbf{I}) = 0$. From the latter expression λ_1 must be an eigenvalue of $\mathbf{A}$, and $\mathbf{V}^T\mathbf{A} = \lambda_1\mathbf{V}^T$. Premultiplying the controllability matrix $\mathscr{C}$ by $\mathbf{V}^T$ yields

$$\mathbf{V}^T\mathscr{C} = \mathbf{V}^T\left[\mathbf{B}, \ \mathbf{A}\mathbf{B}, \ \cdots \mathbf{A}^{n-1}\mathbf{B}\right] = \left[\mathbf{V}^T\mathbf{B}, \ \lambda_1\mathbf{V}^T\mathbf{B}, \ \cdots \lambda_1^{n-1}\mathbf{V}^T\mathbf{B}\right] = 0.$$

Therefore the rows of $\mathscr{C}$ are not linearly independent, and rank $\mathscr{C} < n$, a contradiction.

From the above, a possible test for controllability is to verify that the rank of $[\mathbf{A} - \lambda\mathbf{I}, \ \mathbf{B}]$ is full at each of the eigenvalues of $\mathbf{A}$. This requires accurate computation of the eigenvalues and a reliable rank test. In practice the BPH tests are most useful when the system matrices have special forms that allow exact computation.

Example 10
Example BPH
test

Let a system be described by matrices

$$\mathbf{A} = \begin{bmatrix} \mathbf{A}_{11} & \mathbf{A}_{12} \\ 0 & \mathbf{A}_{22} \end{bmatrix}, \quad \mathbf{B} = \begin{bmatrix} \mathbf{B}_1 \\ 0 \end{bmatrix}, \quad \mathbf{C} = [0, \ \mathbf{C}_2].$$

Then the lower block rows of

$$\text{rank}\left[\mathbf{A} - \lambda\mathbf{I}, \ \mathbf{B}\right] = \begin{bmatrix} \mathbf{A}_{11} - \lambda\mathbf{I} & \mathbf{A}_{12} & \mathbf{B}_1 \\ 0 & \mathbf{A}_{22} - \lambda\mathbf{I} & 0 \end{bmatrix}$$

are linearly dependent when λ has the value of an eigenvalue of $\mathbf{A}_{22}$, and the system is not completely controllable. Similarly the left block columns of

$$\begin{bmatrix} \mathbf{C} \\ \mathbf{A} - \lambda\mathbf{I} \end{bmatrix} = \begin{bmatrix} 0 & \mathbf{C}_2 \\ \mathbf{A}_{11} - \lambda\mathbf{I} & \mathbf{A}_{12} \\ 0 & \mathbf{A}_{22} - \lambda\mathbf{I} \end{bmatrix}$$

are linearly dependent when λ has the value of an eigenvalue of $\mathbf{A}_{11}$, and the system is not completely observable.

The Kalman canonical decomposition

By a sequence of similarity transformations based on the controllability and observability transformations described previously, a system can be transformed to show by inspection four parts in general, one that is controllable and observable, denoted CO; a controllable, unobservable part C$\bar{\text{O}}$; an uncontrollable, observable part $\bar{\text{C}}$O; and an uncontrollable, unobservable part $\bar{\text{C}}\bar{\text{O}}$, as shown in Figure 9.5. The matrices of the transformed system have the form shown:

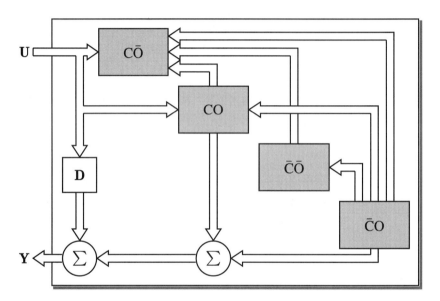

Fig. 9.5   The Kalman canonical decomposition.

$$(9.44) \quad \mathbf{A} = \begin{bmatrix} \mathbf{A}_{\text{C}\bar{\text{O}}} & \mathbf{A}_{12} & \mathbf{A}_{13} & \mathbf{A}_{14} \\ 0 & \mathbf{A}_{\text{CO}} & 0 & \mathbf{A}_{24} \\ 0 & 0 & \mathbf{A}_{\bar{\text{C}}\bar{\text{O}}} & \mathbf{A}_{34} \\ 0 & 0 & 0 & \mathbf{A}_{\bar{\text{C}}\text{O}} \end{bmatrix}, \quad \mathbf{B} = \begin{bmatrix} \mathbf{B}_{\text{C}\bar{\text{O}}} \\ \mathbf{B}_{\text{CO}} \\ 0 \\ 0 \end{bmatrix}, \quad \mathbf{C} = \begin{bmatrix} 0 & \mathbf{C}_{\text{CO}} & 0 & \mathbf{C}_{\bar{\text{C}}\text{O}} \end{bmatrix}.$$

The transfer matrix for this system is $\hat{\mathbf{H}}(s) = \mathbf{D} + \mathbf{C}_2(s\mathbf{I} - \mathbf{A}_{\text{CO}})^{-1}\mathbf{B}_2$, showing that the external behavior is determined exclusively by the controllable, observable part.

The Kalman canonical decomposition provides useful insight into fundamental issues related to controllability, observability, and minimality, but its use is primarily for theoretical understanding rather than for computation.

5 | **Further study**

Minimality theory is the kernel of LTI state-space analysis, and can be found in standard references, including Zadeh and Desoer [58], Antsaklis and Michel [2], Brogan [6], DeCarlo [12], Friedland [16], Kailath [25], and Rugh [46]. The original paper by Kalman [26] is one of the most cited in systems theory.

The Achilles heel of the theory is that it relies on the computation of matrix rank to determine the order of the minimal part, for which the SVD is a useful tool (see Golub and Van Loan [19]) but which in practice must always be justified by knowledge of the system dynamics.

6 | **Problems**

1   Several sets of matrices $(\mathbf{A}, \mathbf{B}, \mathbf{C}, \mathbf{D})$ are given below.

(a) A pair of matrices $(\mathbf{A}, \mathbf{B})$ of conformable dimension for which $\mathscr{C}$ has full row rank is said to be a controllable pair. In each case, determine if the given $(\mathbf{A}, \mathbf{D})$ are controllable pairs.

(b) A pair of matrices $(\mathbf{C}, \mathbf{A})$ of conformable dimension for which $\mathscr{O}$ has full column rank is said to be an observable pair. In each case, determine if the given $(\mathbf{C}, \mathbf{A})$ are observable pairs.

(c) Find an externally equivalent system of minimal order for each case, and calculate the transfer matrix $\hat{\mathbf{H}}(s)$.

(i) $\mathbf{A} = \begin{bmatrix} -11 & -36 & 0 \\ 2 & 8 & 1 \\ -4 & -16 & -2 \end{bmatrix}$, $\quad \mathbf{B} = \begin{bmatrix} 4 \\ -1 \\ 2 \end{bmatrix}$,

$\mathbf{C} - [1, 2, -1]$, $\quad \mathbf{D} = 3$.

(ii) $\mathbf{A} = \begin{bmatrix} -2 & 1 \\ -1 & -2 \end{bmatrix}$, $\quad \mathbf{B} = \begin{bmatrix} 1 & 3 & -1 \\ 0 & 3 & 1 \end{bmatrix}$,

$\mathbf{C} = \begin{bmatrix} 1 & 3 \\ -1 & 3 \end{bmatrix}$, $\quad \mathbf{D} = \begin{bmatrix} 0 & 0 & 0 \\ 0 & 0 & 0 \end{bmatrix}$.

(iii) $\mathbf{A} = \begin{bmatrix} -3 & 0 & 0 \\ 0 & -3 & 0 \\ 0 & 0 & -3 \end{bmatrix}$, $\quad \mathbf{B} = \begin{bmatrix} 0 & 1 \\ 1 & 0 \\ 0 & 0 \end{bmatrix}$,

$\mathbf{C} = \begin{bmatrix} 0 & 1 & 0 \\ 0 & 0 & 1 \end{bmatrix}$, $\quad \mathbf{D} = \begin{bmatrix} 0 & 0 \\ 0 & 0 \end{bmatrix}$.

(iv) $\mathbf{A}$ as in (iii), $\quad \mathbf{B} = \begin{bmatrix} 0 & 1 \\ 1 & 0 \\ 0 & 1 \end{bmatrix}$, $\quad \mathbf{C} = \begin{bmatrix} 0 & 1 & 0 \\ 1 & 0 & 1 \end{bmatrix}$, $\quad \mathbf{D}$ as in (ii).

2   Verify that a state-space model in single-input, single-output "controllable" form (4.43) is controllable, and that the "observable" form (4.46) is observable.

3   If $\mathbf{A}$ is a diagonal matrix, show that if any row of $\mathbf{B}$ is zero, then $(\mathbf{A}, \mathbf{B})$ is not a controllable pair, and if any column of $\mathbf{C}$ is zero, then $(\mathbf{C}, \mathbf{A})$ is not an observable pair.

4   By any method, find the matrices of a minimal state-space system with the impulse-response matrix shown:
$$\mathbf{H}(t) = \begin{bmatrix} te^{-2t} & 3\delta(t) \\ 2te^{-2t} & 1 \end{bmatrix}.$$

5   Compute $\mathbf{W}_{\mathrm{C}}$, and find its rank to test the controllability of the discrete-time pair $\mathbf{A} = \begin{bmatrix} -1 & 6 \\ 1 & 0 \end{bmatrix}$, $\mathbf{B} = \begin{bmatrix} 2 \\ 1 \end{bmatrix}$.

6   Define *elementary similarity transformations* as follows:

(a) Operation O_{ij} interchanges rows i, j in $\mathbf{A}$ and $\mathbf{B}$, then columns i, j in the resulting $\mathbf{A}$ and in $\mathbf{C}$.

(b) Operation $O_i(\alpha)$, for $\alpha \neq 0$, multiplies row i of $\mathbf{A}$ and $\mathbf{B}$ by α, then divides column i of $\mathbf{A}$ and $\mathbf{C}$ by α.

(c) Operation $O_{ij}(\alpha)$ adds the product of (row j) $\times$ α to row i in $\mathbf{A}$ and $\mathbf{B}$, then subtracts the product of (column i) $\times$ α from column j in $\mathbf{A}$ and $\mathbf{C}$.

Note that each of the above operations corresponds to a similarity transformation using an elementary matrix. Instead of computing the controllability transformation $\mathbf{S}$ of Equation (9.3) directly, it can be shown by careful housekeeping that using a sequence of the above elementary operations to transform the matrix $[\mathbf{A}, \mathbf{B}]$ to lower-left row echelon form is equivalent to testing the columns of the matrix $[\mathbf{A}^{n-1}\mathbf{B}, \cdots \mathbf{AB}, \mathbf{B}]$ for independence, from the right. If, in column i, say, of the transformed $\mathbf{A}$, no pivot can be found above row i, the sequence stops, having transformed the system into the form shown in Equation (9.3), with $\mathbf{A}_{11}$ of dimension $i-1 \times i-1$, and the system is not completely controllable. When elementary transformations are replaced by the Householder transformations of Problem 30 in Chapter 5, the result is one of a class of *staircase algorithms* for testing controllability. Test the controllability of the system of Problem 5 by using elementary similarities.

10

Poles and Zeros

The definition and computation of the poles of a transfer function, or transfer function matrix, are of importance in determining system stability or relative stability. The zeros determine the stability of the system when it is subjected to high-gain negative feedback, and also determine the signals that are blocked by the system, as when band-suppression filters are being designed.

An extension of the normal-form computation of Chapter 5 will be used to define the poles and zeros of a rational matrix. Polynomial numerator and denominator matrices will also be required in order to generalize the concept of the scalar numerator and denominator of a transfer function. Such matrices will be obtained, corresponding to left division by the denominator matrix, and a similar right factorization will also be found.

Let the Laplace-transformed output equation for an LTI system with zero initial conditions be

(10.1) $\mathbf{Y}(s) = \mathbf{H}(s)\,\mathbf{U}(s), \quad \mathbf{H}(s) \in \mathbb{R}^{p \times m}(s),$

where the notation $\mathbb{R}^{p \times m}(s)$ denotes the matrices of dimension $p \times m$ of rational functions of s with real coefficients. In this chapter, the notation $\hat{\mathbf{Y}}(s)$, $\hat{\mathbf{H}}(s)$, and $\hat{\mathbf{U}}(s)$ of Chapter 3 has been simplified as above. When $m = p = 1$, then the transfer function matrix can be written $\mathbf{H}(s) = b(s)/a(s)$, where $a(s)$ and $b(s)$ are relatively prime polynomials; that is, they have no common factors. Then the poles of the transfer function are the roots of $a(s)$ and the zeros of the transfer function are the roots of $b(s)$.

Consider a diagonal transfer matrix of the form

(10.2) $\mathbf{M}(s) = \mathrm{diag}\,[\,\epsilon_i(s)/\psi_i(s)\,], \quad i = 1, \cdots r,$

where the pairs of polynomials $\epsilon_i(s)$, $\psi_i(s)$ are relatively prime. Then the poles of $\mathbf{M}(s)$ are defined as the roots of the polynomial $\psi_1(s) \times \psi_2(s) \times \cdots \psi_r(s)$ and the zeros are the roots of $\epsilon_1(s) \times \epsilon_2(s) \times \cdots \epsilon_r(s)$. It will be shown for (10.1) that

231

with nonsingular transformations $\mathbf{U}(s) = \mathbf{Q}(s)\,\tilde{\mathbf{U}}(s)$ and $\tilde{\mathbf{Y}}(s) = \mathbf{P}(s)\,\mathbf{Y}(s)$, the transfer matrix from $\tilde{\mathbf{U}}(s)$ to $\tilde{\mathbf{Y}}(s)$ is of the form (10.2), for which the poles and zeros are known.

Certain symmetries of the pole and zero calculations for a scalar system are exhibited through the symmetric model

$$(10.3) \quad a(s)\,y(s) = b(s)\,u(s),$$

where $\mathbf{H}(s)$ has scalar denominator $a(s)$ and scalar numerator $b(s)$. Then a pole is a value of s at which (10.3) is satisfied with a nonzero output transform $y(s)$ and zero input $u(s)$. Similarly, a zero is a value of s at which (10.3) is satisfied with nonzero input transform $u(s)$ but zero output $y(s)$. When $a(s)$ and $b(s)$ are not assumed to be relatively prime their roots are sometimes called system poles and zeros respectively, whereas the roots after extraction of common factors are called *transmission* poles and zeros to emphasize that they are properties of the transfer matrix.

For vector inputs and outputs, it will be shown that (10.3) can be written as

$$(10.4) \quad \mathbf{D}(s)\,\mathbf{Y}(s) = \mathbf{N}(s)\,\mathbf{U}(s),$$

where the denominator matrix $\mathbf{D}(s)$ and the numerator matrix $\mathbf{N}(s)$ contain polynomial entries. Then the system poles can be defined to be values of s for which $\mathbf{D}(s)$ has less than full column rank, implying that nonzero output transforms $\mathbf{Y}(s)$ exist for zero input transform. Similarly, the zeros can be defined to be values of s for which $\mathbf{N}(s)$ has less than full column rank, implying that some nonzero input transforms $\mathbf{U}(s)$ produce zero output transform. However, care must be taken to distinguish the transmission poles and zeros from those that do not appear in $\mathbf{H}(s)$. The Smith-McMillan form of the next section leads to definitions that precisely account for cancelations.

The final section of this chapter introduces techniques suitable for numerical computation of poles and zeros.

Example 1
Common factor

The system with transfer function $h(s) = (5\,s^2 + 20\,s + 15)/(s^2 + 5\,s + 6)$ has system poles at $s = -2, -3$, the roots of the denominator, and system zeros at $s = -1, -3$, the roots of the numerator. Because of the common factor $s + 3$, the only transmission pole is at $s = -2$, and the only transmission zero is at $s = -1$.

1   The Smith-McMillan form

Using a common denominator polynomial for the entries of $\mathbf{H}(s)$, (10.1) can be rewritten as

(10.5) $\mathbf{Y}(s) = \dfrac{1}{d(s)} \mathbf{G}(s)\, \mathbf{U}(s),$

where $d(s)$ has no common factor with all entries of polynomial matrix $\mathbf{G}(s)$.

The normal form of Section 3.5 of Chapter 5 will be generalized, using polynomial matrices with nonzero constant determinants. Such a matrix is said to be *unimodular.* A unimodular polynomial matrix $\mathbf{P}(s)$ has inverse $\mathbf{P}^{-1}(s) = (1/\det \mathbf{P}(s))\, \mathrm{adj}\, \mathbf{P}(s)$, which is also polynomial and unimodular. Both $\mathbf{P}(s)$ and $\mathbf{P}^{-1}(s)$ have full rank for all finite s.

It will be shown that there exist unimodular $\mathbf{P}(s)$, $\mathbf{Q}(s)$ for which

(10.6) $\mathbf{P}(s)\,\mathbf{G}(s)\,\mathbf{Q}(s) = \mathbf{S}(s)$

$$= \begin{bmatrix} \alpha_1(s) & 0 & \cdots & & & 0 \\ 0 & \alpha_2(s) & & & & \vdots \\ \vdots & & \ddots & & & \\ & & & \alpha_r(s) & & \\ & & & & 0 & \\ & & & & & \ddots \\ 0 & \cdots & & & & 0 \end{bmatrix} = \begin{bmatrix} \mathrm{diag}\,[\,\alpha_i(s)\,] & 0 \\ 0 & 0 \end{bmatrix},$$

where r is called the normal rank of $\mathbf{G}(s)$, and not all of the rows or columns of zeros shown always occur, depending on p, m, and r. Every $\alpha_i(s)$ is a monic polynomial; that is, the coefficient of its highest-degree term is 1. These polynomials also have the property

(10.7) $\alpha_1(s)\mid \alpha_2(s)\mid \alpha_3(s)\mid \cdots \alpha_r(s),$

where $x \mid y$ means that x divides into y without remainder. The matrix $\mathbf{S}(s)$ is unique and is called the Smith form of $\mathbf{G}(s)$.

Premultiplying $\mathbf{H}(s)$ by $\mathbf{P}(s)$, postmultiplying by $\mathbf{Q}(s)$, and performing all possible cancelations of common factors between the $\alpha_i(s)$ and $d(s)$ give

(10.8) $\mathbf{P}(s)\,\mathbf{H}(s)\,\mathbf{Q}(s) = \dfrac{1}{d(s)}\mathbf{P}(s)\,\mathbf{G}(s)\,\mathbf{Q}(s) = \dfrac{1}{d(s)}\mathbf{S}(s)$

$$= \begin{bmatrix} \mathrm{diag}\,[\,\epsilon_i(s)/\psi_i(s)\,] & 0 \\ 0 & 0 \end{bmatrix} = \mathbf{M}(s),$$

where $\mathbf{M}(s)$ defined in the above equation is called the Smith-McMillan form of $\mathbf{H}(s)$.

Example 2
Smith-McMillan
form

Extracting the least common denominator of the transfer matrix shown gives

$$\mathbf{H}(s) = \begin{bmatrix} \dfrac{s}{(s+1)^2(s+2)^2} & \dfrac{s}{(s+2)^2} \\ \dfrac{-s}{(s+2)^2} & \dfrac{-s}{(s+2)^2} \end{bmatrix}$$

$$= \dfrac{1}{(s+1)^2(s+2)^2} \begin{bmatrix} s & s(s+1)^2 \\ -s(s+1)^2 & -s(s+1)^2 \end{bmatrix} = \dfrac{1}{d(s)}\mathbf{G}(s).$$

Then multiplying by the matrices $\mathbf{P}(s)$ and $\mathbf{Q}(s)$ below to produce the Smith form yields

$$\mathbf{P}(s)\,\mathbf{G}(s)\,\mathbf{Q}(s) = \begin{bmatrix} 1 & 0 \\ (s+1)^2 & 1 \end{bmatrix}\begin{bmatrix} s & s(s+1)^2 \\ -s(s+1)^2 & -s(s+1)^2 \end{bmatrix}\begin{bmatrix} 1 & -(s+1)^2 \\ 0 & 1 \end{bmatrix}$$

$$= \begin{bmatrix} s & 0 \\ 0 & s^2(s+1)^2(s+2) \end{bmatrix} = \mathbf{S}(s),$$

so that dividing entries by the common denominator gives the Smith-McMillan form

$$\mathbf{M}(s) = \begin{bmatrix} \dfrac{s}{(s+1)^2(s+2)^2} & 0 \\ 0 & \dfrac{s^2}{(s+2)} \end{bmatrix}.$$

**Left
factorization**

Premultiplying (10.8) by $\mathbf{P}^{-1}(s)$ and postmultiplying by $\mathbf{Q}^{-1}(s)$ show that $\mathbf{H}(s)$ can be factored as

(10.9) $$\mathbf{H}(s) = \mathbf{P}^{-1}(s)\,\mathbf{M}(s)\,\mathbf{Q}^{-1}(s)$$

$$= \mathbf{P}^{-1}(s)\begin{bmatrix} \text{diag}\,[\,\psi_i(s)\,] & 0 \\ 0 & \mathbf{I}_{p-r} \end{bmatrix}^{-1}\begin{bmatrix} \text{diag}\,[\,\epsilon_i(s)\,] & 0 \\ 0 & 0 \end{bmatrix}\mathbf{Q}^{-1}(s)$$

$$= \left(\begin{bmatrix} \text{diag}\,[\,\psi_i(s)\,] & 0 \\ 0 & \mathbf{I}_{p-r} \end{bmatrix}\mathbf{P}(s)\right)^{-1}\left(\begin{bmatrix} \text{diag}\,[\,\epsilon_i(s)\,] & 0 \\ 0 & 0 \end{bmatrix}\mathbf{Q}^{-1}(s)\right)$$

$$= \mathbf{D}^{-1}(s)\,\mathbf{N}(s).$$

The polynomial matrix $\mathbf{N}(s)$ is the numerator matrix, and polynomial $\mathbf{D}(s)$ is the denominator matrix in this left-factored form. Equation (10.4) can be written directly from this equation. Alternatively, on premultiplying by $\mathbf{D}(s)$, (10.4) becomes

(10.10) $$[-\mathbf{D}(s)\ \ \mathbf{N}(s)]\begin{bmatrix} \mathbf{Y}(s) \\ \mathbf{U}(s) \end{bmatrix} = 0,$$

where the entries of $\mathbf{D}(s)$ and $\mathbf{N}(s)$ can be assumed to have denominators equal to 1, and hence these entries are in the field $\mathbb{R}(s)$ of rational functions. Therefore the set $\left\{\begin{bmatrix} \mathbf{Y}(s) \\ \mathbf{U}(s) \end{bmatrix}\right\}$ of input-output vector transforms satisfying (10.10) is a vector space, the kernel of $[\,-\mathbf{D}(s)\ \ \mathbf{N}(s)\,]$. Although $\mathbf{D}(s)$, $\mathbf{N}(s)$ are rational in s, the vectors $\mathbf{Y}(s)$, $\mathbf{U}(s)$ need not be rational functions of s.

Right factorization Right factorizations can also be found, of the form $\mathbf{H}(s) = \tilde{\mathbf{N}}(s)\,\tilde{\mathbf{D}}^{-1}(s)$, with polynomial numerator matrix $\tilde{\mathbf{N}}(s)$ and polynomial denominator matrix $\tilde{\mathbf{D}}(s)$. Rewrite (10.9) as

(10.11)

$$\mathbf{H}(s) = \mathbf{P}^{-1}(s)\,\mathbf{M}(s)\,\mathbf{Q}^{-1}(s)$$

$$= \mathbf{P}^{-1}(s) \begin{bmatrix} \operatorname{diag}[\,\epsilon_i(s)\,] & 0 \\ 0 & 0 \end{bmatrix} \begin{bmatrix} \operatorname{diag}[\,\psi_i(s)\,] & 0 \\ 0 & \mathbf{I}_{m-r} \end{bmatrix}^{-1} \mathbf{Q}^{-1}(s)$$

$$= \left(\mathbf{P}^{-1}(s) \begin{bmatrix} \operatorname{diag}[\,\epsilon_i(s)\,] & 0 \\ 0 & 0 \end{bmatrix} \right) \left(\mathbf{Q}(s) \begin{bmatrix} \operatorname{diag}[\,\psi_i(s)\,] & 0 \\ 0 & \mathbf{I}_{m-r} \end{bmatrix} \right)^{-1}$$

$$= \tilde{\mathbf{N}}(s)\,\tilde{\mathbf{D}}^{-1}(s),$$

and define the vector $\mathbf{V}(s) = \tilde{\mathbf{D}}^{-1}(s)\,\mathbf{U}(s)$, so (10.1) becomes

(10.12)

$$\begin{bmatrix} \mathbf{Y}(s) \\ \mathbf{U}(s) \end{bmatrix} = \begin{bmatrix} \tilde{\mathbf{N}}(s) \\ \tilde{\mathbf{D}}(s) \end{bmatrix} \mathbf{V}(s).$$

The vector space of input-output vector transforms $\left\{ \begin{bmatrix} \mathbf{Y}(s) \\ \mathbf{U}(s) \end{bmatrix} \right\}$ is the range of the rational matrix in (10.12), in addition to being the kernel of the rational matrix in (10.10).

Example 3 Left and right factored forms For the matrix

$$\mathbf{H}(s) = \frac{1}{s+2} \begin{bmatrix} s^2+s & -s^3 \\ 2s & -2s^2 \end{bmatrix} = \frac{1}{d(s)}\mathbf{G}(s),$$

the Smith-McMillan form is obtained as

$$\mathbf{M}(s) = \frac{1}{d(s)}\mathbf{P}(s)\,\mathbf{G}(s)\,\mathbf{Q}(s)$$

$$= \frac{1}{s+2} \begin{bmatrix} 0 & 1/2 \\ 1 & -(s+1)/2 \end{bmatrix} \begin{bmatrix} s^2+s & -s^3 \\ 2s & -2s^2 \end{bmatrix} \begin{bmatrix} 1 & s \\ 0 & 1 \end{bmatrix}$$

$$= \begin{bmatrix} s/(s+2) & 0 \\ 0 & s^2/(s+2) \end{bmatrix},$$

which has zeros at 0, 0, 0 and poles at -2, -2.

The matrix $\mathbf{H}(s)$ can then be written in left-factored form (10.9) as

$$\mathbf{H}(s) = \left(\begin{bmatrix} s+2 & 0 \\ 0 & s+2 \end{bmatrix} \begin{bmatrix} 0 & 1/2 \\ 1 & -(s+1)/2 \end{bmatrix} \right)^{-1} \left(\begin{bmatrix} s & 0 \\ 0 & s^2 \end{bmatrix} \begin{bmatrix} 1 & -s \\ 0 & 1 \end{bmatrix} \right)$$

$$= \begin{bmatrix} 0 & (s+2)/2 \\ s+2 & -(s+1)(s+2)/2 \end{bmatrix}^{-1} \begin{bmatrix} s & -s^2 \\ 0 & s^2 \end{bmatrix},$$

and in right-factored form from (10.11) as

$$
\mathbf{H}(s) = \left(\begin{bmatrix} s+1 & 1 \\ -2 & 0 \end{bmatrix} \begin{bmatrix} s & 0 \\ 0 & s^2 \end{bmatrix} \right) \left(\begin{bmatrix} 1 & s \\ 0 & 1 \end{bmatrix} \begin{bmatrix} s+2 & 0 \\ 0 & s+2 \end{bmatrix} \right)^{-1}
$$

$$
= \begin{bmatrix} s^2+s & s^2 \\ -2s & 0 \end{bmatrix} \begin{bmatrix} s+2 & s^2+2s \\ 0 & s+2 \end{bmatrix}^{-1}.
$$

Zeros With the Smith-McMillan form and the left-factorization (10.9) of $\mathbf{H}(s)$, (10.4) becomes

(10.13) $$
\left(\begin{bmatrix} \text{diag}\,[\,\psi_i(s)\,] & 0 \\ 0 & \mathbf{I}_{p-r} \end{bmatrix} \mathbf{P}(s) \right) \mathbf{Y}(s) = \left(\begin{bmatrix} \text{diag}\,[\,\epsilon_i(s)\,] & 0 \\ 0 & 0 \end{bmatrix} \mathbf{Q}^{-1}(s) \right) \mathbf{U}(s).
$$

To find the finite zeros of the system, investigate the possibility of nonzero vectors $\mathbf{U}(s)$ satisfying (10.13) for zero $\mathbf{Y}(s)$. From the properties of (10.2) and because $\mathbf{Q}^{-1}(s)$ is nonsingular for all finite s, the transmission zeros of $\mathbf{H}(s) \in \mathbb{R}^{p \times m}$ are the roots of the polynomials $\epsilon_i(s)$, together with the infinity of points in $\mathbb{C}$ when the normal rank r is less than the number of columns m. In the latter case the zeros of the $\epsilon_i(s)$ are called the *isolated* transmission zeros of $\mathbf{H}(s)$.

Poles Similarly, the poles of $\mathbf{H}(s)$ are found by investigating the possibility of nonzero $\mathbf{Y}(s)$ for zero $\mathbf{U}(s)$ in (10.13). Because $\mathbf{P}(s)$ is nonsingular for all finite s, such a solution is possible at the points where $\mathbf{D}(s)$ has less than full rank; that is, the poles of $\mathbf{H}(s)$ are the roots of the $\psi_i(s)$.

Summary The finite transmission poles and zeros of a rational matrix $\mathbf{H}(s)$ are the roots of the denominators and numerators respectively of the diagonal elements of the Smith-McMillan form of $\mathbf{H}(s)$. If the normal rank of $\mathbf{G}(s)$ is less than m then the zeros include all points in $\mathbb{C}$.

Example 4
Poles and zeros In Example 2, the poles are at $-1, -1, -2, -2, -2$, and the zeros are at $0, 0, 0$.

Example 5
Identical poles and zeros The following illustrates that identical transmission poles and zeros may not cancel when the transfer matrix is larger than 1×1:

$$
\frac{1}{d(s)} \mathbf{S}(s) = \frac{1}{s(s+1)} \begin{bmatrix} 1 & 0 \\ 0 & (s+1)^2 \end{bmatrix} = \begin{bmatrix} \dfrac{1}{s(s+1)} & 0 \\ 0 & \dfrac{s+1}{s} \end{bmatrix}.
$$

The poles have values $0, -1, 0$, and there is one zero at -1, which does *not* cancel the pole at -1. Therefore, in general, as in this example, the transmission zeros cannot be defined simply as the values of s for which the rank of $\mathbf{H}(s)$ is less than full, since the rank of $\mathbf{H}(s)$ may not be defined for all values of s.

Example 6
Proper $\mathbf{H}(s)$,
improper $\mathbf{M}(s)$

For the transfer matrix $\mathbf{H}(s)$ transformed to Smith-McMillan form $\mathbf{M}(s)$ as shown,

$$\mathbf{H}(s) = \frac{1}{s+1}\begin{bmatrix} 1 & s+1 \\ -s-1 & s+1 \end{bmatrix} \xrightarrow[K_{21}(-s-1)]{H_{21}(s+1)} \frac{1}{s+1}\begin{bmatrix} 1 & 0 \\ 0 & s^2+3s+2 \end{bmatrix}$$

$$= \begin{bmatrix} \frac{1}{s+1} & 0 \\ 0 & \frac{s+2}{1} \end{bmatrix} = \mathbf{M}(s),$$

the result $\mathbf{M}(s)$ is improper, whereas $\mathbf{H}(s)$ was proper. Example 2 contains matrices with the same properties. These examples illustrate the general result that multiplication by unimodular matrices may introduce poles or zeros at infinity.

1.1   Construction of the Smith form

Replacing $\mathbf{G}(s)$ in (10.6) by $\mathbf{A}(s) = [a_{ij}]$ for notational convenience, we can show that unimodular matrices $\mathbf{P}(s)$, $\mathbf{Q}(s)$ exist such that

(10.14) $\mathbf{P}(s)\,\mathbf{A}(s)\,\mathbf{Q}(s) = \mathbf{S}(s)$

is in the form of (10.6).

The following elementary row operations are used, similar to those in Section 3.1 of Chapter 5:

1. Interchange rows i and j, denoting the operation as H_{ij}.
2. Multiply row i by a nonzero constant α, denoting the operation as $H_i(\alpha)$.
3. Add row j multiplied by polynomial $\beta(s)$ to another row i, denoting the operation as $H_{ij}(\beta(s))$.

The analogous column operations K_{ij}, $K_i(\alpha)$, $K_{ij}(\beta(s))$ are also defined.

Each elementary row operation corresponds to premultiplication by an elementary unimodular matrix. Therefore a sequence of such row operations corresponds to premultiplication by a matrix $\mathbf{P}(s)$ that is a product of unimodular matrices, and is therefore unimodular. Similarly, a sequence of such column operations corresponds to postmultiplication by a unimodular matrix $\mathbf{Q}(s)$.

Provided it can be shown that a sequence of row and column operations transforms $\mathbf{A}(s)$ to $\mathbf{S}(s)$, the transformation illustrated below is obtained by performing row operations on the top block row and column operations on the right block column:

(10.15) $\begin{bmatrix} \mathbf{I} & \mathbf{A}(s) \\ 0 & \mathbf{I} \end{bmatrix} \rightarrow \begin{bmatrix} \mathbf{P}(s) & \mathbf{S}(s) \\ 0 & \mathbf{Q}(s) \end{bmatrix}.$

The required sequence of operations on $\mathbf{A}(s)$ is not unique, but one which produces the desired result can be given as follows.

1. Stop if all matrix entries are zero; otherwise, by row and column interchanges, move a nonzero entry of lowest degree to position 1, 1. Multiply row 1 by a nonzero constant to make (the current) entry a_{11} monic.

   If the matrix has more than one row, subtract polynomial multiples of row 1 from lower rows to give all entries below entry 1, 1 in column 1 lower degree than entry 1, 1, as follows. Divide (the current) a_{21} by (the current) a_{11} to find q_{21} and r_{21} such that $a_{21} = q_{21}a_{11} + r_{21}$ where degree $r_{21} <$ degree a_{11}. Subtract q_{21} times row 1 from row 2 to transform entry 2, 1 into r_{21}. Transform lower rows similarly.

   If the matrix contains more than one column, subtract multiples of column 1 from succeeding columns so that all entries in row 1 to the right of entry 1, 1 have lower degree than entry 1, 1, using column operations corresponding to the above row operations.

2. If any element other than entry a_{11} is nonzero in row 1 or column 1, go to step 1.

3. After a finite sequence of operations, the above steps produce a zero first row and first column except for entry a_{11}, because the matrix contains polynomials of finite degree, and because at the start of each repetition, entry a_{11} is brought to lowest degree, after which the other entries of row 1 and column 1 are reduced to lower degree than entry a_{11}.

   Go to step 1, transforming only the submatrix below and to the right of the previous top-left entry, so that references to positions 1, 1 and 2, 1, and to rows or columns 1 or 2 now refer to the submatrix currently being transformed.

4. The above operations must result in a diagonal matrix. If now any a_{ii} does not divide into $a_{i+1,i+1}$ without remainder, add row $i+1$ to row i and go to step 1.

Example 7
Smith form
computation

The Smith form of the following polynomial matrix will be found, using the operations shown:

$$\begin{bmatrix} s+2 & s+1 & s+3 \\ s^3+2s^2+s & s^3+s^2+s & 2s^3+3s^2+s \\ s^2+3s+2 & s^2+2s+1 & 3s^2+6s+3 \end{bmatrix} \begin{matrix} H_{21}(-s^2-1) \\ H_{31}(-s-1) \\ \longrightarrow \end{matrix} \begin{bmatrix} s+2 & s+1 & s+3 \\ -2 & -1 & s^3-3 \\ 0 & 0 & 2s^2+2s \end{bmatrix}$$

$$\begin{matrix} K_{21}(-1) \\ K_{31}(-1) \\ \longrightarrow \end{matrix} \begin{bmatrix} s+2 & -1 & 1 \\ -2 & 1 & s^3-1 \\ 0 & 0 & 2s^2+2s \end{bmatrix} \begin{matrix} K_{12} \\ H_{21}(1) \\ \longrightarrow \end{matrix} \begin{bmatrix} -1 & s+2 & 1 \\ 0 & s & s^3 \\ 0 & 0 & 2s^2+2s \end{bmatrix}$$

$$\begin{matrix} K_{21}(s+2) \\ K_{31}(1) \\ H_1(-1) \\ \longrightarrow \end{matrix} \begin{bmatrix} 1 & 0 & 0 \\ 0 & s & s^3 \\ 0 & 0 & 2s^2+2s \end{bmatrix} \begin{matrix} K_{32}(-s^2) \\ H_3(1/2) \\ \longrightarrow \end{matrix} \begin{bmatrix} 1 & 0 & 0 \\ 0 & s & 0 \\ 0 & 0 & s^2+s \end{bmatrix}.$$

In this example the unimodular operation matrices are

$$\mathbf{P}(s) = \mathbf{H}_3(1/2)\,\mathbf{H}_1(-1)\,\mathbf{H}_{21}(1)\,\mathbf{H}_{31}(-s-1)\,\mathbf{H}_{21}(-s^2-1),$$

and

$$\mathbf{Q}(s) = \mathbf{K}_{21}(-1)\,\mathbf{K}_{31}(-1)\,\mathbf{K}_{12}\mathbf{K}_{21}(s+2)\,\mathbf{K}_{31}(1)\,\mathbf{K}_{32}(-s^2).$$

2 | Computation of poles and zeros

The Smith-form calculation in the previous development is suitable for definition, but is often unsuitable for floating-point calculation, which is inexact. This unsuitability arises from the sensitivity of the roots of a polynomial to perturbations of its coefficients, and from the difficulty of calculating the term of highest degree of the required remainders. An alternative which is often used for transfer matrices with state-space realizations is to write these equations as

$$(10.16) \qquad \begin{bmatrix} \mathbf{A} - s\mathbf{I} & 0 & \mathbf{B} \\ \mathbf{C} & -\mathbf{I} & \mathbf{D} \end{bmatrix} \begin{bmatrix} \mathbf{X}(s) \\ \mathbf{Y}(s) \\ \mathbf{U}(s) \end{bmatrix} = 0.$$

Let the *system* poles be values of s for which (10.16) is satisfied with zero input $\mathbf{U}(s)$ but nonzero $\mathbf{X}(s)$, $\mathbf{Y}(s)$, or both, that is, values of s for which the equation

$$(10.17) \qquad \begin{bmatrix} \mathbf{A} - s\mathbf{I} & 0 \\ \mathbf{C} & -\mathbf{I} \end{bmatrix} \begin{bmatrix} \mathbf{X}(s) \\ \mathbf{Y}(s) \end{bmatrix} = 0$$

may have nonzero solution vector $\begin{bmatrix} \mathbf{X}(s) \\ \mathbf{Y}(s) \end{bmatrix}$. By inspection, these are the values of s for which $\mathbf{A} - s\mathbf{I}$ has less than full rank, that is, the eigenvalues of $\mathbf{A}$ as expected, and if the system is a minimal representation of $\mathbf{H}(s)$, the system poles are all transmission poles.

Similarly, the system zeros are values of s for which

$$(10.18) \qquad \begin{bmatrix} \mathbf{A} - s\mathbf{I} & \mathbf{B} \\ \mathbf{C} & \mathbf{D} \end{bmatrix} \begin{bmatrix} \mathbf{X}(s) \\ \mathbf{U}(s) \end{bmatrix} = \left(\begin{bmatrix} \mathbf{A} & \mathbf{B} \\ \mathbf{C} & \mathbf{D} \end{bmatrix} - s \begin{bmatrix} \mathbf{I} & 0 \\ 0 & 0 \end{bmatrix} \right) \begin{bmatrix} \mathbf{X}(s) \\ \mathbf{U}(s) \end{bmatrix} = 0$$

may have nonzero solution vector $\begin{bmatrix} \mathbf{X}(s) \\ \mathbf{U}(s) \end{bmatrix}$, and it can be shown that if the system is minimal, these values will be transmission zeros. The computation of the required values of s is sometimes called a *generalized* eigenvalue problem in the literature of numerical analysis. Reliable algorithms exist when the matrix in parentheses in (10.18) is square and has a determinant which is not identically zero.

Example 8
Poles and zeros
of state-space
model

For the system $\mathbf{A} = -1$, $\mathbf{B} = 1$, $\mathbf{C} = -2$, $\mathbf{D} = 1$, the coefficient matrices of (10.17) and (10.18) are respectively

$$\begin{bmatrix} \mathbf{A} - s\mathbf{I} & 0 \\ \mathbf{C} & -\mathbf{I} \end{bmatrix} = \begin{bmatrix} -1-s & 0 \\ -2 & -1 \end{bmatrix}, \quad \begin{bmatrix} \mathbf{A} - s\mathbf{I} & \mathbf{B} \\ \mathbf{C} & \mathbf{D} \end{bmatrix} = \begin{bmatrix} -1-s & 1 \\ -2 & 1 \end{bmatrix}.$$

The determinant of the first matrix is $(s+1)$, so the pole is at -1, and the determinant of the second matrix is $1-s$, so the zero is at 1. These results can be confirmed by comparison with the system transfer function, which is $(s-1)/(s+1)$.

Example 9
Scalar minimal
system

The relationship between system poles and transmission poles, and between system zeros and transmission zeros, is easy to illustrate for single-input, single-output systems. For example, let a scalar transfer matrix be given by

$$\mathbf{H}(s) = b(s)/a(s) = (b_0 s^3 + b_1 s^2 + b_2 s + b_3)/(s^3 + a_1 s^2 + a_2 s + a_3),$$

with $b_0 \neq 0$. Assume that there are no cancelations, so that the equations of a minimal state-space realization of order $n = 3$ can be written in the observable form of Chapter 4, Section 7, as

$$\begin{bmatrix} -s & 0 & -a_3 & 0 & b_3 - a_3 b_0 \\ 1 & -s & -a_2 & 0 & b_2 - a_2 b_0 \\ 0 & 1 & -a_1 - s & 0 & b_1 - a_1 b_0 \\ 0 & 0 & 1 & -1 & b_0 \end{bmatrix} \begin{bmatrix} \mathbf{X}(s) \\ \mathbf{Y}(s) \\ \mathbf{U}(s) \end{bmatrix} = 0.$$

To compute the system poles, find the values of s for which nonzero solutions exist for $\mathbf{U}(s) = 0$. These are the roots of $a(s)$ since the realization utilizes a companion form for its $\mathbf{A}$ matrix. To find the zeros, set $\mathbf{Y}(s) = 0$, and perform the operations $H_{34}(a_1 - b_1/b_0)$, $H_{24}(a_2 - b_2/b_0)$, $H_{14}(a_3 - b_3/b_0)$, resulting in the equations

$$\begin{bmatrix} -s & 0 & -b_3/b_0 & 0 \\ 1 & -s & -b_2/b_0 & 0 \\ 0 & 1 & -b_1/b_0 - s & 0 \\ 0 & 0 & 1 & b_0 \end{bmatrix} \begin{bmatrix} \mathbf{X}(s) \\ \mathbf{U}(s) \end{bmatrix} = 0.$$

Because $b_0 \neq 0$, these equations have nonzero solutions for values of s at which the top-left 3×3 block loses rank, that is, since the matrix is in a companion form, at the roots of the polynomial $s^3 + (b_1/b_0)s^2 + (b_2/b_0)s + (b_3/b_0)$, which are the roots of $b(s)$. Since the system is minimal, the system poles are all transmission poles, and the system zeros are all transmission zeros.

Example 10
Generalized
eigenvalues

A realization of the transfer matrix of Example 5 is given by

$$
\mathbf{A} = \begin{bmatrix} 0 & 1 & 0 \\ 0 & -1 & 0 \\ 0 & 0 & 0 \end{bmatrix}, \quad
\mathbf{B} = \begin{bmatrix} 0 & 0 \\ 1 & 0 \\ 0 & 1 \end{bmatrix}, \quad
\mathbf{C} = \begin{bmatrix} 1 & 0 & 0 \\ 0 & 0 & 1 \end{bmatrix}, \quad
\mathbf{D} = \begin{bmatrix} 0 & 0 \\ 0 & 1 \end{bmatrix},
$$

for which the eigenvalues of $\mathbf{A}$, or poles, are at $0, 0, -1$, and for which, from (10.18), the zeros are the generalized eigenvalues of

$$
\left[\begin{array}{ccc|cc}
-s & 1 & 0 & 0 & 0 \\
0 & -1-s & 0 & 1 & 0 \\
0 & 0 & -s & 0 & 1 \\
\hline
1 & 0 & 0 & 0 & 0 \\
0 & 0 & 1 & 0 & 1
\end{array}\right].
$$

The above matrix has determinant $-(s+1)$, showing that the zero is at -1, and as before, the zero does not cancel the pole at the same location.

Example 11
Nonsquare
system

For the minimal system with matrices

$$
\mathbf{A} = \begin{bmatrix} 0 & -6 \\ 1 & -5 \end{bmatrix}, \quad
\mathbf{B} = \begin{bmatrix} 1 & -6 \\ 1 & -4 \end{bmatrix}, \quad
\mathbf{C} = \begin{bmatrix} 0 & 1 \end{bmatrix}, \quad
\mathbf{D} = \begin{bmatrix} 0 & 1 \end{bmatrix},
$$

the transmission zeros are the values of s for which the nonsquare matrix

$$
\begin{bmatrix}
-s & -6 & 1 & -6 \\
1 & -s-5 & 1 & -4 \\
0 & 1 & 0 & 1
\end{bmatrix}
$$

loses rank. The Smith form of this matrix can be calculated as

$$
\begin{bmatrix}
-s & -6 & 1 & -6 \\
1 & -s-5 & 1 & -4 \\
0 & 1 & 0 & 1
\end{bmatrix}
\begin{array}{c} H_{12} \\ H_{21}(s) \\ \longrightarrow \end{array}
\begin{bmatrix}
1 & -s-5 & 1 & -4 \\
0 & -s^2-5s-6 & s+1 & -4s-6 \\
0 & 1 & 0 & 1
\end{bmatrix}
$$

$$
\begin{array}{c} K_{21}(s+5) \\ K_{31}(-1) \\ K_{41}(4) \\ \longrightarrow \end{array}
\begin{bmatrix}
1 & 0 & 0 & 0 \\
0 & -s^2-5s-6 & s+1 & -4s-6 \\
0 & 1 & 0 & 1
\end{bmatrix}
$$

$$
\begin{array}{c} H_{23} \\ H_{32}(s^2+5s+6) \\ \longrightarrow \end{array}
\begin{bmatrix}
1 & 0 & 0 & 0 \\
0 & 1 & 0 & 1 \\
0 & 0 & s+1 & s^2+s
\end{bmatrix}
\begin{array}{c} K_{42}(-1) \\ K_{43}(-s) \\ \longrightarrow \end{array}
\begin{bmatrix}
1 & 0 & 0 & 0 \\
0 & 1 & 0 & 0 \\
0 & 0 & s+1 & 0
\end{bmatrix}.
$$

The normal rank of this matrix is less than the number of columns; therefore the zeros include all values of the complex plane, but there is also an isolated zero at -1.

3 | **Further study**

Further analysis of the Smith-McMillan form is in Kailath [25], and, for example, Antsaklis and Michel [2]. Under high-gain feedback the closed-loop system eigenvalues typically approach either infinity or the open-loop zeros, so the location of the zeros becomes important for feedback design; see Maciejowski [33] or Patel and Munro [41].

4 | **Problems**

1   Calculate the Smith form for the following matrices:

(a) $\mathbf{G}(s) = \begin{bmatrix} s+1 & 0 \\ s+2 & s+2 \end{bmatrix}$,   (b) $\mathbf{G}(s) = \begin{bmatrix} s+1 & 0 \\ 0 & s+2 \end{bmatrix}$,

(c) $\mathbf{G}(s) = \begin{bmatrix} s^3+3s^2+5s+3 & s^3+3s^2+6s+4 \\ s^2+2s+1 & s^2+s \end{bmatrix}$.

2   Recall that polynomials over a field, with ordinary arithmetic, obey the same axioms as the integers. Then, along with interchanges and multiplication by ± 1, the basic row operation in the Smith-form computation for integer matrices is to add $-q$ times row 1 to row i, where $a_{i1} = a_{11}q + r$, with $r < a_{11}$, thus changing entry a_{i1} to r. The basic column operations are analogous. Calculate the Smith form for the following integer matrices:

(a) $\mathbf{G} = \begin{bmatrix} 10 & 3 & -7 \\ 34 & 10 & -24 \end{bmatrix}$,   (b) $\mathbf{G} = \begin{bmatrix} 10 & -10 & -6 \\ 10 & -11 & -5 \\ 12 & -10 & -8 \end{bmatrix}$.

3   For the systems with matrices given below, find the transmission poles and transmission zeros of the system, (i) by finding the Smith-McMillan form of the transfer matrix $\mathbf{H}(s)$, and (ii) by the method of equations (10.17) and (10.18).

(a) $\mathbf{A} = 3$, $\mathbf{B} = 1$, $\mathbf{C} = 8$, $\mathbf{D} = 2$,

(b) $\mathbf{A} = \begin{bmatrix} 0 & -2 & 0 \\ 1 & -3 & 0 \\ 0 & 0 & -2 \end{bmatrix}$, $\mathbf{B} = \begin{bmatrix} 0 & 0 \\ 1 & 0 \\ 0 & -2 \end{bmatrix}$,

$\mathbf{C} = \begin{bmatrix} 0 & 1 & 0 \\ 0 & 0 & 1 \end{bmatrix}$, $\mathbf{D} = \begin{bmatrix} 0 & 0 \\ 0 & 0 \end{bmatrix}$,

$$(c)\ \mathbf{A} = \begin{bmatrix} 0 & -4 & 0 \\ 1 & -5 & 0 \\ 0 & 0 & -4 \end{bmatrix}, \quad \mathbf{B} = \begin{bmatrix} 0 & 0 \\ 1 & 1 \\ -1 & 2 \end{bmatrix},$$

$$\mathbf{C} = \begin{bmatrix} 0 & 1 & 0 \\ 0 & 0 & 1 \end{bmatrix}, \quad \mathbf{D} = \begin{bmatrix} 0 & 0 \\ 0 & 0 \end{bmatrix}.$$

4   As an extreme example of a polynomial with roots which are sensitive to coef-ficient inaccuracies, consider the polynomial $s^{16} - \epsilon$. Calculate the ratio of the change in root magnitude to perturbation magnitude for a perturbation of 10^{-16}, as might be caused by storage or round-off error, from the value of $\epsilon = 0$.

5   For the system described by the matrices

$$\mathbf{A} = \begin{bmatrix} 0 & -4 & 0 \\ 1 & -5 & 0 \\ 0 & 0 & -4 \end{bmatrix}, \quad \mathbf{B} = \begin{bmatrix} 0 & 0 \\ 1 & 1 \\ -1 & 2 \end{bmatrix},$$

$$\mathbf{C} = \begin{bmatrix} 0 & 1 & 0 \\ 0 & 0 & 1 \end{bmatrix}, \quad \mathbf{D} = \begin{bmatrix} 0 & 0 \\ 0 & 0 \end{bmatrix}:$$

(a) Find the transfer matrix in the form $\hat{\mathbf{H}}(s) = \frac{1}{d(s)} \mathbf{G}(s)$, where $d(s)$ is a polynomial and $\mathbf{G}(s)$ is a polynomial matrix.

(b) Find the Smith form of $\mathbf{G}(s)$ and the Smith-McMillan form of $\hat{\mathbf{H}}(s)$.

(c) Find the transmission poles and transmission zeros of the system from the Smith-McMillan form.

(d) Find the transmission poles and transmission zeros of the system by using the methods of Section 2.

6   Solve Problem 5 for the system described by the matrices

$$\mathbf{A} = \begin{bmatrix} 0 & 1 & 0 \\ -3 & -4 & 0 \\ 0 & 0 & -3 \end{bmatrix}, \quad \mathbf{B} = \begin{bmatrix} 0 \\ 1 \\ 1 \end{bmatrix}, \quad \mathbf{C} = \begin{bmatrix} 0 & 1 & 0 \\ 0 & 0 & -3 \end{bmatrix}, \quad \mathbf{D} = \begin{bmatrix} 0 \\ 0 \end{bmatrix}.$$

7   Using the Smith-McMillan form, find polynomial matrices $\mathbf{D}(s)$, $\mathbf{N}(s)$ of a left factorization $\mathbf{H}(s) = \mathbf{D}^{-1}(s) \mathbf{N}(s)$ of the transfer matrix of the following sys-tems:

(a) the systems of Problem 3,

(b) the system of Problem 6.

8   Find polynomial matrices $\tilde{\mathbf{D}}$, $\tilde{\mathbf{N}}(s)$ of a right-factorization $\mathbf{H}(s) = \tilde{\mathbf{N}}(s)\,\tilde{\mathbf{D}}^{-1}(s)$ of the transfer matrix of the following systems:

(a) the system of Problem 5,

(b) the system of Problem 6.

References

[1] E. Anderson, Z. Bai, C. Bischof, J. Demmel, J. Dongarra, J. Du Croz, A. Greenbaum, S. Hammarling, A. McKenney, S. Ostrouchov, and D. Sorensen. *LAPACK Users' Guide*. SIAM, Philadelphia, 1992.

[2] P. J. Antsaklis and A. N. Michel. *Linear Systems*. McGraw-Hill, New York, 1997.

[3] K. J. Åström and B. Wittenmark. *Computer-Controlled Systems, Theory and Design*. Prentice Hall, Upper Saddle River, N.J., 1997.

[4] F. Ayres, Jr. *Schaum's Outline of Theory and Problems of Matrices*. McGraw-Hill, New York, 1962.

[5] G. Birkhoff and S. MacLane. *A Survey of Modern Algebra*. Macmillan, New York, 1953.

[6] W. L. Brogan. *Modern Control Theory*. Prentice Hall, Englewood Cliffs, N.J., 1991.

[7] R. T. Byerly and W. Kimbark, editors. *Stability of Large Electric Power Systems*. IEEE Press, New York, 1974.

[8] F. Cellier. *Continuous system modeling*. Springer-Verlag, New York, 1991.

[9] C.-T. Chen. *Linear System Theory and Design*. Holt, Rinehart and Winston, New York, 1984.

[10] L. O. Chua. *Linear and Nonlinear Circuits*. McGraw-Hill, New York, 1987.

[11] E. A. Coddington and N. Levinson. *Theory of Ordinary Differential Equations*. McGraw-Hill, New York, 1955.

[12] R. A. DeCarlo. *Linear Systems, a State Variable Approach with Numerical Application*. Prentice Hall, Englewood Cliffs, N.J., 1989.

[13] N. Deo. *Graph Theory with Applications to Engineering and Computer Science*. Prentice Hall, Englewood Cliffs, N.J., 1974.

[14] A. Dholakia. *Introduction to Convolutional Codes with Applications.* Kluwer, Boston, 1994.

[15] C. N. Dorny. *Understanding Dynamic Systems: Approaches to Modeling, Analysis, and Design.* Prentice Hall, Englewood Cliffs, N.J., 1993.

[16] B. Friedland. *Control System Design, an Introduction to State-Space Methods.* McGraw-Hill, New York, 1986.

[17] F. R. Gantmacher. *The Theory of Matrices,* Volume 2. Chelsea, New York, 1960.

[18] H. Goldstein. *Classical Mechanics.* Addison-Wesley, Reading, Mass., 1980.

[19] G. H. Golub and C. F. Van Loan, editors. *Matrix Computations,* Third Edition. The Johns Hopkins University Press, Baltimore, 1996.

[20] P. B. Guest. *Laplace Transforms and an Introduction to Distributions.* Ellis Horwood, New York, 1991.

[21] B. L. Ho and R. E. Kalman. Effective construction of linear state-variable models from input/output functions. *Regelungstechnik*, 12(14):545–548, 1966.

[22] C. H. Houpis and G. B. Lamont. *Digital Control Systems.* McGraw-Hill, New York, 1992.

[23] Integrated Systems, Inc. *MATRIX$_X$ User's Guide.* Integrated Systems, Inc., Santa Clara, Cal., 1984.

[24] J.-N. Juang. *Applied System Identification.* Prentice Hall, Englewood Cliffs, N.J., 1994.

[25] T. Kailath. *Linear Systems.* Prentice Hall, Englewood Cliffs, N.J., 1980.

[26] R. E. Kalman. Mathematical description of linear dynamical systems. *SIAM J. Control, Ser. A*, 1(2):152–192, 1963.

[27] R. King, M. Ahmadi, R. Gorgui-Naguib, A. Kwabwe, and M. Azimi-Sadjadi. *Digital Filtering in One and Two Dimensions.* Plenum Press, New York, 1989.

[28] H. E. Koenig, Y. Tokad, H. K. Kesavan, and H. G. Hedges. *Analysis of Discrete Physical Systems.* McGraw-Hill, New York, 1967.

[29] E. Kreyszig. *Advanced Engineering Mathematics.* Wiley, New York, 1988.

[30] B. P. Lathi. *Linear Systems and Signals.* Berkeley-Cambridge Press, Carmichael, Cal., 1992.

[31] C. L. Lawson and R. J. Hanson. *Solving Least Squares Problems.* Prentice Hall, Englewood Cliffs, N.J., 1974.

[32] D. G. Luenberger. *Optimization by Vector Space Methods*. Wiley, New York, 1969.

[33] J. M. Maciejowski. *Multivariable Feedback Design*. Addison-Wesley, New York, 1989.

[34] S. MacLane and G. Birkhoff. *Algebra*. Macmillan, New York, 1967.

[35] The MathWorks, Inc. *MATLAB User's Guide*. The MathWorks Inc., South Natick, Mass., 1990.

[36] R. M. M. Mattheij and J. Molenaar. *Ordinary Differential Equations in Theory and Practice*. Wiley, New York, 1996.

[37] N. Minorsky. *Nonlinear Oscillations*. D. Van Nostrand Company, Inc., Princeton, N.J., 1962.

[38] N. Minorsky. *Theory of Nonlinear Control Systems*. McGraw-Hill, New York, 1969.

[39] C. Moler and C. Van Loan. Nineteen dubious ways to compute the exponential of a matrix. *SIAM Review*, 20(4):801–836, 1978.

[40] A. V. Oppenheim and A. S. Willsky. *Signals and Systems*. Prentice Hall, Englewood Cliffs, N.J., 1983.

[41] R. V. Patel and N. Munro. *Multivariable System Theory and Design*. Pergamon Press, Oxford, 1982.

[42] R. P. Paul. *Robot Manipulators: Mathematics, Programming, and Control*. MIT Press, Cambridge, Mass., 1981.

[43] W. H. Press, B. P. Flannery, S. A. Teukolsky, and W. T. Vetterling. *Numerical Recipes, The Art of Scientific Computing*. Cambridge University Press, Cambridge, 1989.

[44] C. E. Roberts. *Ordinary Differential Equations*. Prentice Hall, Englewood Cliffs, N.J., 1979.

[45] D. Rowell and D. N. Wormley. *System Dynamics: An Introduction*. Prentice Hall, Englewood Cliffs, N.J., 1997.

[46] W. J. Rugh. *Linear system theory*. Prentice Hall, Upper Saddle River, N.J., 1996.

[47] G. M. Sandquist. *Introduction to System Science*. Prentice Hall, Englewood Cliffs, N.J., 1985.

[48] M. W. Spong and M. Vidyasagar. *Robot Dynamics and Control*. Wiley, New York, 1989.

[49] T. E. Stern. *Theory of Nonlinear Networks and Systems*. Addison-Wesley, Reading, Mass., 1965.

[50] G. W. Stewart. *Introduction to Matrix Computations.* Academic Press, New York, 1973.

[51] G. Strang. *Linear Algebra and Its Applications.* Academic Press, New York, 1988.

[52] F. Szidarovszky and A. T. Bahill. *Linear Systems Theory.* CRC Press, Boca Raton, Fla., 1991.

[53] J. Thoma. *Simulation by Bondgraphs: Introduction to a Graphical Method.* Springer-Verlag, Berlin, 1990.

[54] M. Vidyasagar. *Nonlinear systems analysis.* Prentice Hall, Englewood Cliffs, N.J., 1993.

[55] J. Vlach and K. Singhal. *Computer Methods for Circuit Analysis and Design.* Van Nostrand Reinhold, New York, 1983.

[56] J. C. Willems. Models for dynamics. In U. Kirchgraber, editor, *Dynamics Reported,* Volume 2, pp. 171–269. Wiley, New York, 1988.

[57] J. C. Willems. Paradigms and puzzles in the theory of dynamical systems. *IEEE Transactions on Automatic Control*, AC-36(3):259–294, 1991.

[58] L. A. Zadeh and C. A. Desoer. *Linear System Theory, the State Space Approach.* McGraw-Hill, New York, 1963.

Appendix

The following sections contain solutions of odd-numbered problems.

<table>
<tr><td>Chapter 1</td><td>Solutions</td></tr>
</table>

1 See Table S1.1.

3 Since the system is 2nd order in v, let $x_1 = v$, $x_2 = \dot{v}$, giving the state-space equations,

$$\begin{bmatrix} \dot{x}_1 \\ \dot{x}_2 \end{bmatrix} = \begin{bmatrix} x_2 \\ -(\sin x_1)\, x_2 - u_2\, x_1 + u_1 + u_2 \end{bmatrix} = \begin{bmatrix} f_1 \\ f_2 \end{bmatrix},$$
$$y = (\cos x_1)\, u_2 = G.$$

(a) Since $\frac{\partial f_1}{\partial t} = 0$, $\frac{\partial f_2}{\partial t} = 0$, $\frac{\partial G}{\partial t} = 0$, the system is time-invariant. The equations are nonlinear because of the terms $(\sin x_1)\, x_2$ and $u_2 x_1$.

(b) The operating point is defined by $u_1^o = 0$, $u_2^o = 1$, and is *constant;* that is, $\dot{x}_1 = 0$, $\dot{x}_2 = 0$. Substituting these values in the state-space equations, we get

$$0 = x_2, \quad \Rightarrow x_2^o = 0,$$
$$0 = -(\sin x_1)\, 0 - 1\, x_1 + 0 + u_2 = -x_1 + 1 \quad \Rightarrow x_1^o = 1,$$
$$y^o = (\cos x_1^o)\, u_2 = \cos(1).$$

Then

$$\mathbf{A} = \begin{bmatrix} \frac{\partial f_1}{\partial x_1} & \frac{\partial f_1}{\partial x_2} \\ \frac{\partial f_2}{\partial x_1} & \frac{\partial f_2}{\partial x_2} \end{bmatrix}_o = \begin{bmatrix} 0 & 1 \\ -(\cos x_1)\, x_2 - u_2 & -\sin x_1 \end{bmatrix}_o$$
$$= \begin{bmatrix} 0 & 1 \\ -1 & -\sin(1) \end{bmatrix},$$

$$\mathbf{B} = \begin{bmatrix} \frac{\partial f_1}{\partial u_1} & \frac{\partial f_1}{\partial u_2} \\ \frac{\partial f_2}{\partial u_1} & \frac{\partial f_2}{\partial u_2} \end{bmatrix}_o = \begin{bmatrix} 0 & 0 \\ 1 & 1 - x_1 \end{bmatrix}_o = \begin{bmatrix} 0 & 0 \\ 1 & 0 \end{bmatrix},$$

$$\mathbf{C} = [\, \frac{\partial G}{\partial x_1} \quad \frac{\partial G}{\partial x_2} \,]_o = [\, (-\sin x_1)\, u_2 \quad 0 \,]_o = [\, -\sin(1), \, 0 \,],$$

249

Table S1.1   Systems, variables, equations, matrices.

Device	Input	Output	State	Equations	A, B, C, D
	i	v	q	$\dot{q} = i$ $v = q/C$	0, 1, 1/C, 0
	i	v	v	$\dot{v} = i/C$ $v = v$	0, 1/C, 1, 0
	$\begin{bmatrix} i_1 \\ i_2 \end{bmatrix}$	v_1	$\begin{bmatrix} v_1 \\ v_2 \end{bmatrix}$	$\begin{bmatrix} \dot{v}_1 \\ \dot{v}_2 \end{bmatrix} = \begin{bmatrix} i_1/C_1 \\ i_2/C_2 \end{bmatrix}$ $v_1 = v_1$	$\begin{bmatrix} 0 & 0 \\ 0 & 0 \end{bmatrix}$, $\begin{bmatrix} 1/C_1 & 0 \\ 0 & 1/C_2 \end{bmatrix}$, $[1, 0]$, $[0, 0]$
	i	v	v	$\dot{v} = -\dfrac{v}{RC} + \dfrac{i}{C}$ $v = v$	$-1/(RC)$, $1/C$, 1, 0
	u	y	x	$x(t+1) = u(t)$ $y(t) = x(t)$	0, 1, 1, 0
	u	y	x	$x(t+1) = x(t) + u(t)$ $y(t) = x(t) + u(t)$	1, 1, 1, 1

$$\mathbf{D} = [\, \tfrac{\partial G}{\partial u_1} \quad \tfrac{\partial G}{\partial u_2} \,]_o = [\,0,\ \cos x_1\,]_o = [\,0,\ \cos(1)\,].$$

5   Writing the state equations for the composite system, using the given state and the equations for the block components, assuming continuous time, we get

$$\dot{\mathbf{X}} = \begin{bmatrix} \dot{\mathbf{X}}_1 \\ \dot{\mathbf{X}}_2 \end{bmatrix} = \begin{bmatrix} \mathbf{A}_1\mathbf{X}_1 + \mathbf{B}_1\mathbf{Y}_2 \\ \mathbf{A}_2\mathbf{X}_2 + \mathbf{B}_2\mathbf{U}_2 \end{bmatrix} = \begin{bmatrix} \mathbf{A}_1 & \mathbf{B}_1\mathbf{C}_2 \\ 0 & \mathbf{A}_2 \end{bmatrix}\begin{bmatrix} \mathbf{X}_1 \\ \mathbf{X}_2 \end{bmatrix} + \begin{bmatrix} \mathbf{B}_1\mathbf{D}_2 \\ \mathbf{B}_2 \end{bmatrix}\mathbf{U}_2,$$

$$\mathbf{Y} = \mathbf{C}_1\mathbf{X}_1 + \mathbf{D}_1\mathbf{Y}_2 = [\,\mathbf{C}_1 \quad \mathbf{D}_1\mathbf{C}_2\,]\begin{bmatrix} \mathbf{X}_1 \\ \mathbf{X}_2 \end{bmatrix} + \mathbf{D}_1\mathbf{D}_2\mathbf{U}_2,$$

so that

$$\mathbf{A} = \begin{bmatrix} \mathbf{A}_1 & \mathbf{B}_1\mathbf{C}_2 \\ 0 & \mathbf{A}_2 \end{bmatrix}, \quad \mathbf{B} = \begin{bmatrix} \mathbf{B}_1\mathbf{D}_2 \\ \mathbf{B}_2 \end{bmatrix}, \quad \mathbf{C} = [\, \mathbf{C}_1 \quad \mathbf{D}_1\mathbf{C}_2 \,], \quad \mathbf{D} = \mathbf{D}_1\mathbf{D}_2,$$

which would also be the case had discrete-time systems been assumed.

7 The equations can be written in state-space form as

$$\frac{d}{dt} \begin{bmatrix} y \\ y' \\ m \\ v \end{bmatrix} = \begin{bmatrix} 0 & 1 & 0 & 0 \\ 0 & 0 & 1/(EI) & 0 \\ 0 & 0 & 0 & 1 \\ 0 & 0 & 0 & 0 \end{bmatrix} \begin{bmatrix} y \\ y' \\ m \\ v \end{bmatrix} + \begin{bmatrix} 0 \\ 0 \\ 0 \\ 1 \end{bmatrix} u,$$

$$y = [\, 1,\, 0,\, 0,\, 0 \,] \begin{bmatrix} y \\ y' \\ m \\ v \end{bmatrix} + 0\, u.$$

(a) The matrices $\mathbf{A}$, $\mathbf{B}$, $\mathbf{C}$, $\mathbf{D}$ are shown in the equations.

(b) The given boundary conditions are not all *initial values,* and thus do not satisfy the requirement for the state. Furthermore, the values at $t = 0$ are not independent of the inputs to the right, that is, for $t > 0$, which implies that the system is not causal with respect to t.

Chapter 2   **Solutions**

1 In the first two cases, $\mathbf{A}$ is a scalar, and in the third, $\mathbf{A}$ is diagonal, and the required powers are the powers of the scalar diagonal entries:

(a) $\mathbf{A} = 1/2 \Rightarrow \mathbf{A}^k = 2^{-k}$.

(b) $\mathbf{A} = 2 \Rightarrow \mathbf{A}^k = 2^k$.

(c) $\mathbf{A} = \begin{bmatrix} 2^{-1} & 0 \\ 0 & 3^{-1} \end{bmatrix} \Rightarrow \mathbf{A}^k = \begin{bmatrix} 2^{-k} & 0 \\ 0 & 3^{-k} \end{bmatrix}$.

3 Use formula 2.5:

(a) $\mathbf{X}(7) = \mathbf{A}^{7-4}\mathbf{X}(4) = \begin{bmatrix} 1 & 2 + 2\alpha + 2\alpha^2 \\ 0 & \alpha^3 \end{bmatrix} \begin{bmatrix} 1 \\ 0 \end{bmatrix} = \begin{bmatrix} 1 \\ 0 \end{bmatrix}$.

(b) $\begin{bmatrix} 1 \\ 0 \end{bmatrix} = \mathbf{X}(4) = \mathbf{A}^{4-2}\mathbf{X}(2) = \begin{bmatrix} 1 & 2 + 2\alpha \\ 0 & \alpha^2 \end{bmatrix} \mathbf{X}(2)$, or

$\mathbf{X}(2) = \begin{bmatrix} 1 & 2 + 2\alpha \\ 0 & \alpha^2 \end{bmatrix}^{-1} \begin{bmatrix} 1 \\ 0 \end{bmatrix} = \begin{bmatrix} 1 \\ 0 \end{bmatrix}$.

(c) If $\alpha = 0$, then $\mathbf{X}(7) = \mathbf{A}^3\mathbf{X}(4)$ as before, but $\mathbf{X}(4) = \mathbf{A}^2\mathbf{X}(2)$ has singular coefficient matrix $\mathbf{A}^2$ with unknown $\mathbf{X}(2)$. In this case solutions exist, but are not unique, with $\mathbf{X}(2) = \begin{bmatrix} 1 - 2\beta \\ \beta \end{bmatrix}$, for arbitrary β.

---

5 The power $\mathbf{A}^k$ requires $(k-1)n^3$ multiplications when calculated directly, so the powers of $\mathbf{A}$ separately calculated in the formula

$$\mathbf{X}(\ell) = \mathbf{A}^\ell\mathbf{X}(0) + [\,\mathbf{A}^{\ell-1}\mathbf{B}, \ \mathbf{A}^{\ell-2}\mathbf{B}, \cdots \mathbf{AB}, \ \mathbf{B}\,] \begin{bmatrix} \mathbf{U}(0) \\ \mathbf{U}(1) \\ \vdots \\ \mathbf{U}(\ell-2) \\ \mathbf{U}(\ell-1) \end{bmatrix}$$

require $((\ell-1) + (\ell-2) + \cdots)n^3 = (\ell-1)(\ell-2)n^3/2$ multiplications, and postmultiplying requires $n^2 + (\ell-1)n^2m + \ell n^2$ multiplications. The total becomes $(\ell-1)(\ell-2)n^3/2 + n^2 + (\ell-1)n^2m + \ell n^2$, which is dominated by the first term for large n. Each recursion of $\mathbf{X}(k+1) = \mathbf{AX}(k) + \mathbf{BU}(k)$ requires $n^2 + nm$ multiplications, totaling $\ell(n^2 + nm)$, of order n^2 rather than n^3 for the naïve calculation. For the dimensions given, these totals are 4.86×10^{12} and 1.05×10^8 respectively.

---

7 The weighting sequence is

$$\{0, \ \mathbf{CB}, \ \mathbf{CAB}, \ \mathbf{CA}^2\mathbf{B}, \cdots\},$$

and the free response for the given initial state is the sequence

$$\{\mathbf{CB}, \ \mathbf{CAB}, \mathbf{CA}^2\mathbf{B}, \cdots\},$$

which is the weighting sequence shifted left by one step.

---

9 In each case, use formula (2.7), with binary arithmetic for the binary system:

(a)

$$\mathbf{H}_0 = \mathbf{D} = 1,$$

$$\mathbf{H}_1 = \mathbf{CB} = 2, \quad \mathbf{AB} = \begin{bmatrix} 2^{-1} \\ 0 \end{bmatrix},$$

$$\mathbf{H}_2 = \mathbf{C}(\mathbf{AB}) = 1, \quad \mathbf{A}^2\mathbf{B} = \begin{bmatrix} 2^{-2} \\ 0 \end{bmatrix},$$

$$\mathbf{H}_3 = \mathbf{C}(\mathbf{A}^2\mathbf{B}) = 2^{-1}, \quad \mathbf{A}^3\mathbf{B} = \begin{bmatrix} 2^{-3} \\ 0 \end{bmatrix},$$

$$\vdots$$

$$\mathbf{H}_k = \mathbf{C}(\mathbf{A}^{k-1}\mathbf{B}) = 2^{-(k-2)},$$

(b)

$$\mathbf{H}_0 = \mathbf{D} = [\,0, 0\,],$$
$$\mathbf{H}_1 = \mathbf{CB} = [\,1, 0\,], \quad \mathbf{CA} = [\,1, 1\,],$$
$$\mathbf{H}_2 = (\mathbf{CA})\mathbf{B} = [\,0, 1\,], \quad \mathbf{CA}^2 = [\,0, 1\,],$$
$$\mathbf{H}_3 = (\mathbf{CA}^2)\mathbf{B} = [\,1, 0\,], \quad \mathbf{CA}^3 = [\,1, 1\,] = \mathbf{CA}^1,$$
$$\mathbf{H}_4 = (\mathbf{CA}^3)\mathbf{B} = [\,0, 1\,] = \mathbf{H}_2,$$

$$\vdots$$

$$\mathbf{H}_k = \mathbf{H}_{k-2}.$$

---

11 The sequence $\{\mathbf{H}_k\}$ is calculated as

$$\mathbf{H}_0 = \mathbf{D} = 1,$$
$$\mathbf{H}_1 = \mathbf{CB} = 2, \quad \mathbf{AB} = \begin{bmatrix} 0 \\ 2^{-1} \end{bmatrix},$$
$$\mathbf{H}_2 = \mathbf{C}(\mathbf{AB}) = 1, \quad \mathbf{A}^2\mathbf{B} = \begin{bmatrix} 0 \\ 2^{-2} \end{bmatrix},$$

$$\vdots$$

$$\mathbf{H}_k = \mathbf{C}(\mathbf{A}^{k-1})\mathbf{B} = 2^{-(k-2)},$$

which is identical to the sequence in question 9(a); consequently the two systems have identical forced response for identical inputs.

---

13 See Section 2:

(a) By definition, $e^{t\mathbf{A}} =$

$$\begin{bmatrix} 1 & 0 \\ 0 & 1 \end{bmatrix} + \frac{1}{1!}\begin{bmatrix} -2t & t \\ 0 & -2t \end{bmatrix} + \frac{1}{2!}\begin{bmatrix} (-2t)^2 & -4t^2 \\ 0 & (-2t)^2 \end{bmatrix} + \frac{1}{3!}\begin{bmatrix} (-2t)^3 & 12t^3 \\ 0 & (-2t)^3 \end{bmatrix}$$

$$+ \cdots = \begin{bmatrix} e^{-2t} & 0 + \frac{t}{1!} + \frac{-4t^2}{2!} + \cdots \\ 0 & e^{-2t} \end{bmatrix},$$

in which the upper-right entry is not easy to recognize, but will be shown by the methods of Chapter 3 and Chapter 7 to be te^{-2t}. Another method is given in the following.

(b) Because the matrix $\mathbf{A}$ is triangular, the solution can be obtained in closed form as follows.

$$\dot{x}_2 = -2x_2 \Rightarrow x_2(t) = e^{-2t}x_2(0),$$

$$\dot{x}_1 = -2x_1 + x_2(t) \Rightarrow$$

$$x_1(t) = e^{-2t}x_1(0)$$

$$+ \int_0^t e^{-2(t-\tau)}x_2(\tau)d\tau = e^{-2t}x_1(0) + \int_0^t e^{-2(t-\tau)}e^{-2\tau}x_2(0)\,d\tau$$

$$= e^{-2t}x_1(0) + te^{-2t}x_2(0),$$

or

$$\mathbf{X}(t) = \begin{bmatrix} e^{-2t} & te^{-2t} \\ 0 & e^{-2t} \end{bmatrix}\mathbf{X}(0) = e^{t\mathbf{A}}\mathbf{X}(0),$$

confirming the previously calculated value of $e^{t\mathbf{A}}$. For the given $\mathbf{X}(0)$,

$$\mathbf{X}(t) = \begin{bmatrix} e^{-2t} + te^{-2t} \\ e^{-2t} \end{bmatrix}.$$

(c) The free solution is $\mathbf{X}(10) = e^{(10-9)\mathbf{A}}\mathbf{X}(9)$, so that

$$\mathbf{X}(9) = \begin{bmatrix} e^{-2} & e^{-2} \\ 0 & e^{-2} \end{bmatrix}^{-1}\mathbf{X}(10) = \begin{bmatrix} e^{2} & -e^{2} \\ 0 & e^{2} \end{bmatrix}\begin{bmatrix} 1 \\ 2 \end{bmatrix} = \begin{bmatrix} -e^{2} \\ 2e^{2} \end{bmatrix}.$$

15 If $\mathbf{AB} = \mathbf{BA}$ then the binomial expansion can be used for each power in the series

$$e^{\mathbf{A}+\mathbf{B}} = \sum_{r=0}^{\infty} \frac{1}{r!}(\mathbf{A}+\mathbf{B})^r = \sum_{r=0}^{\infty} \frac{1}{r!} \sum_{k=0}^{r} \binom{r}{k} \mathbf{A}^{r-k}\mathbf{B}^k$$

$$= \sum_{r=0}^{\infty} \sum_{k=0}^{r} \frac{1}{(r-k)!k!} \mathbf{A}^{r-k}\mathbf{B}^k.$$

The product $e^{\mathbf{A}}e^{\mathbf{B}}$ is

$$e^{\mathbf{A}}e^{\mathbf{B}} = \sum_{r=0}^{\infty} \frac{1}{r!}\mathbf{A}^r \sum_{k=0}^{\infty} \frac{1}{k!}\mathbf{B}^k = \sum_{r=0}^{\infty} \sum_{k=0}^{\infty} \frac{1}{r!k!}\mathbf{A}^r\mathbf{B}^k,$$

which is a sum of terms taken at all grid points in the first quadrant $r \geq 0, k \geq 0$. If the formula is changed to sum along diagonals as shown in Figure S2.15,

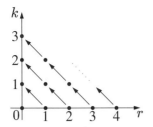

Fig. S2.15 Illustrating summation along diagonals

the formula becomes

$$e^{\mathbf{A}}e^{\mathbf{B}} = \sum_{r=0}^{\infty}\sum_{k=0}^{r}\frac{1}{(r-k)!k!}\mathbf{A}^{r-k}\mathbf{B}^{k},$$

which is the same as the formula for $e^{\mathbf{A}+\mathbf{B}}$.

---

17 For the second of the two equations, which is

$$\dot{x}_2 = 2x_2,$$

the general solution is

$$x_2(t) = e^{2t}x_2(t_0),$$

so that $x_2(1) = 1$ implies that $x_2(0) = e^{-2}$.
   The first equation is now

$$\dot{x}_1 = x_1 + e^{2t}e^{-2} + 1,$$

which has general solution

$$x_1(t) = e^t x_1(0) + \int_0^t e^{t-\tau}(e^{-2}e^{2\tau} + 1)\,d\tau$$

$$= e^t x_1(0) + e^t\left(e^{-2}\int_0^t e^{\tau}\,d\tau + \int_0^t e^{-\tau}\,d\tau\right)$$

$$= e^t x_1(0) + e^t\left(e^{-2}(e^t - 1) - (e^{-t} - 1)\right),$$

and setting $x_1(1) = 1$ gives $x_0(0) = e^{-1} + e^{-2} - 1$.

---

19 For the first system, the Markov sequence is

$$\{\mathbf{H}_k\} = \left\{\begin{array}{l}\mathbf{D},\ k = 0 \\ [\,\mathbf{C}_1,\ \mathbf{C}_2\,]\begin{bmatrix}\mathbf{A}_1 & \mathbf{A}_2 \\ 0 & \mathbf{A}_3\end{bmatrix}^{k-1}\begin{bmatrix}\mathbf{B}_1 \\ 0\end{bmatrix} = \mathbf{C}_1\mathbf{A}_1^{k-1}\mathbf{B}_1,\ k > 0\end{array}\right\},$$

which is the Markov sequence of the second system, showing that the two systems have identical forced behavior.

21 Because the initial states are identical, at any sample time kT the continuous-time system has free response

$$\mathbf{Y}_{\text{free}}(t) = \mathbf{C}e^{kT\mathbf{A}}\mathbf{X}(0) = \mathbf{C}\left(e^{T\mathbf{A}}\right)^k\mathbf{X}(0),$$

which is the response of the discrete-time system since the initial states are identical. With Equation (2.41), the forced responses have identical values at the sample times since $\mathbf{U}(kT) = \mathbf{U}(0) = $ constant.

23 For the discretized system, $\mathbf{C}$ and $\mathbf{D}$ remain unchanged. The state-update equation is $\mathbf{X}(k+1) = \mathbf{A}'\mathbf{X}(k) + \mathbf{B}'\mathbf{U}(k)$, where

$$\mathbf{A}' = e^{\alpha T}\begin{bmatrix} \cos\omega T & \sin\omega T \\ -\sin\omega T & \cos\omega T \end{bmatrix}, \quad \mathbf{B}' = \int_0^T e^{\alpha\tau}\begin{bmatrix} \cos\omega\tau & \sin\omega\tau \\ -\sin\omega\tau & \cos\omega\tau \end{bmatrix}d\tau\begin{bmatrix} 0 \\ 1 \end{bmatrix}.$$

By integrating,

$$\int_0^T e^{\alpha\tau}\cos\omega\tau\, d\tau = \int_0^T \operatorname{Re} e^{(\alpha+j\omega)\tau}d\tau = \operatorname{Re}\frac{1}{\alpha+j\omega}(e^{(\alpha+j\omega)T}-1)$$

$$= \operatorname{Re}\frac{\alpha-j\omega}{\alpha^2+\omega^2}e^{\alpha T}(\cos\omega T - 1 + j\sin\omega T)$$

$$= \frac{e^{\alpha T}}{\alpha^2+\omega^2}(\alpha(\cos\omega T - 1) + \omega\sin\omega T),$$

$$\int_0^T e^{\alpha\tau}\sin\omega\tau\, d\tau = \int_0^T \operatorname{Im} e^{(\alpha+j\omega)\tau}d\tau$$

$$= \frac{e^{\alpha T}}{\alpha^2+\omega^2}(\alpha\sin\omega T + \omega(1 - \cos\omega T)),$$

so that

$$\mathbf{B}' = \frac{e^{\alpha T}}{\alpha^2+\omega^2}\begin{bmatrix} \alpha\sin\omega T + \omega(1 - \cos\omega T) \\ \alpha(\cos\omega T - 1) + \omega\sin\omega T \end{bmatrix}.$$

Chapter 3 **Solutions**

1 In all cases:
$$\mathbf{Y}(t) = \mathbf{Y}_{\text{free}}(t) + \mathbf{Y}_{\text{forced}}(t)$$
$$= \mathcal{L}^{-1}\left(\mathbf{C}(s\mathbf{I}-\mathbf{A})^{-1}\mathbf{X}(0)\right) + \mathcal{L}^{-1}\left(\mathbf{D} + \mathbf{C}(s\mathbf{I}-\mathbf{A})^{-1}\mathbf{B}\right)\hat{\mathbf{U}}(s).$$

(a)

$$\mathbf{Y}_{\text{free}}(t) = \mathcal{L}^{-1}\begin{bmatrix} 1 \\ 0 \\ 1 \end{bmatrix}(s-1/4)^{-1}3 = \begin{bmatrix} 3e^{t/4} \\ 0 \\ 3e^{t/4} \end{bmatrix},$$

$$\mathbf{Y}_{\text{forced}}(t) = \mathcal{L}^{-1}\left(\begin{bmatrix} 0 & 0 \\ 2 & 0 \\ 0 & 1 \end{bmatrix} + \begin{bmatrix} 1 \\ 0 \\ 1 \end{bmatrix}(s-1/4)^{-1}[1, 2]\right)\begin{bmatrix} 1 \\ 0 \end{bmatrix}$$

$$= \mathcal{L}^{-1}\begin{bmatrix} (s-1/4)^{-1} \\ 2 \\ (s-1/4)^{-1} \end{bmatrix} = \begin{bmatrix} e^{t/4} \\ 2\delta(t) \\ e^{t/4} \end{bmatrix},$$

$$\mathbf{Y}(t) = \mathbf{Y}_{\text{free}} + \mathbf{Y}_{\text{forced}} = \begin{bmatrix} 4e^{t/4} \\ 2\delta(t) \\ 4e^{t/4} \end{bmatrix}.$$

(b)

$$\mathbf{Y}_{\text{forced}}(t) = \mathcal{L}^{-1}\left([0, 0] + [2, 1]\begin{bmatrix} s-1/4 & 0 \\ 0 & s+1 \end{bmatrix}^{-1}\begin{bmatrix} 1 & 0 \\ 0 & 1 \end{bmatrix}\right)\begin{bmatrix} 1/s \\ 0 \end{bmatrix}$$

$$= \mathcal{L}^{-1}[2(s-1/4)^{-1} \quad (s+1)^{-1}]\begin{bmatrix} 1/s \\ 0 \end{bmatrix} = \mathcal{L}^{-1}\frac{2}{s(s-1/4)}$$

$$= \mathcal{L}^{-1}\left(\frac{-8}{s} + \frac{8}{s-1/4}\right) = -8 + 8e^{t/4}$$

$$\mathbf{Y}_{\text{free}}(t) = \mathcal{L}^{-1}[2, 1]\begin{bmatrix} (s-1/4)^{-1} & 0 \\ 0 & (s+1)^{-1} \end{bmatrix}\begin{bmatrix} 0 \\ 2 \end{bmatrix} = \mathcal{L}^{-1}\frac{2}{s+1} = 2e^{-t}$$

$$\mathbf{Y}(t) = 2e^{-t} - 8 + 8e^{t/4}.$$

(c) $(s\mathbf{I}-\mathbf{A})^{-1} = \begin{bmatrix} s-1/4 & -1 \\ 0 & s-1/4 \end{bmatrix}^{-1} = \begin{bmatrix} (s-1/4)^{-1} & (s-1/4)^{-2} \\ 0 & (s-1/4)^{-1} \end{bmatrix}$

$$\mathbf{Y}_{\text{forced}}(t) = \mathcal{L}^{-1}\left(0 + [2, 0]\begin{bmatrix} (s-1/4)^{-1} & (s-1/4)^{-2} \\ 0 & (s-1/4)^{-1} \end{bmatrix}\begin{bmatrix} 1 \\ 0 \end{bmatrix}\right)\frac{1}{s+1}$$

$$= \mathcal{L}^{-1}\frac{2}{(s-1/4)(s+1)} = \frac{8}{5}e^{t/4} - \frac{8}{5}e^{-t}$$

$$\mathbf{Y}_{\text{free}}(t) = \mathcal{L}^{-1}[2, 0]\begin{bmatrix} (s-1/4)^{-1} & (s-1/4)^{-2} \\ 0 & (s-1/4)^{-1} \end{bmatrix}\begin{bmatrix} 0 \\ 2 \end{bmatrix} = \mathcal{L}^{-1}\frac{4}{(s-1/4)^2}$$

$$= 4te^{t/4}$$

$$\mathbf{Y}(t) = 4te^{t/4} + \frac{8}{5}e^{t/4} - \frac{8}{5}e^{-t}.$$

(d)

$$\mathbf{Y}_{\text{forced}}(t) = \mathcal{L}^{-1}\left(0 + [1, 0, 1]\begin{bmatrix} s-1/4 & 0 & 0 \\ -1 & s-1/3 & 1 \\ 0 & 0 & s-2 \end{bmatrix}^{-1}\begin{bmatrix} 1 \\ 0 \\ 0 \end{bmatrix}\right)\frac{1}{s+1}$$

$$= \mathcal{L}^{-1}[1, 0, 1]\frac{1}{(s-1/4)(s-1/3)(s-2)}\begin{bmatrix} (s-1/3)(s-2) \\ \times \\ 0 \end{bmatrix}\frac{1}{s+1}$$

$$= \mathcal{L}^{-1}\frac{1}{s-1/4}\frac{1}{s+1} = \frac{4}{5}e^{t/4} - \frac{4}{5}e^{-t}$$

$$\mathbf{Y}_{\text{free}}(t) = \mathcal{L}^{-1}\frac{1}{(s-1/4)(s-1/3)(s-2)} \times$$

$$[1, 0, 1]\begin{bmatrix} (s-1/3)(s-2) & 0 & 0 \\ (s-2) & (s-1/4)(s-2) & -(s-1/4) \\ 0 & 0 & (s-1/4)(s-1/3) \end{bmatrix}\begin{bmatrix} 0 \\ 2 \\ -1 \end{bmatrix}$$

$$= \mathcal{L}^{-1}\frac{[(s-1/3)(s-2) \quad 0 \quad (s-1/4)(s-1/3)]}{(s-1/4)(s-1/3)(s-2)}\begin{bmatrix} 0 \\ 2 \\ -1 \end{bmatrix}$$

$$= \mathcal{L}^{-1}\frac{-1}{s-2} = -e^{2t}$$

$$\mathbf{Y}(t) = -e^{2t} + \frac{4}{5}e^{t/4} - \frac{4}{5}e^{-t}.$$

3 In each case, the detailed computations are modified from those of Problem 1:

(a)

$$\{\mathbf{Y}_{\text{free}}(k)\} = \mathcal{Z}^{-1}\begin{bmatrix} 1 \\ 0 \\ 1 \end{bmatrix}(z-1/4)^{-1}z\,3 = \left\{\begin{bmatrix} 3(1/4)^k \\ 0 \\ 3(1/4)^k \end{bmatrix}\right\}$$

$$\{\mathbf{Y}_{\text{forced}}(k)\} = \mathcal{Z}^{-1}\left(\begin{bmatrix} 0 & 0 \\ 2 & 0 \\ 0 & 1 \end{bmatrix} + \begin{bmatrix} 1 \\ 0 \\ 1 \end{bmatrix}(z-1/4)^{-1}[1, 2]\right)\begin{bmatrix} 1 \\ 0 \end{bmatrix}$$

$$= \mathcal{Z}^{-1}\begin{bmatrix} (z-1/4)^{-1} \\ 2 \\ (z-1/4)^{-1} \end{bmatrix} = \left\{\begin{bmatrix} 0 \\ 2 \\ 0 \end{bmatrix}, \begin{bmatrix} 1 \\ 0 \\ 1 \end{bmatrix}, \cdots \begin{bmatrix} (1/4)^{k-1} \\ 0 \\ (1/4)^{k-1} \end{bmatrix}\right\},$$

$$\{\mathbf{Y}(k)\} = \left\{\begin{bmatrix} 3(1/4)^k \\ 0 \\ 3(1/4)^k \end{bmatrix}\right\} + \left\{\begin{bmatrix} 0 \\ 2 \\ 0 \end{bmatrix}, \begin{bmatrix} 1 \\ 0 \\ 1 \end{bmatrix}, \cdots \begin{bmatrix} (1/4)^{k-1} \\ 0 \\ (1/4)^{k-1} \end{bmatrix}\right\}.$$

(b)

$$\{\mathbf{Y}_{\text{forced}}(k)\} = \mathcal{Z}^{-1}[2(z-1/4)^{-1} \quad (z+1)^{-1}]\begin{bmatrix} \frac{z}{z-1} \\ 0 \end{bmatrix} = \mathcal{Z}^{-1}\frac{2}{z-1/4}\frac{z}{z-1}$$

$$= \mathcal{Z}^{-1}2z\left(\frac{-4/3}{z-1/4} + \frac{4/3}{z-1}\right) = \frac{8}{3}\{1\} - \frac{8}{3}\{(1/4)^k\}$$

$$\{\mathbf{Y}_{\text{free}}(k)\} = \mathcal{Z}^{-1}[2, 1]\begin{bmatrix} (z-1/4)^{-1} & 0 \\ 0 & (z+1)^{-1} \end{bmatrix}z\begin{bmatrix} 0 \\ 2 \end{bmatrix} = \mathcal{Z}^{-1}\frac{2z}{z+1}$$

$$= 2\{(-1)^k\}$$

$$\{\mathbf{Y}(k)\} = 2\{(-1)^k\} + \frac{8}{3}\{1\} - \frac{8}{3}\{4^{-k}\}.$$

(c)

$$\{\mathbf{Y}_{\text{forced}}(k)\} = \mathcal{Z}^{-1}\frac{2z}{(z-1/4)(z-1/3)} = \mathcal{Z}^{-1}2z\left(\frac{-12}{z-1/4} + \frac{12}{z-1/3}\right)$$
$$= -24\{4^{-k}\} + 24\{3^{-k}\}$$
$$\{\mathbf{Y}_{\text{free}}(k)\} = \mathcal{Z}^{-1}[2,\,0]\begin{bmatrix}(z-1/4)^{-1} & (z-1/4)^{-2}\\ 0 & (z-1/4)^{-1}\end{bmatrix}z\begin{bmatrix}0\\2\end{bmatrix}$$
$$= \mathcal{Z}^{-1}\frac{4z}{(z-1/4)^2}$$
$$= \mathcal{Z}^{-1}\{0 + 4z^{-1} + 2z^{-2} + \frac{3}{4}z^{-3} + \frac{1}{4}z^{-4} + \cdots\}$$
$$= \{0,\,4,\,2,\,\frac{3}{4},\,\frac{1}{4},\cdots\}$$
$$\{\mathbf{Y}(k)\} = \{0,\,4,\,2,\,\frac{3}{4},\,\frac{1}{4},\cdots\} - 24\{4^{-k}\} + 24\{3^{-k}\}.$$

(d)

$$\{\mathbf{Y}_{\text{forced}}(k)\} = \mathcal{Z}^{-1}\frac{1}{z-1/4}\frac{z}{z-1/2} = \mathcal{Z}^{-1}z\left(\frac{-4}{z-1/4} + \frac{4}{z-1/2}\right)$$
$$= -4\{4^{-k}\} + 4\{2^{-k}\}$$
$$\{\mathbf{Y}_{\text{free}}(k)\} = \mathcal{Z}^{-1}\frac{-z}{z-2} = -\{2^k\}$$
$$\{\mathbf{Y}(k)\} = -\{2^k\} - 4\{4^{-k}\} + 4\{2^{-k}\}$$

5 If we assign x_1 and x_2 as the output of the right and left delay elements respectively, the equations are
$$\begin{bmatrix}x_1(k+1)\\x_2(k+1)\end{bmatrix} = \begin{bmatrix}0 & 1\\1 & 1\end{bmatrix}\begin{bmatrix}x_1\\x_2\end{bmatrix} + \begin{bmatrix}1\\1\end{bmatrix}u(k)$$
$$y(k) = [1,\,1]\begin{bmatrix}x_1\\x_2\end{bmatrix} + [1]u(k).$$

(a)

$$\hat{\mathbf{H}}(z) = 1 + [1,\,1]\begin{bmatrix}z & 1\\1 & z+1\end{bmatrix}^{-1}\begin{bmatrix}1\\1\end{bmatrix} = 1 + [1,\,1]\frac{1}{z^2+z+1}\begin{bmatrix}z+1 & 1\\1 & z\end{bmatrix}\begin{bmatrix}1\\1\end{bmatrix}$$
$$= \frac{z^2+z}{z^2+z+1}.$$

(b)

$$\mathbf{Y}_{\text{free}}(z) = [1, 1]\frac{1}{z^2+z+1}\begin{bmatrix} z+1 & 1 \\ 1 & z \end{bmatrix} z \begin{bmatrix} x_1(0) \\ x_2(0) \end{bmatrix}$$

$$= \frac{z^2 x_1(0) + z(z+1)x_2(0)}{z^2+z+1}.$$

(c) By long division,

$$\hat{\mathbf{H}}(z) = \frac{z^2+z}{z^2+z+1} = 1 + 0z^{-1} + z^{-2} + z^{-3} + 0z^{-4} + z^{-5} + \cdots,$$

hence

$$\{\mathbf{H}(k)\} = \{1, 0, 1, 1, 0, 1, 1, 0, \cdots\}.$$

(d)

$$\hat{\mathbf{Y}}_{\text{forced}}(z) = \frac{z^2+z}{z^2+z+1}\frac{z^3+z^2+z+1}{z^3} = \frac{z^4+1}{z^4+z^3+z^2}$$

$$= 1 + z^{-1} + z^{-3} + z^{-5} + z^{-6} + z^{-8} + z^{-9} + \cdots$$

by long division, so that

$$\{\mathbf{Y}_{\text{forced}}(k)\} = \{1, 1, 0, 1, 0, 1, 1, 0, 1, 1, 0, \cdots\}.$$

(e) Computing the convolutions yields

1	0	1	1	0	1	1	0	1	1	$\cdots$
1	1	1	1	0	0	0	0	0	0	$\cdots$
1	0	1	1	0	1	1	0	1	1	$\cdots$
	1	0	1	1	0	1	1	0	1	$\cdots$
		1	0	1	1	0	1	1	0	$\cdots$
			1	0	1	1	0	1	1	$\cdots$
				0						$\cdots$
1	1	0	1	0	1	1	0	1	1	$\cdots$

7 Because of the the form of $\mathbf{C}$, the computation of the expression

$$\hat{\mathbf{H}}(s) = \mathbf{D} + \mathbf{C}(s\mathbf{I}-\mathbf{A})^{-1}\mathbf{B} = 0 + [0, 0, 1]\begin{bmatrix} s & 0 & 4 \\ -1 & s & -7 \\ 0 & -1 & s-11 \end{bmatrix}^{-1}\begin{bmatrix} 1 \\ 3 \\ -2 \end{bmatrix},$$

requires only the last row of $(s\mathbf{I}-\mathbf{A})^{-1}$. The inverse requires the determinant and the adjoint. Expanding along the third column gives

$$\det(s\mathbf{I}-\mathbf{A}) = (4)(1) + (-7)(s) + (s-11)(s^2) = s^3 - 11s^2 - 7s + 4,$$

and the cofactors computed in this determinant are the quantities required in the last row of $\text{adj}(s\mathbf{I}-\mathbf{A})$. Thus

$$\hat{\mathbf{H}}(s) = 0 + \frac{1}{s^3-11s^2-7s+4}[0,\,0,\,1]\begin{bmatrix} \times & \times & \times \\ \times & \times & \times \\ 1 & s & s^2 \end{bmatrix}\begin{bmatrix} 1 \\ 3 \\ -2 \end{bmatrix}$$

$$= 0 + \frac{-2s^2+3s+1}{s^3-11s^2-7s+4},$$

where, because of the form of $\mathbf{C}$ and $\mathbf{A}$, the coefficients in $\mathbf{A}$ and $\mathbf{B}$ are those of the transfer function, with changes of sign in $\mathbf{A}$.

Chapter 4   Solutions

1   In each case, it is necessary to choose the state variables.

(a) With the tree and state variables shown in Figure S4.1(a),

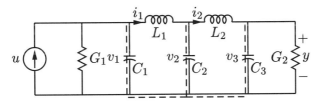

Fig. S4.1(a)   Linear circuit with tree and state variables.

the equations are

$$\frac{d}{dt}\begin{bmatrix} v_1 \\ v_2 \\ v_3 \\ i_1 \\ i_2 \end{bmatrix} = \begin{bmatrix} -G_1/C_1 & 0 & 0 & 1/C_1 & 0 \\ 0 & 0 & 0 & 1/C_2 & -1/C_2 \\ 0 & 0 & -G_2/C_3 & 0 & 1/C_3 \\ 1/L_1 & -1/L_1 & 0 & 0 & 0 \\ 0 & 1/L_2 & -1/L_2 & 0 & 0 \end{bmatrix}\begin{bmatrix} v_1 \\ v_2 \\ v_3 \\ i_1 \\ i_2 \end{bmatrix}$$

$$+ \begin{bmatrix} 1/C_1 \\ 0 \\ 0 \\ 0 \\ 0 \end{bmatrix}u$$

$$y = [0 \quad 0 \quad 1 \quad 0 \quad 0]\begin{bmatrix} v_1 \\ v_2 \\ v_3 \\ i_1 \\ i_2 \end{bmatrix} + 0\,u.$$

(b) The tree is shown in Figure S4.1(b).

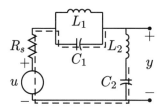

Fig. S4.1(b)   The second linear circuit with tree and state variables.

Choose capacitor voltages and inductor currents as state variables, to get the equations

$$\frac{d}{dt}\begin{bmatrix} v_1 \\ v_2 \\ i_1 \\ i_2 \end{bmatrix} = \begin{bmatrix} 0 & 0 & -1/C_1 & 1/C_1 \\ 0 & 0 & 0 & 1/C_2 \\ 1/L_1 & 0 & 0 & 0 \\ -1/L_2 & -1/L_2 & 0 & -R_s/L_2 \end{bmatrix} \begin{bmatrix} v_1 \\ v_2 \\ i_1 \\ i_2 \end{bmatrix} + \begin{bmatrix} 0 \\ 0 \\ 0 \\ 1/L_2 \end{bmatrix} u$$

$$y = \begin{bmatrix} -1 & 0 & 0 & -R_s \end{bmatrix} \begin{bmatrix} v_1 \\ v_2 \\ i_1 \\ i_2 \end{bmatrix} + 1\,u.$$

(c) Choosing λ, q as state variables yields

$$\frac{d}{dt}\begin{bmatrix} \lambda \\ q \end{bmatrix} = \begin{bmatrix} 2q + q^3 - R(3\lambda - \lambda^3) \\ -G(2q + q^3) - (3\lambda - \lambda^3) \end{bmatrix}$$

(d) Choosing state variables x_1, x_2 from right to left yields

$$\begin{bmatrix} x_1(k+1) \\ x_2(k+1) \end{bmatrix} = \begin{bmatrix} 0 & 1 \\ 0 & 0 \end{bmatrix} \begin{bmatrix} x_1(k) \\ x_2(k) \end{bmatrix} + \begin{bmatrix} 0 \\ 1 \end{bmatrix} u(k)$$

$$y = \begin{bmatrix} 1, & 0 \end{bmatrix} \begin{bmatrix} x_1(k) \\ x_2(k) \end{bmatrix} + 0u(k).$$

(e) Choosing state variables x_1, x_2 from right to left yields

$$\begin{bmatrix} x_1(k+1) \\ x_2(k+1) \end{bmatrix} = \begin{bmatrix} 0 & 1 \\ 2 & -3 \end{bmatrix} \begin{bmatrix} x_1(k) \\ x_2(k) \end{bmatrix} + \begin{bmatrix} 0 \\ 1 \end{bmatrix} u(k)$$

$$y = \begin{bmatrix} 2, & -3 \end{bmatrix} \begin{bmatrix} x_1(k) \\ x_2(k) \end{bmatrix} + 1u(k).$$

(f) Since the operational diagram is equivalent to the previous one with integrators replacing delays, the equations are identical, except that derivatives replace the shift operation:

$$\frac{d}{dt}\begin{bmatrix} x_1 \\ x_2 \end{bmatrix} = \begin{bmatrix} 0 & 1 \\ 2 & -3 \end{bmatrix} \begin{bmatrix} x_1(t) \\ x_2(t) \end{bmatrix} + \begin{bmatrix} 0 \\ 1 \end{bmatrix} u(t)$$

$$y = [\,2, \ -3\,] \begin{bmatrix} x_1(t) \\ x_2(t) \end{bmatrix} + 1u(t).$$

3 The coordinates of the mass with respect to equilibrium are

$$x_m = x - \ell \sin(\theta), \quad y_m = \ell(1 - \cos(\theta)),$$

from which it follows that the squared velocity of the mass is

$$\dot{x}_m^2 + \dot{y}_m^2 = \dot{x}^2 + \ell^2\dot{\theta}^2 - 2\ell\dot{x}\dot{\theta}\cos(\theta),$$

so that the system kinetic energy is

$$T = \frac{m}{2}\dot{x}^2 + \frac{m}{2}\ell^2\dot{\theta}^2 - m\ell\dot{x}\dot{\theta}\cos(\theta).$$

The potential energy is

$$V = mg\ell(1 - \cos(\theta)) + \frac{k}{2}\ell_1^2\theta^2,$$

the dissipation term is

$$D = \frac{b}{2}\ell_1^2\dot{\theta}^2,$$

and the Lagrangian is

$$L = \frac{m}{2}\dot{x}^2 + \frac{m}{2}\ell^2\dot{\theta}^2 - m\ell\dot{x}\dot{\theta}\cos(\theta) + mg\ell\cos(\theta) - \frac{k}{2}\ell_1^2\theta^2 - mg\ell.$$

Performing the differentiations with respect to x gives the Euler-Lagrange equation

$$m\ddot{x} - m\ell\ddot{\theta}\cos(\theta) + m\ell\dot{\theta}^2\sin(\theta) = f,$$

and similarly, with respect to θ,

$$m\ell^2\ddot{\theta} - m\ell\ddot{x}\cos(\theta) + b\ell_1^2\dot{\theta} + k\ell_1^2\theta + mg\ell\sin(\theta) = 0.$$

5 In each case, divide to make each transfer-matrix entry a constant plus a strictly proper function. Then the matrices are written by inspection.

(a) $\mathbf{A} = \begin{bmatrix} 0 & 1 & 0 & 0 \\ 0 & 0 & 1 & 0 \\ 0 & 0 & 0 & 1 \\ -5 & -8 & 2 & -6 \end{bmatrix}$, $\mathbf{B} = \begin{bmatrix} 0 \\ 0 \\ 0 \\ 1 \end{bmatrix}$, $\mathbf{C} = [\,3 \ \ 4 \ \ 0 \ \ 7\,]$, $\mathbf{D} = 4$.

(b) $\hat{\mathbf{H}}(s) = 2 + \dfrac{-6}{s+4}$, $\mathbf{A} = -4$, $\mathbf{B} = 1$, $\mathbf{C} = -6$, $\mathbf{D} = 2$.

(c) $\mathbf{A} = 0$, $\mathbf{B} = 1$, $\mathbf{C} = 1$, $\mathbf{D} = 0$.

(d) As for the previous question: $\mathbf{A} = 0$, $\mathbf{B} = 1$, $\mathbf{C} = 1$, $\mathbf{D} = 0$.

(e) Because the transfer function is improper, there is no state-space realization.

(f) $\hat{\mathbf{H}}(s) = \begin{bmatrix} 2 + \dfrac{-2s + 0}{s^2 + s + 1} \\ 7 \\ 2/s \end{bmatrix}$

$$\mathbf{A} = \left[\begin{array}{cc|c} 0 & 1 & 0 \\ -1 & -1 & 0 \\ 0 & 0 & 0 \end{array}\right], \quad \mathbf{B} = \begin{bmatrix} 0 \\ 1 \\ 1 \end{bmatrix}, \quad \mathbf{C} = \left[\begin{array}{cc|c} 0 & -2 & 0 \\ 0 & 0 & 0 \\ 0 & 0 & 2 \end{array}\right], \quad \mathbf{D} = \begin{bmatrix} 2 \\ 7 \\ 0 \end{bmatrix}.$$

---

7   Refer to Figure P4.7.

(a) If

$$\hat{\mathbf{Y}}(s) = \left(\hat{\mathbf{H}}_1(s) + \hat{\mathbf{H}}_2(s)\right) \hat{\mathbf{U}}(s)$$
$$= \left(\mathbf{D}_1 + \mathbf{C}_1(s\mathbf{I} - \mathbf{A}_1)^{-1}\mathbf{B}_1 + \mathbf{D}_2 + \mathbf{C}_2(s\mathbf{I} - \mathbf{A}_2)^{-1}\mathbf{B}_2\right) \hat{\mathbf{U}}(s)$$
$$= \left(\mathbf{D}_1 + \mathbf{D}_2\right) \hat{\mathbf{U}}(s) + \left(\mathbf{C}_1(s\mathbf{I} - \mathbf{A}_1)^{-1}\mathbf{B}_1 + \mathbf{C}_2(s\mathbf{I} - \mathbf{A}_2)^{-1}\mathbf{B}_2\right) \hat{\mathbf{U}}(s),$$

then let $\mathbf{D} = \mathbf{D}_1 + \mathbf{D}_2$, and as for (4.49), let

$$\mathbf{A} = \begin{bmatrix} \mathbf{A}_1 & 0 \\ 0 & \mathbf{A}_2 \end{bmatrix}, \quad \mathbf{B} = \begin{bmatrix} \mathbf{B}_1 \\ \mathbf{B}_2 \end{bmatrix}, \quad \mathbf{C} = [\mathbf{C}_1, \mathbf{C}_2].$$

Confirming, we see that the transfer matrix for this system is

$$\hat{\mathbf{H}}(s) = \mathbf{D}_1 + \mathbf{D}_2 + [\mathbf{C}_1, \mathbf{C}_2] \begin{bmatrix} s\mathbf{I} - \mathbf{A}_1 & 0 \\ 0 & s\mathbf{I} - \mathbf{A}_2 \end{bmatrix}^{-1} \begin{bmatrix} \mathbf{B}_1 \\ \mathbf{B}_2 \end{bmatrix}$$
$$= \mathbf{D}_1 + \mathbf{D}_2 + \mathbf{C}_1(s\mathbf{I} - \mathbf{A}_1)^{-1}\mathbf{B}_1 + \mathbf{C}_2(s\mathbf{I} - \mathbf{A}_2)^{-1}\mathbf{B}_2$$

as required.

(b) Write $\hat{\mathbf{Y}}(s) = \hat{\mathbf{H}}_2(s)\hat{\mathbf{H}}_1(s) \hat{\mathbf{U}}(s)$ as

$$\hat{\mathbf{Y}}_1(s) = \hat{\mathbf{H}}_1(s) \hat{\mathbf{U}}(s),$$
$$\hat{\mathbf{Y}}(s) = \hat{\mathbf{H}}_2(s) \hat{\mathbf{Y}}_1(s),$$

corresponding to the time-domain equations

$$\frac{d}{dt}\mathbf{X}_1 = \mathbf{A}_1\mathbf{X}_1 + \mathbf{B}_1\mathbf{U}$$
$$\mathbf{Y}_1 = \mathbf{C}_1\mathbf{X}_1 + \mathbf{D}_1\mathbf{U}$$
$$\frac{d}{dt}\mathbf{X}_2 = \mathbf{A}_2\mathbf{X}_2 + \mathbf{B}_2\mathbf{Y}_1 = \mathbf{A}_2\mathbf{X}_2 + \mathbf{B}_2(\mathbf{C}_1\mathbf{X}_1 + \mathbf{D}_1\mathbf{U})$$
$$\mathbf{Y}_2 = \mathbf{C}_2\mathbf{X}_2 + \mathbf{D}_2\mathbf{Y}_1 = \mathbf{C}_2\mathbf{X}_2 + \mathbf{D}_2(\mathbf{C}_1\mathbf{X}_1 + \mathbf{D}_1\mathbf{U})$$

corresponding to state vector $\mathbf{X} = \begin{bmatrix} \mathbf{X}_1 \\ \mathbf{X}_2 \end{bmatrix}$ with matrices

$$\mathbf{A} = \begin{bmatrix} \mathbf{A}_1 & 0 \\ \mathbf{B}_2\mathbf{C}_1 & \mathbf{A}_2 \end{bmatrix}, \quad \mathbf{B} = \begin{bmatrix} \mathbf{B}_1 \\ \mathbf{B}_2\mathbf{D}_1 \end{bmatrix}$$
$$\mathbf{C} = [\,\mathbf{D}_2\mathbf{C}_1 \quad \mathbf{C}_2\,], \quad \mathbf{D} = \mathbf{D}_2\mathbf{D}_1.$$

---

9 Refer to Figure P4.9 in the following:

(a) From the diagram, the input of the right integrator is dy/dt and the input to the left integrator is

$$\frac{d^2y}{dt^2} = \frac{2}{\tau^2}\left(u - \tau\frac{dy}{dt} - y\right),$$

which simplifies to

$$u = \frac{\tau^2}{2}\frac{d^2y}{dt^2} + \tau\frac{dy}{dt} + y$$

as required.

(b) Taking Laplace transforms yields

$$\frac{\hat{y}(s)}{\hat{u}(s)} = \frac{1}{1 + \tau s + (\tau^2/2!)s^2} = \frac{2/\tau^2}{s^2 + (2/\tau)s + (2/\tau^2)}.$$

(c) Multiplying yields

$$b_0 s^2 + b_1 s + b_2 =$$
$$(s^2 + a_1 s + a_2)\left(1 - \frac{s\tau}{1} + \frac{s^2\tau^2}{2} - \frac{s^3\tau^3}{6} + \frac{s^4\tau^4}{24} - \cdots\right),$$

and equating terms of degree 0 to 4 gives

$$b_2 = a_2$$
$$b_1 = a_1 - \tau a_2$$
$$b_0 = 1 - a_1\tau + a_2(\tau^2/2)$$
$$0 = -\tau + a_1(\tau^2/2) - a_2(\tau^3/6)$$
$$0 = (\tau^2/2) - a_1(\tau^3/6) + a_2(\tau^4/24).$$

The bottom two equations can be solved for a_1, a_2, giving $a_1 - 6/\tau$ and $a_2 = 12/\tau^2$, so that, using the top three equations, the $(2, 2)$ Padé approximation of $e^{-s\tau}$ is

$$\frac{b(s)}{a(s)} = \frac{s^2 - (6/\tau)s + 12/\tau^2}{s^2 + (6/\tau)s + 12/\tau^2}.$$

11 For the mod-2 "T" flip-flop equation $y(k+1) = y(k) + u(k)$,

(a) The operational diagram is shown in Figure S4.11(a):

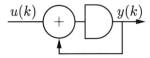

Fig. S4.11(a) A "T" flip-flop.

(b) The required equation is $y(k+1) = u(k)$, which is obtained by adding $y(k)$ to the right-hand side of the equation for the "T" flip-flop, which results in Figure S4.11(b).

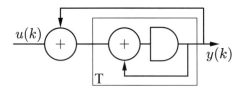

Fig. S4.11(b) The equivalent "D" flip-flop.

Chapter 5   Solutions

1 The multiplication table for $\mathbb{Z}_6$ is shown:

×	0	1	2	3	4	5
0	0	0	0	0	0	0
1	0	1	2	3	4	5
2	0	2	4	0	2	4
3	0	3	0	3	0	3
4	0	4	2	0	4	2
5	0	5	4	3	2	1

Because there is no entry equal to 1 in rows with indices 2, 3, 4, there is no multiplicative inverse for these nonzero values, violating axiom M_5.

3 The field axioms on Problem 1.1 have to be checked.

(a) The 4-byte signed integers can only represent values within a finite range, as for the previous question, so that sums (or products) that leave this range are not allowed, violating axioms A_1 and M_1.

(b) The floating-point numbers can represent values only within a finite range, violating axioms A_1 and M_1.

5   If $x^2 + x + 1$ is not prime, then it must have factors. The products of all possible factors of degree 0 and 1 are shown in the table:

$\times$	0	1	x	$x+1$
0	0	0	0	0
1	0	1	x	$x+1$
x	0	x	x^2	x^2+x
$x+1$	0	$x+1$	x^2+x	x^2+1

and because no product equals x^2+x+1, this polynomial is prime.

The multiplication table is obtained by replacing the entries of the above table by their remainders on division by x^2+x+1, and the addition table is similarly obtained, giving

$\times$	0	1	x	$x+1$
0	0	0	0	0
1	0	1	x	$x+1$,
x	0	x	$x+1$	1
$x+1$	0	$x+1$	1	x

$+$	0	1	x	$x+1$
0	0	1	x	$x+1$
1	1	0	$x+1$	x
x	x	$x+1$	0	1
$x+1$	$x+1$	x	1	0

These tables are symmetric and contain unique units of addition and multiplication as required.

7   Each matrix $\mathbf{A}_j \mathbf{B}_j$ is the product of an $m \times 1$ and a $1 \times n$ matrix, producing an $m \times n$ matrix, of which the entry in row i and column k is $a_{ij}b_{jk}$. The sum of these will be an $m \times n$ matrix with entry ik equal to $\sum_{j=1}^{n} a_{ij}b_{jk}$, which is the formula for the ik entry of the product $\mathbf{AB}$.

9   The sequence of operations on $[\mathbf{I}, \mathbf{A}]$ that transforms $\mathbf{A}$ to upper-right reduced echelon form is not unique, but one sequence is

$$\begin{bmatrix} 1 & 0 & 0 & 0 & 2 \\ 0 & 1 & 0 & -2 & 3 \\ 0 & 0 & 1 & -1 & -1 \end{bmatrix} \xrightarrow[H_1(-1)]{H_{13}} \begin{bmatrix} 0 & 0 & -1 & 1 & 1 \\ 0 & 1 & 0 & -2 & 3 \\ 1 & 0 & 0 & 0 & 2 \end{bmatrix} \xrightarrow{H_{21}(2)} \begin{bmatrix} 0 & 0 & -1 & 1 & 1 \\ 0 & 1 & -2 & 0 & 5 \\ 1 & 0 & 0 & 0 & 2 \end{bmatrix}$$

$$\xrightarrow{H_2(1/5)} \begin{bmatrix} 0 & 0 & -1 & 1 & 1 \\ 0 & 1/5 & -2/5 & 0 & 1 \\ 1 & 0 & 0 & 0 & 2 \end{bmatrix} \xrightarrow[H_{32}(-2)]{H_{12}(-1)} \begin{bmatrix} 0 & -1/5 & -3/5 & 1 & 0 \\ 0 & 1/5 & -2/5 & 0 & 1 \\ 1 & -2/5 & 4/5 & 0 & 0 \end{bmatrix}$$

so that a left inverse is given by

$$\mathbf{A}^L = \begin{bmatrix} 0 & -1/5 & -3/5 \\ 0 & 1/5 & -2/5 \end{bmatrix} + \alpha [1 \quad -2/5 \quad 4/5],$$

where α is arbitrary.

11   The determinant is computed as follows.

(a) Expanding along row 1 gives

$$\det \mathbf{A} = (1)(-1)^{1+1}(-1) + (1)(-1)^{1+2}(0) + (-2)(-1)^{1+3}(-1) = 1.$$

(b) Expanding along column 3 gives

$$\det \mathbf{A} = (-2)(-1)^{1+3}(-1) + (1)(-1)^{2+3}(-1) + (-2)(-1)^{3+3}(1) = 1.$$

(c) By permutations:

$$\det \mathbf{A} = (-1)^0 a_{11}a_{22}a_{33} + (-1)^1 a_{12}a_{21}a_{33} + (-1)^2 a_{12}a_{23}a_{31}$$
$$+(-1)^3 a_{13}a_{22}a_{31} + (-1)^4 a_{13}a_{21}a_{32} + (-1)^5 a_{11}a_{23}a_{32}$$
$$= (1)(0)(-2) - (1)(-1)(-2) + (1)(1)(2) - (-2)(0)(2)$$
$$+(-2)(-1)(1) - (1)(1)(1) = 1.$$

13   From the definition of orthogonal matrices, $\mathbf{A}^T\mathbf{A} = \mathbf{I}$. Therefore

$$\det(\mathbf{A}^T\mathbf{A}) = \det(\mathbf{A}^T)(\det \mathbf{A}) = (\det \mathbf{A})^2 = 1,$$

or $\det \mathbf{A} = \pm 1$.

15   The sequence of operations is not unique. One possible sequence is shown:

$$\begin{bmatrix} \mathbf{A} & \mathbf{I}_3 \\ \mathbf{I}_4 & \end{bmatrix} = \left[\begin{array}{rrrr|rrr} 0 & 1 & -1 & 1 & 1 & 0 & 0 \\ -2 & 3 & 0 & 4 & 0 & 1 & 0 \\ -2 & 4 & -1 & 5 & 0 & 0 & 1 \\ 1 & 0 & 0 & 0 & & & \\ 0 & 1 & 0 & 0 & & & \\ 0 & 0 & 1 & 0 & & & \\ 0 & 0 & 0 & 1 & & & \end{array}\right] \xrightarrow{H_{12}} \left[\begin{array}{rrrr|rrr} -2 & 3 & 0 & 4 & 0 & 1 & 0 \\ 0 & 1 & -1 & 1 & 1 & 0 & 0 \\ -2 & 4 & -1 & 5 & 0 & 0 & 1 \\ 1 & 0 & 0 & 0 & & & \\ 0 & 1 & 0 & 0 & & & \\ 0 & 0 & 1 & 0 & & & \\ 0 & 0 & 0 & 1 & & & \end{array}\right]$$

$$\xrightarrow{H_{31}(-1)} \left[\begin{array}{rrrr|rrr} -2 & 3 & 0 & 4 & 0 & 1 & 0 \\ 0 & 1 & -1 & 1 & 1 & 0 & 0 \\ 0 & 1 & -1 & 1 & 0 & -1 & 1 \\ 1 & 0 & 0 & 0 & & & \\ 0 & 1 & 0 & 0 & & & \\ 0 & 0 & 1 & 0 & & & \\ 0 & 0 & 0 & 1 & & & \end{array}\right] \xrightarrow{H_{32}(-1)} \left[\begin{array}{rrrr|rrr} -2 & 3 & 0 & 4 & 0 & 1 & 0 \\ 0 & 1 & -1 & 1 & 1 & 0 & 0 \\ 0 & 0 & 0 & 0 & -1 & -1 & 1 \\ 1 & 0 & 0 & 0 & & & \\ 0 & 1 & 0 & 0 & & & \\ 0 & 0 & 1 & 0 & & & \\ 0 & 0 & 0 & 1 & & & \end{array}\right]$$

$$\xrightarrow{K_1(-.5)} \left[\begin{array}{rrrr|rrr} 1 & 3 & 0 & 4 & 0 & 1 & 0 \\ 0 & 1 & -1 & 1 & 1 & 0 & 0 \\ 0 & 0 & 0 & 0 & -1 & -1 & 1 \\ -.5 & 0 & 0 & 0 & & & \\ 0 & 1 & 0 & 0 & & & \\ 0 & 0 & 1 & 0 & & & \\ 0 & 0 & 0 & 1 & & & \end{array}\right]$$

$$
\begin{array}{c}
K_{21}(-3) \\
K_{41}(-4) \\
\longrightarrow
\end{array}
\left[\begin{array}{cccc|ccc}
1 & 0 & 0 & 0 & 0 & 1 & 0 \\
0 & 1 & -1 & 1 & 1 & 0 & 0 \\
0 & 0 & 0 & 0 & -1 & -1 & 1 \\
-.5 & 1.5 & 0 & 2 & & & \\
0 & 1 & 0 & 0 & & & \\
0 & 0 & 1 & 0 & & & \\
0 & 0 & 0 & 1 & & &
\end{array}\right]
$$

$$
\begin{array}{c}
K_{32}(1) \\
K_{42}(-1) \\
\longrightarrow
\end{array}
\left[\begin{array}{cccc|ccc}
1 & 0 & 0 & 0 & 0 & 1 & 0 \\
0 & 1 & 0 & 0 & 1 & 0 & 0 \\
0 & 0 & 0 & 0 & -1 & -1 & 1 \\
-.5 & 1.5 & 1.5 & .5 & & & \\
0 & 1 & 1 & -1 & & & \\
0 & 0 & 1 & 0 & & & \\
0 & 0 & 0 & 1 & & &
\end{array}\right]
=\left[\begin{array}{cc} \mathbf{N} & \mathbf{P} \\ \mathbf{Q} & \end{array}\right].
$$

17 One possible sequence is

$$
\left[\begin{array}{c} \mathbf{A} \quad \mathbf{I}_3 \\ \mathbf{I}_4 \end{array}\right]=
\left[\begin{array}{cccc|ccc}
0 & 1 & 1 & 1 & 1 & 0 & 0 \\
0 & 1 & 0 & 0 & 0 & 1 & 0 \\
0 & 0 & 1 & 1 & 0 & 0 & 1 \\
1 & 0 & 0 & 0 & & & \\
0 & 1 & 0 & 0 & & & \\
0 & 0 & 1 & 0 & & & \\
0 & 0 & 0 & 1 & & &
\end{array}\right]
\begin{array}{c}
K_{12} \\
K_{31}(1) \\
K_{41}(1) \\
\longrightarrow
\end{array}
\left[\begin{array}{cccc|ccc}
1 & 0 & 0 & 0 & 1 & 0 & 0 \\
1 & 0 & 1 & 1 & 0 & 1 & 0 \\
0 & 0 & 1 & 1 & 0 & 0 & 1 \\
0 & 1 & 0 & 0 & & & \\
1 & 0 & 1 & 1 & & & \\
0 & 0 & 1 & 0 & & & \\
0 & 0 & 0 & 1 & & &
\end{array}\right]
$$

$$
\begin{array}{c}
K_{23} \\
K_{42}(1) \\
\longrightarrow
\end{array}
\left[\begin{array}{cccc|ccc}
1 & 0 & 0 & 0 & 1 & 0 & 0 \\
1 & 1 & 0 & 0 & 0 & 1 & 0 \\
0 & 1 & 0 & 0 & 0 & 0 & 1 \\
0 & 0 & 1 & 0 & & & \\
1 & 1 & 0 & 0 & & & \\
0 & 1 & 0 & 1 & & & \\
0 & 0 & 0 & 1 & & &
\end{array}\right]
\begin{array}{c}
H_{21}(1) \\
H_{32}(1) \\
\longrightarrow
\end{array}
\left[\begin{array}{cccc|ccc}
1 & 0 & 0 & 0 & 1 & 0 & 0 \\
0 & 1 & 0 & 0 & 1 & 1 & 0 \\
0 & 0 & 0 & 0 & 1 & 1 & 1 \\
0 & 0 & 1 & 0 & & & \\
1 & 1 & 0 & 0 & & & \\
0 & 1 & 0 & 1 & & & \\
0 & 0 & 0 & 1 & & &
\end{array}\right]
=\left[\begin{array}{cc} \mathbf{N} & \mathbf{P} \\ \mathbf{Q} & \end{array}\right].
$$

19 Let $\mathbf{A} \in \mathbb{R}^{m \times n}$, and suppose $\mathbf{A}\mathbf{Q}_1 = \left[\begin{array}{c} \mathbf{I}_r \\ \mathbf{V} \end{array}\right]$. Then a suitable matrix $\mathbf{P}$ such that $\mathbf{P}\mathbf{A}\mathbf{Q}_1 = \left[\begin{array}{c} \mathbf{I}_r \\ \mathbf{0} \end{array}\right]$ is given by $\mathbf{P} = \left[\begin{array}{cc} \mathbf{I}_r & 0 \\ -\mathbf{V} & \mathbf{I}_{m-r} \end{array}\right]$, which can be written by simple inspection of $\mathbf{A}\mathbf{Q}_1$. But in general the reduced column echelon form is a row permutation of the above, of the form $\mathbf{J} \left[\begin{array}{c} \mathbf{I}_r \\ \mathbf{V} \end{array}\right]$, where $\mathbf{J}$ is a permutation matrix, an identity matrix $\mathbf{I}_m$ with permuted rows, and for which $\mathbf{J}^T \mathbf{J} = \mathbf{I}$. Premultiplication by $\mathbf{J}$ corresponds to a permutation of rows, and postmultiplication by $\mathbf{J}^T$ is the corresponding column permutation. Therefore let $\mathbf{P}$ be $\left[\begin{array}{cc} \mathbf{I}_r & 0 \\ -\mathbf{V} & \mathbf{I}_{m-r} \end{array}\right]\mathbf{J}^T$

so that

$$\mathbf{PAQ}_1 = \left(\begin{bmatrix} \mathbf{I}_r & 0 \\ -\mathbf{V} & \mathbf{I}_{m-r} \end{bmatrix} \mathbf{J}^T \right) \left(\mathbf{J} \begin{bmatrix} \mathbf{I}_r \\ \mathbf{V} \end{bmatrix} \right) = \begin{bmatrix} \mathbf{I}_r \\ 0 \end{bmatrix},$$

as required, and $\mathbf{P}$ can be written by inspection.

---

21 Setting λ to 0 yields $\phi(0) = a_n = \prod_{i=1}^{n}(\lambda - \lambda_i)|_{\lambda=0} = (-1)^n \prod_{i=1}^{n} \lambda_i$.

Now we need to prove that $\prod_{i=1}^{n}(\lambda - \lambda_i) = \lambda^n - \lambda^{n-1}\sum_{i=1}^{n}\lambda_i +$ lower-order terms, which is true by inspection for $n = 1$. Suppose this statement is true for the product of k terms. Then the product of $k+1$ terms is

$$\left(\lambda^k - \lambda^{k-1}\sum_{i=1}^{k}\lambda_i + \text{lower-order terms} \right)(\lambda - \lambda_{k+1})$$

$$= \lambda^{k+1} - \lambda^k \left(\lambda_{k+1} + \sum_{i=1}^{k}\lambda_i \right) + \text{lower-order terms}$$

$$= \lambda^{k+1} - \lambda^k \sum_{i=1}^{k+1}\lambda_i + \text{lower-order terms},$$

so the statement is true by induction for n terms.

Again, $\phi(\lambda)|_{\lambda=0} = a_n = \det(\mathbf{I} - \mathbf{A})|_{\lambda=0} = \det(-\mathbf{A}) = (-1)^n \det(\mathbf{A})$.

For $n = 3$, and by expanding along the bottom row,

$$\det \begin{bmatrix} \lambda - a_{11} & -a_{12} & -a_{13} \\ -a_{21} & \lambda - a_{22} & -a_{23} \\ -a_{31} & -a_{32} & \lambda - a_{33} \end{bmatrix}$$

$$= (\lambda - a_{33})(-1)^{3+3}(\lambda^2 - (a_{11}+a_{22})\lambda - a_{12}a_{21})$$
$$\quad - a_{32}(-1)^{3+2}(\text{terms of degree} \leq 1)$$
$$\quad - a_{31}(-1)^{3+1}(\text{terms of degree} \leq 1)$$
$$= \lambda^3 + \lambda^2(a_{11}+a_{22}+a_{33}) + \text{lower-order terms},$$

so the coefficient of λ^2 is $-\operatorname{trace}(\mathbf{A})$.

---

23 Calculating $\{\mathbf{H}_k\}$ yields

$$\mathbf{H}_0 = 0,$$

$$\mathbf{H}_1 = 3te^{-2t}\big|_{t=0} = 0,$$

$$\mathbf{H}_2 = 3e^{-2t} - 6te^{-2t}\big|_{t=0} = 3,$$

$$\mathbf{H}_3 = -12e^{-2t} + 12te^{-2t}\big|_{t=0} = -12,$$

$$\mathbf{H}_4 = 36e^{-2t} - 24te^{-2t}\big|_{t=0} = 36,$$

$$\mathbf{H}_5 = -96e^{-2t} + 48te^{-2t}\big|_{t=0} = -96,$$

$$\vdots$$

Since the form te^{-2t} is the solution of a second-order differential equation, $\mathbf{S}_r$ must have rank 2, and since rank $\mathbf{S}_2 = \mathrm{rank}\begin{bmatrix} 0 & 3 \\ 3 & -12 \end{bmatrix} = 2$, let $r = 2$. Calculating $\mathbf{P}, \mathbf{Q}$ yields

$$\begin{array}{cc|cc}
1 & 0 & 0 & 3 \\
0 & 1 & 3 & -12 \\
\hline
 & & 1 & 0 \\
 & & 0 & 1
\end{array}
\longrightarrow
\begin{array}{cc|cc}
4/3 & 1/3 & 1 & 0 \\
1/3 & 0 & 0 & 1 \\
\hline
 & & 1 & 0 \\
 & & 0 & 1
\end{array}$$

so that $\mathbf{P}_1 = \begin{bmatrix} 4/3 & 1/3 \\ 1/3 & 0 \end{bmatrix}$ and $\mathbf{Q}_1 = \mathbf{I}_2$. Then

$$\mathbf{A} = \mathbf{P}_1\begin{bmatrix} 3 & -12 \\ -12 & 36 \end{bmatrix}\mathbf{Q}_1 = \begin{bmatrix} 0 & -4 \\ 1 & -4 \end{bmatrix}, \quad \mathbf{B} = \mathbf{P}_1\begin{bmatrix} 0 \\ 3 \end{bmatrix} = \begin{bmatrix} 1 \\ 0 \end{bmatrix},$$

$$\mathbf{C} = [0, 3]\mathbf{Q}_1 = [0, 3], \quad \mathbf{D} = 0.$$

---

25 Step 0: $\mathbf{D} = 0$.

Step 1:

$$\mathbf{S}_1 = 1, \quad \mathrm{rank}\,\mathbf{S}_1 = 1$$

$$\mathbf{S}_2 = \begin{bmatrix} 1 & 1+\beta \\ 1+\beta & 1+\beta^2 \end{bmatrix}, \quad \mathrm{rank}\,\mathbf{S}_2 = 2,$$

Use $\mathbf{S}_2$ and check that all terms are computed correctly by the resulting $\mathbf{A}$, $\mathbf{B}$, $\mathbf{C}$, $\mathbf{D}$.

$$\begin{array}{cc|cc}
1 & 0 & 1 & 1+\beta \\
0 & 1 & 1+\beta & 2\beta+\beta^2
\end{array}
\longrightarrow
\begin{array}{cc|cc}
-(2\beta+\beta^2) & 1+\beta & 1 & 0 \\
1+\beta & -1 & 0 & 1
\end{array}$$

so that $\mathbf{P}_1 =$ as shown, and $\mathbf{Q}_1 = \mathbf{I}_2$.

Step 2:

$$\mathbf{A} = \mathbf{P}_1\begin{bmatrix} 1+\beta & 2\beta+\beta^2 \\ 2\beta+\beta^2 & 3\beta^2+\beta^3 \end{bmatrix}\mathbf{Q}_1 = \begin{bmatrix} 0 & -\beta^2 \\ 1 & 2\beta \end{bmatrix}, \quad \mathbf{B} = \mathbf{P}_1\begin{bmatrix} 1 \\ 1+\beta \end{bmatrix} = \begin{bmatrix} 1 \\ 0 \end{bmatrix},$$

$$\mathbf{C} = [1 \quad 1+\beta]\mathbf{Q}_1 = [1 \quad 1+\beta].$$

We must check that $r = 2$ is adequate. For the computed $\mathbf{A}, \mathbf{B}, \mathbf{C}, \mathbf{D}$, the characteristic polynomial is $\lambda^2 - 2\beta\lambda + \beta^2$, so if $\mathbf{H}_j = j\beta^{j-1} + \beta^j$ up to $\mathbf{H}_{k-1}$, then $\mathbf{H}_k = 2\beta\mathbf{H}_{k-1} - \beta^2\mathbf{H}_{k-2} = 2\beta((k-1)\beta^{k-2} + \beta^{k-1}) - \beta^2((k-2)\beta^{k-3} + \beta^{k-2}) = k\beta^{k-1} + \beta^k$, as required.

27 The general solution is given by Equation (5.81).

(a) Calculating the normal form yields

$$
\begin{array}{cc}
\mathbf{I}_2 & \mathbf{A} \quad \mathbf{B} \\
 & \mathbf{I}_2
\end{array}
\longrightarrow
\frac{\left.\begin{array}{cc|cc|c} 1/2 & 1/2 & 1 & 0 & 1 \\ 1/2 & -1/2 & 0 & 1 & 1 \end{array}\right.}{\begin{array}{cc} & 1 & 0 \\ & 0 & 1 \end{array}}
=
\begin{array}{c|c|c}
\mathbf{P}_1 & \mathbf{I} & \mathbf{P}_1\mathbf{B} \\
 & \mathbf{Q}_1 &
\end{array}
,
$$

so that $\mathbf{P}_2$ and $\mathbf{Q}_2$ do not appear, hence the solution exists and is unique, equal to $\mathbf{P}_1\mathbf{B} = \begin{bmatrix} 1 \\ 1 \end{bmatrix}$. The matrices $\mathbf{P}_1$ and $\mathbf{Q}_1$ are not unique.

(b) Calculating the normal form yields

$$
\begin{array}{cc}
\mathbf{I}_2 & \mathbf{A} \quad \mathbf{B} \\
 & \mathbf{I}_3
\end{array}
\longrightarrow
\frac{\left.\begin{array}{cc|ccc|c} 0 & 1 & 1 & 0 & 0 & 0 \\ 1 & 0 & 0 & 1 & 0 & 2 \end{array}\right.}{\begin{array}{ccc} .5 & -.5 & 1 \\ 0 & 1 & -1 \\ 0 & 0 & 1 \end{array}}
=
\begin{array}{c|cc|c}
\mathbf{P}_1 & \mathbf{I} & 0 & \mathbf{P}_1\mathbf{B} \\
 & \mathbf{Q}_1 & \mathbf{Q}_2 &
\end{array}
,
$$

where $\mathbf{P}_2$ does not appear, so that the solution exists, and is given by

$$
\mathbf{X} = \mathbf{Q}_1\mathbf{P}_1\mathbf{B} + \mathbf{Q}_2\mathbf{Y}_2 = \begin{bmatrix} .5 & -.5 \\ 0 & 1 \\ 0 & 0 \end{bmatrix} \begin{bmatrix} 0 \\ 2 \end{bmatrix} + \begin{bmatrix} 1 \\ -1 \\ 1 \end{bmatrix} \mathbf{Y}_2
$$

$$
= \begin{bmatrix} -1 \\ 2 \\ 0 \end{bmatrix} + \begin{bmatrix} 1 \\ -1 \\ 1 \end{bmatrix} \mathbf{Y}_2,
$$

where $\mathbf{Y}_2$ is an arbitrary scalar.

(c) Calculating the normal form yields

$$
\begin{array}{cc}
\mathbf{I}_3 & \mathbf{A} \quad \mathbf{B} \\
 & \mathbf{I}_3
\end{array}
\longrightarrow
\frac{\left.\begin{array}{ccc|ccc|c} 0 & 1 & 0 & 1 & 0 & 0 & 2 \\ 1 & 0 & 0 & 0 & 1 & 0 & 2 \\ -1 & -1 & 1 & 0 & 0 & 0 & 0 \end{array}\right.}{\begin{array}{ccc} .5 & -.5 & 1 \\ 0 & 1 & -1 \\ 0 & 0 & 1 \end{array}}
=
\begin{array}{c|cc|c}
\mathbf{P}_1 & \mathbf{I} & 0 & \mathbf{P}_1\mathbf{B} \\
\mathbf{P}_2 & 0 & 0 & \mathbf{P}_2\mathbf{B} \\
 & \mathbf{Q}_1 & \mathbf{Q}_2 &
\end{array}
,
$$

where $\mathbf{P}_2\mathbf{B} = 0$, so a solution exists, and is

$$
\mathbf{X} = \mathbf{Q}_1\mathbf{P}_1\mathbf{B} + \mathbf{Q}_2\mathbf{Y}_2 = \begin{bmatrix} .5 & -.5 \\ 0 & 1 \\ 0 & 0 \end{bmatrix} \begin{bmatrix} 2 \\ 2 \end{bmatrix} + \begin{bmatrix} 1 \\ -1 \\ 1 \end{bmatrix} \mathbf{Y}_2 = \begin{bmatrix} 0 \\ 2 \\ 0 \end{bmatrix} + \begin{bmatrix} 1 \\ -1 \\ 1 \end{bmatrix} \mathbf{Y}_2,
$$

where $\mathbf{Y}_2$ is an arbitrary scalar.

(d) Calculating the normal form yields

$$
\begin{array}{c}
\mathbf{I}_3 \quad \mathbf{A} \quad \mathbf{B} \\
\mathbf{I}_3
\end{array}
\longrightarrow
\left[
\begin{array}{ccc|ccc|c}
0 & 1 & 0 & 1 & 0 & 0 & 2 \\
1 & 0 & 0 & 0 & 1 & 0 & 2 \\
-1 & -1 & 1 & 0 & 0 & 0 & -8 \\
\hline
 & & & .5 & -.5 & 1 & \\
 & & & 0 & 1 & -1 & \\
 & & & 0 & 0 & 1 &
\end{array}
\right]
=
\begin{array}{ccc}
\mathbf{P}_1 & \mathbf{I} & 0 & \mathbf{P}_1\mathbf{B} \\
\mathbf{P}_2 & 0 & 0 & \mathbf{P}_2\mathbf{B} \\
\mathbf{Q}_1 & \mathbf{Q}_2 & &
\end{array};
$$

in this case $\mathbf{P}_2\mathbf{B} = -8 \neq 0$, so no solution exists.

29 The sequence of row operations is

$$
\mathbf{A} =
\begin{bmatrix}
1 & 4 & -2 \\
-1 & 0 & 1 \\
2 & 1 & -2
\end{bmatrix}
\xrightarrow{H_{13}}
\begin{bmatrix}
2 & 1 & -2 \\
-1 & 0 & 1 \\
1 & 4 & -2
\end{bmatrix}
\xrightarrow{\substack{H_{21}(1/2) \\ H_{31}(-1/2)}}
\begin{bmatrix}
2 & 1 & -2 \\
0 & .5 & 0 \\
0 & 3.5 & -1
\end{bmatrix}
$$

$$
\xrightarrow{H_{23}}
\begin{bmatrix}
2 & 1 & -2 \\
0 & 3.5 & -1 \\
0 & .5 & 0
\end{bmatrix}
\xrightarrow{H_{32}(-1/7)}
\begin{bmatrix}
2 & 1 & -2 \\
0 & 3.5 & -1 \\
0 & 0 & 1/7
\end{bmatrix}.
$$

Chapter 6   Solutions

1 We shall show that there is no zero element, and hence axiom A_4 is not satisfied. Let $\mathbf{X} = \begin{bmatrix} \alpha \\ m\alpha + b \end{bmatrix} \in \mathcal{V}$, and let $\mathbf{Z} = \begin{bmatrix} \beta \\ m\beta + b \end{bmatrix}$ denote the zero element, which must satisfy $\mathbf{X} + \mathbf{Z} = \mathbf{X}$; that is,

$$
\begin{bmatrix} \alpha \\ m\alpha + b \end{bmatrix} + \begin{bmatrix} \beta \\ m\beta + b \end{bmatrix} = \begin{bmatrix} \alpha \\ m\alpha + b \end{bmatrix}.
$$

Then β must simultaneously satisfy $\beta = 0$ and $\beta = -b/m$, which is impossible for finite m and nonzero b.

3 With the given definition of piecewise continuous functions, the sum of any pair of such functions is piecewise continuous, and the product by a scalar is piecewise continuous. The axioms are then true as for the periodic functions, with the zero vector given by the function that is identically zero for all $t \in [a, b]$.

5 The bases for the range and null space are calculated as in Equation (6.5):

(a) $\begin{bmatrix} \mathbf{A} \\ \hline \mathbf{I} \end{bmatrix} = \begin{bmatrix} 1 & 1 \\ 1 & -1 \\ 1 & 0 \\ \hline 0 & 1 \end{bmatrix} \overset{\text{col}}{\sim} \begin{bmatrix} 1 & 0 \\ 0 & -2 \\ 1 & -1 \\ \hline 0 & 1 \end{bmatrix} = \begin{bmatrix} \mathbf{AQ} \\ \hline \mathbf{Q} \end{bmatrix}.$

Since $\mathbf{A}$ has full column rank of 2, a basis for $\mathcal{R}(\mathbf{A})$ is $\mathbf{AQ}$, or $\mathbf{A}$ itself. Since $\mathcal{N}(\mathbf{A})$ has dimension 0, a basis for $\mathcal{N}(\mathbf{A})$ is $\begin{bmatrix} 0 \\ 0 \end{bmatrix}$.

(b) $\begin{bmatrix} \mathbf{A} \\ \mathbf{I} \end{bmatrix} = \begin{bmatrix} 0 & 1 & 1 \\ 2 & 1 & -1 \\ 1 & 0 & 0 \\ 0 & 1 & 0 \\ 0 & 0 & 1 \end{bmatrix} \overset{\text{col}}{\sim} \begin{bmatrix} 1 & 0 & 0 \\ 1 & 2 & -2 \\ 0 & 1 & 0 \\ 1 & 0 & -1 \\ 0 & 0 & 1 \end{bmatrix} \overset{\text{col}}{\sim} \begin{bmatrix} 1 & 0 & 0 \\ 1 & 2 & 0 \\ 0 & 1 & 1 \\ 1 & 0 & -1 \\ 0 & 0 & 1 \end{bmatrix} = \begin{bmatrix} \mathbf{AQ_1} & 0 \\ \mathbf{Q_1} & \mathbf{Q_2} \end{bmatrix}.$

A basis for $\mathcal{R}(\mathbf{A})$ is $\mathbf{AQ_1} = \begin{bmatrix} 1 & 0 \\ 1 & 2 \end{bmatrix}$. A basis for $\mathcal{N}(\mathbf{A})$ is $\mathbf{Q_2} = \begin{bmatrix} 1 \\ -1 \\ 1 \end{bmatrix}$.

(c) $\begin{bmatrix} \mathbf{A} \\ \mathbf{I} \end{bmatrix} = \begin{bmatrix} 0 & 1 & 1 \\ 2 & 1 & -1 \\ 2 & 2 & 0 \\ 1 & 0 & 0 \\ 0 & 1 & 0 \\ 0 & 0 & 1 \end{bmatrix} \overset{\text{col}}{\sim} \begin{bmatrix} 1 & 0 & 0 \\ 1 & 2 & -2 \\ 2 & 2 & -2 \\ 0 & 1 & 0 \\ 1 & 0 & -1 \\ 0 & 0 & 1 \end{bmatrix} \overset{\text{col}}{\sim} \begin{bmatrix} 1 & 0 & 0 \\ 1 & 2 & 0 \\ 2 & 2 & 0 \\ 0 & 1 & 1 \\ 1 & 0 & -1 \\ 0 & 0 & 1 \end{bmatrix} = \begin{bmatrix} \mathbf{AQ_1} & 0 \\ \mathbf{Q_1} & \mathbf{Q_2} \end{bmatrix}.$

A basis for $\mathcal{R}(\mathbf{A})$ is $\mathbf{AQ_1} = \begin{bmatrix} 1 & 0 \\ 1 & 2 \\ 2 & 2 \end{bmatrix}$. A basis for $\mathcal{N}(\mathbf{A})$ is $\mathbf{Q_2} = \begin{bmatrix} 1 \\ -1 \\ 1 \end{bmatrix}$.

---

7 See Example 14 in Section 4.2.

(a) Performed as a sequence of elementary column operations, the Gram-Schmidt process produces

$$\begin{bmatrix} 0 & 1 & 2 \\ -3 & 2 & 0 \\ 0 & 2 & 4 \end{bmatrix} \overset{K_{13}}{\longrightarrow} \begin{bmatrix} 2 & 1 & 0 \\ 0 & 2 & -3 \\ 4 & 2 & 0 \end{bmatrix} \overset{\substack{K_{21}(-10/20) \\ K_{31}(0)}}{\longrightarrow} \begin{bmatrix} 2 & 0 & 0 \\ 0 & 2 & -3 \\ 4 & 0 & 0 \end{bmatrix}$$

$$\overset{K_{23}}{\longrightarrow} \begin{bmatrix} 2 & 0 & 0 \\ 0 & -3 & 2 \\ 4 & 0 & 0 \end{bmatrix} \overset{K_{32}(6/9)}{\longrightarrow} \begin{bmatrix} 2 & 0 & 0 \\ 0 & -3 & 0 \\ 4 & 0 & 0 \end{bmatrix}.$$

The first operation moves the largest vector to the left, which becomes $\mathbf{Z_1}$. The next two operations subtract $(\mathbf{X}_i^T \mathbf{Z_1})/(\mathbf{Z_1^T Z_1})$ times $\mathbf{Z_1}$ from vectors on its right, to replace them with vectors orthogonal to $\mathbf{Z_1}$. Then a similar process is performed on the rightmost two vectors, and the process terminates when the last vector is zero, leaving two orthogonal vectors spanning a two-dimensional subspace of $\mathbb{R}^3$.

(b) Performing the column operations as in the previous case gives

$$\begin{bmatrix} 1 & 0 & 2 \\ 2 & -3 & 0 \\ 0 & 1 & 2 \end{bmatrix} \overset{K_{12}}{\longrightarrow} \begin{bmatrix} 0 & 1 & 2 \\ -3 & 2 & 0 \\ 1 & 0 & 2 \end{bmatrix} \overset{\substack{K_{21}(6/10) \\ K_{31}(-2/10)}}{\longrightarrow} \begin{bmatrix} 0 & 1 & 2 \\ -3 & 0.2 & 0.6 \\ 1 & 0.6 & 1.8 \end{bmatrix}$$

$$K_{23} \begin{bmatrix} 0 & 2 & 1 \\ -3 & 0.6 & 0.2 \\ 1 & 1.8 & 0.6 \end{bmatrix} K_{32}(-3.2/7.6) \begin{bmatrix} 0 & 2 & 0.1579 \\ -3 & 0.6 & -0.0526 \\ 1 & 1.8 & -0.1579 \end{bmatrix},$$

where, in this case, an orthogonal basis for $\mathbb{R}^3$ results.

9 Let the length of any given $\mathbf{X} = \begin{bmatrix} x_1 \\ x_2 \end{bmatrix}$ be m. Then $\mathbf{X} = m \begin{bmatrix} \cos\phi \\ \sin\phi \end{bmatrix}$, where

$\phi = \tan^{-1}(x_2/x_1)$, and

$$\mathbf{HX} = m \begin{bmatrix} \cos\theta & -\sin\theta \\ \sin\theta & \cos\theta \end{bmatrix} \begin{bmatrix} \cos\theta \\ \sin\theta \end{bmatrix} = m \begin{bmatrix} \cos\theta\cos\phi - \sin\theta\sin\phi \\ \sin\theta\cos\phi + \cos\theta\sin\phi \end{bmatrix}$$

$$= m \begin{bmatrix} \cos(\theta+\phi) \\ \sin(\theta+\phi) \end{bmatrix},$$

which is the counter-clockwise rotation of $\mathbf{X}$ through the angle θ.

11 Using the notation of Example 17, a vector $\mathbf{X}_1$ in the new frame has coordinates $\mathbf{X}$ in the original frame given by

$$\mathbf{X} = \mathbf{R}_{z,\theta}\mathbf{R}_{y,-\phi}\mathbf{X}_1.$$

The inverse transformation is

$$\mathbf{X}_1 = (\mathbf{R}_{y,-\phi})^{-1}(\mathbf{R}_{z,\theta})^{-1}\mathbf{X} = \mathbf{R}_{y,\phi}\mathbf{R}_{z,-\theta}\mathbf{X}$$

$$= \begin{bmatrix} \cos\phi & 0 & \sin\phi \\ 0 & 1 & 0 \\ -\sin\phi & 0 & \cos\phi \end{bmatrix} \begin{bmatrix} \cos\theta & \sin\theta & 0 \\ -\sin\theta & \cos\theta & 0 \\ 0 & 0 & 1 \end{bmatrix} \mathbf{X}$$

$$= \begin{bmatrix} \cos\phi\cos\theta & \cos\phi\sin\theta & \sin\phi \\ -\sin\theta & \cos\theta & 0 \\ -\sin\phi\cos\theta & -\sin\phi\sin\theta & \cos\phi \end{bmatrix} \mathbf{X}.$$

The y_1, z_1 coordinates are obtained by multiplying $\mathbf{X}$ by the bottom two rows of the above transformation matrix; that is,

$$\mathbf{P} = \begin{bmatrix} -\sin\theta & \cos\theta & 0 \\ -\sin\phi\cos\theta & -\sin\phi\sin\theta & \cos\phi \end{bmatrix}.$$

Chapter 7	Solutions

1 See Chapter 3 for formulas for the transfer matrix, and substitute the original and transformed matrices as required:

(a)

$$\hat{\mathbf{H}}(z) = \mathbf{D} + \mathbf{C}(z\mathbf{I} - \mathbf{A})^{-1}\mathbf{B} = 0 + [5,\ 1]\begin{bmatrix} z & -1 \\ 2 & z+3 \end{bmatrix}^{-1}\begin{bmatrix} 0 \\ 1 \end{bmatrix}$$

$$= [5,\ 1]\frac{1}{z^2+3z+2}\begin{bmatrix} z+3 & 1 \\ -2 & z \end{bmatrix}\begin{bmatrix} 0 \\ 1 \end{bmatrix} = \frac{z+5}{z^2+3z+2}.$$

(b) The new system has matrices $\mathbf{A}'$, $\mathbf{B}'$, $\mathbf{C}'$, $\mathbf{D}'$, where $\mathbf{D}' = \mathbf{D}$, and

$$\mathbf{A}' = \mathbf{S}^{-1}\mathbf{A}\mathbf{S} = \begin{bmatrix} -2 & -1 \\ -1 & -1 \end{bmatrix}\begin{bmatrix} 0 & 1 \\ -2 & -3 \end{bmatrix}\begin{bmatrix} -1 & 1 \\ 1 & -2 \end{bmatrix} = \begin{bmatrix} -1 & 0 \\ 0 & -2 \end{bmatrix},$$

$$\mathbf{B}' = \mathbf{S}^{-1}\mathbf{B} = \begin{bmatrix} -2 & -1 \\ -1 & -1 \end{bmatrix}\begin{bmatrix} 0 \\ 1 \end{bmatrix} = \begin{bmatrix} -1 \\ -1 \end{bmatrix},$$

$$\mathbf{C}' = \mathbf{C}\mathbf{S} = [5,\ 1]\begin{bmatrix} -1 & 1 \\ 1 & -2 \end{bmatrix} = [-4\quad 3].$$

(c) The transfer function $\hat{\mathbf{H}}'(z)$ of the new system is

$$\hat{\mathbf{H}}'(z) = \mathbf{D}' + \mathbf{C}'(z\mathbf{I} - \mathbf{A}')^{-1}\mathbf{B}' = 0 + [-4\quad 3]\begin{bmatrix} z+1 & 0 \\ 0 & z+2 \end{bmatrix}^{-1}\begin{bmatrix} -1 \\ -1 \end{bmatrix}$$

$$= \frac{4}{z+1} - \frac{3}{z+2} = \frac{z+5}{(z+1)(z+2)},$$

which is the same as for the untransformed system.

---

3 By the definition of eigenvalues and eigenvectors,

$$(\mathbf{A} - (\alpha + j\omega)\mathbf{I})\,(\mathbf{U} + j\mathbf{V}) = 0,$$

or

$$\mathbf{A}\mathbf{U} + j\mathbf{A}\mathbf{V} - \alpha\mathbf{U} - \alpha j\mathbf{V} - j\omega\mathbf{U} + \omega\mathbf{V} = 0.$$

Equating the real and imaginary parts to zero,

$$\mathbf{A}\mathbf{U} - \alpha\mathbf{U} + \omega\mathbf{V} = 0,$$

and

$$\mathbf{A}\mathbf{V} - \alpha\mathbf{V} - \omega\mathbf{U} = 0.$$

Now since $\mathbf{A}$ is real, its characteristic polynomial has real coefficients; hence its roots are real or occur in complex-conjugate pairs, so $\alpha - j\omega$ is an eigenvalue. To verify that $\mathbf{U} - j\mathbf{V}$ is an eigenvector, compute

$$(\mathbf{A} - (\alpha - j\omega)\mathbf{I})\,(\mathbf{U} - j\mathbf{V}) = (\mathbf{A}\mathbf{U} - \alpha\mathbf{U} + \omega\mathbf{V}) - j(\mathbf{A}\mathbf{V} - \alpha\mathbf{V} - \omega\mathbf{U}),$$

of which the real and imaginary parts have been shown previously to be zero.

5 The fundamental loop equation is

$$L\frac{di}{dt} + Ri = v$$

and the fundamental cut-set equation is

$$C\frac{dv}{dt} + Gv + i = 0,$$

from which the state equations are

$$\frac{d}{dt}\begin{bmatrix} i \\ v \end{bmatrix}\begin{bmatrix} -R/L & 1/L \\ -1/C & -G/C \end{bmatrix}\begin{bmatrix} i \\ v \end{bmatrix}.$$

The characteristic equation is

$$\lambda^2 - \lambda\left(-\frac{R}{L} - \frac{G}{C}\right) + \left(\frac{RG}{LC} + \frac{1}{LC}\right) = 0$$

so that for small R and G the eigenvalues are complex as shown:

$$\lambda = -\frac{1}{2}\left(\frac{R}{L} + \frac{G}{C}\right) \pm \frac{j}{2}\sqrt{\frac{4}{LC} - \left(\frac{R}{L} - \frac{G}{C}\right)^2}.$$

Computing the first eigenvector by column operations gives

$$\begin{bmatrix} \mathbf{A} - \lambda_1\mathbf{I} \\ \mathbf{I} \end{bmatrix} = \begin{bmatrix} \frac{1}{2}(\frac{G}{C} - \frac{R}{L}) - \frac{j}{2}\sqrt{} & \frac{1}{L} \\ \frac{-1}{C} & \frac{1}{2}(\frac{R}{L} - \frac{G}{C}) - \frac{j}{2}\sqrt{} \\ 1 & 0 \\ 0 & 1 \end{bmatrix}$$

$$\rightarrow \begin{bmatrix} \frac{1}{2}(\frac{G}{C} - \frac{R}{L}) - \frac{j}{2}\sqrt{} & 0 \\ \frac{-1}{C} & 0 \\ 1 & (-1/L)/\left(\frac{1}{2}(\frac{G}{C} - \frac{R}{L}) - \frac{j}{2}\sqrt{}\right) \\ 0 & 1 \end{bmatrix}$$

$$\rightarrow \begin{bmatrix} \frac{1}{2}(\frac{G}{C} - \frac{R}{L}) - \frac{j}{2}\sqrt{} & 0 \\ \frac{-1}{C} & 0 \\ 1 & -1/L \\ 0 & \frac{1}{2}(\frac{G}{C} - \frac{R}{L}) - \frac{j}{2}\sqrt{} \end{bmatrix} = \begin{bmatrix} \times & 0 \\ \times & \mathbf{S}_1 \end{bmatrix},$$

where the contents of the square root are as above. Taking the real part and the negative of the imaginary part of the above computed eigenvector $\mathbf{S}_1$, the similarity transformation matrix is

$$\mathbf{S} = [\operatorname{Re}(\mathbf{S}_1), -\operatorname{Im}(\mathbf{S}_1)] = \begin{bmatrix} -1/L & 0 \\ \frac{1}{2}(\frac{G}{C} - \frac{R}{L}) & \frac{1}{2}\sqrt{\frac{4}{LC} - (\frac{R}{L} - \frac{G}{C})^2} \end{bmatrix},$$

and without computation the resulting system matrix must be

$$\mathbf{S}^{-1}\mathbf{AS} = \begin{bmatrix} -\frac{1}{2}\left(\frac{R}{L}+\frac{G}{C}\right) & -\frac{1}{2}\sqrt{\frac{4}{LC}-\left(\frac{R}{L}-\frac{G}{C}\right)^2} \\ \frac{1}{2}\sqrt{\frac{4}{LC}-\left(\frac{R}{L}-\frac{G}{C}\right)^2} & -\frac{1}{2}\left(\frac{R}{L}+\frac{G}{C}\right) \end{bmatrix}.$$

7 The matrix is block-triangular, the 1×1 block having eigenvalue -4, and the remaining 2×2 block has characteristic polynomial $\phi(\lambda) = \lambda^2 - \lambda - 2$, so the remaining 2 eigenvalues are -1, 2.

9 The eigenvalues are -2, -2, -2, and the formulas for $e^{t\mathbf{A}}$ and $\mathbf{A}^k$ are given by $\alpha_0\mathbf{I}+\alpha_1\mathbf{A}+\alpha_2\mathbf{A}^2$, with coefficients α_0, α_1, α_2 found by solving, respectively,

$$\begin{bmatrix} 1 & (-2) & (-2)^2 \\ 0 & 1 & 2(-2) \\ 0 & 0 & 2 \end{bmatrix} \begin{bmatrix} \alpha_0 \\ \alpha_1 \\ \alpha_2 \end{bmatrix} = \begin{bmatrix} e^{t(-2)} \\ te^{t(-2)} \\ t^2 e^{t(-2)} \end{bmatrix} \text{ or } \begin{bmatrix} (-2)^k \\ k(-2)^{k-1} \\ k(k-1)(-2)^{k-2} \end{bmatrix}.$$

Solving, in the first case gives

$$\begin{bmatrix} \alpha_0 \\ \alpha_1 \\ \alpha_2 \end{bmatrix} = \begin{bmatrix} 1 & 2 & 2 \\ 0 & 1 & 2 \\ 0 & 0 & 0.5 \end{bmatrix} \begin{bmatrix} e^{t(-2)} \\ te^{t(-2)} \\ t^2 e^{t(-2)} \end{bmatrix} = e^{-2t} \begin{bmatrix} 1 + 2t + 2t^2 \\ t + 2t^2 \\ t^2/2 \end{bmatrix},$$

so that, collecting terms in equal powers of t,

$$e^{t\mathbf{A}} = e^{-2t}\begin{bmatrix} 1 & 0 & 0 \\ 0 & 1 & 0 \\ 0 & 0 & 1 \end{bmatrix} + te^{-2t}\begin{bmatrix} 2 & 0 & -8 \\ 1 & 2 & -12 \\ 0 & 1 & -4 \end{bmatrix} + t^2 e^{-2t}\begin{bmatrix} 2 & -4 & 8 \\ 2 & -4 & 8 \\ 0.5 & -1 & 2 \end{bmatrix},$$

and in the second case,

$$\begin{bmatrix} \alpha_0 \\ \alpha_1 \\ \alpha_2 \end{bmatrix} = \begin{bmatrix} 1 & 2 & 2 \\ 0 & 1 & 2 \\ 0 & 0 & 0.5 \end{bmatrix} \begin{bmatrix} (-2)^k \\ k(-2)^{k-1} \\ k(k-1)(-2)^{k-2} \end{bmatrix} = (-2)^{k-2}\begin{bmatrix} 4 - 6k + 2k^2 \\ -4k + 2k^2 \\ k(k-1)/2 \end{bmatrix},$$

so that

$$\mathbf{A}^k = (-2)^k \begin{bmatrix} 1 & 0 & 0 \\ 0 & 1 & 0 \\ 0 & 0 & 1 \end{bmatrix} + k(-2)^{k-1}\begin{bmatrix} 2 & 0 & -8 \\ 1 & 2 & -12 \\ 0 & 1 & -4 \end{bmatrix}$$

$$+k(k-1)(-2)^{k-2}\begin{bmatrix} 2 & -4 & 8 \\ 2 & -4 & 8 \\ .5 & -1 & 2 \end{bmatrix}.$$

11 Given $\mathbf{F}$ and T, compute $T\mathbf{A} = \log \mathbf{F}$, so that

$$\mathbf{A} = \frac{1}{T} \log \mathbf{F}.$$

Since the logarithm is not unique; that is, $\log z = \log |z| + j2n\pi \arg z$, for $n = 0, \pm 1, \pm 2, \cdots$, $\mathbf{A}$ is not unique. This is the phenomenon of aliasing.
From $\mathbf{A}$, compute $\mathbf{M} = \int_0^T e^{\tau \mathbf{A}} d\tau$, and solve $\mathbf{MB} = \mathbf{G}$.

13 Multiplying eigenvector $\mathbf{X}_i$ by $f(\mathbf{A})$ gives
$$\begin{aligned} f(\mathbf{A})\, \mathbf{X}_i &= (\alpha_0 \mathbf{I} + \alpha_1 \mathbf{A} + \cdots \alpha_{n-1} \mathbf{A}^{n-1})\, \mathbf{X}_i \\ &= (\alpha_0 \mathbf{I} + \alpha_1 \lambda_i + \cdots \alpha_{n-1} \lambda_i^{n-1})\, \mathbf{X}_i = f(\lambda_i)\, \mathbf{X}_i, \end{aligned}$$
showing that the matrix $f(\mathbf{A})$ has eigenvector $\mathbf{X}_i$ and corresponding eigenvector $f(\lambda_i)$.

Chapter 8   Solutions

1 The internal stability is determined by calculating the eigenvalues of the $\mathbf{A}$ matrix.

(a) The eigenvalues are $\pm j$, -1, so the system is stable.

(b) The eigenvalues are $\pm j$, 1, so the system is unstable.

(c) The eigenvalues are 0, 0, -1, and since there is a multiple eigenvalue on the imaginary axis, the rank of $\mathbf{A} - 0\mathbf{I}$ must be tested. This rank is 2, so there is only one eigenvalue corresponding to eigenvalue 0, and the corresponding Jordan block is 2×2. Hence the system is unstable.

3 The matrix $\mathbf{A}$ is asymptotically stable, implying $\text{Re}(\lambda_i) < 0$ (or here for simplicity, assuming the λ_i are real, $\lambda_i < 0$). The discrete process is to be stable, and $(\mathbf{I} + h\mathbf{A})$ has eigenvalues $1 + h\lambda_i$, so $|1 + h\lambda_i| < 1$ for all eigenvalues λ_i of $\mathbf{A}$. Therefore the step size is limited by the inequalities $0 < h < -2/\lambda_i$, which is true for all i provided $0 < h < -2/\lambda_M$, where λ_M is the eigenvalue with largest absolute value, that is, corresponding to the fastest time-constant in the continuous-time solution. Thus the step size of the Euler forward method is limited at all times to be a fraction of the fastest time constant, even though the solution may be changing very slowly, the solution components of fast time-constant having decayed to very small values.

5 For the trapezoidal method the discrete equation of the previous problem reduces to

$$\mathbf{X}(k+1) = (\mathbf{I}-h\mathbf{A})^{-1}(\mathbf{I}+h\mathbf{A})\,\mathbf{X},$$

for which the eigenvalues are $(1-h\lambda_i)^{-1}(1+h\lambda_i)$. Thus, for asymptotic stability,

$$1 > \frac{|1+h\alpha_i/2 + jh\omega_i/2|^2}{|1-h\alpha_i/2 - jh\omega_i/2|^2} = \frac{1 + h\alpha_i + h^2\alpha_i^2/4 + h^2\omega_i^2/4}{1 - h\alpha_i + h^2\alpha_i^2/4 + h^2\omega_i^2/4},$$

or

$$-h\alpha_i > h\alpha_i,$$

which is true for any positive step size h since the α_i are all nonzero and negative.

7 See Section 3.3.

(a) Writing the equations in detail as in Example 9 yields

$$\begin{bmatrix} -1/2 & 1 \\ -1 & -1/2 \end{bmatrix}\begin{bmatrix} p_{11} & p_{12} \\ p_{12} & p_{22} \end{bmatrix} + \begin{bmatrix} p_{11} & p_{12} \\ p_{12} & p_{22} \end{bmatrix}\begin{bmatrix} -1/2 & -1 \\ 1 & -1/2 \end{bmatrix} = -\begin{bmatrix} 2 & 1 \\ 1 & 2 \end{bmatrix},$$

in which there are three unknowns. Writing the three corresponding equations yields

$$\begin{bmatrix} -1 & -2 & 0 \\ -1 & -1 & 1 \\ 0 & -2 & -1 \end{bmatrix}\begin{bmatrix} p_{11} \\ p_{12} \\ p_{22} \end{bmatrix} = \begin{bmatrix} -2 \\ -1 \\ -2 \end{bmatrix},$$

and solving gives

$$\mathbf{P} = \begin{bmatrix} 2.4 & 0.2 \\ 0.2 & 1.6 \end{bmatrix}.$$

(b) The equations to be solved are

$$\begin{bmatrix} -2 & 0 & 0 \\ 1 & 0 & 0 \\ 0 & 2 & 2 \end{bmatrix}\begin{bmatrix} p_{11} \\ p_{12} \\ p_{22} \end{bmatrix} = \begin{bmatrix} -2 \\ -1 \\ -2 \end{bmatrix},$$

where the coefficient matrix is singular, and the equations are inconsistent. In fact, by inspection, the two eigenvectors of $\mathbf{A}$ are $\lambda_1 = -1$, $\lambda_2 = 1$, and $\lambda_1 + \lambda_2 = 0$, indicating that the equations are singular.

(c) Since $\mathbf{A}$ is 2×2, the equations can be solved as above, but in this case $\mathbf{A}$ is upper-triangular, so the solution can be obtained by solving smaller equations in sequence. Writing the equations in detail gives

$$\begin{bmatrix} -1 & 0 \\ 1 & 2 \end{bmatrix} \begin{bmatrix} p_{11} & p_{12} \\ p_{12} & p_{22} \end{bmatrix} + \begin{bmatrix} p_{11} & p_{12} \\ p_{12} & p_{22} \end{bmatrix} \begin{bmatrix} -1 & 1 \\ 0 & 2 \end{bmatrix} = -\begin{bmatrix} 1 & 0 \\ 0 & 1 \end{bmatrix},$$

from which,

$$-2p_{11} = -1 \qquad \Rightarrow p_{11} = 0.5$$

$$p_{11} + p_{12} = 0 \qquad \Rightarrow p_{12} = -p_{11} = -0.5$$

$$2p_{12} + 4p_{22} = -1 \Rightarrow p_{22} = (-1 - 2(-0.5))/4 = 0$$

so that

$$\mathbf{P} = \begin{bmatrix} .5 & -.5 \\ -.5 & 0 \end{bmatrix}.$$

In this case $\mathbf{P}$ exists and is symmetric but is not positive-definite, as can be verified by computing its eigenvalues as $\lambda = 0.81, -.31$, or by noting that $\mathbf{A}$ has an eigenvalue in the right-half plane.

9 Writing the partitioned transformed equation yields

$$(\mathbf{A}')^T(\mathbf{S}^T\mathbf{PS}) + (\mathbf{S}^T\mathbf{PS})\,\mathbf{A}' = \begin{bmatrix} \mathbf{A}_{11}^T & 0 & 0 \\ \mathbf{A}_{12}^T & \mathbf{A}_{22}^T & 0 \\ \mathbf{A}_{13}^T & \mathbf{A}_{23}^T & \mathbf{A}_{33}^T \end{bmatrix} \begin{bmatrix} \mathbf{P}_{11} & \mathbf{P}_{12} & \mathbf{P}_{13} \\ \mathbf{P}_{12}^T & \mathbf{P}_{22} & \mathbf{P}_{23} \\ \mathbf{P}_{13}^T & \mathbf{P}_{23}^T & \mathbf{P}_{33} \end{bmatrix}$$

$$+ \begin{bmatrix} \mathbf{P}_{11} & \mathbf{P}_{12} & \mathbf{P}_{13} \\ \mathbf{P}_{12}^T & \mathbf{P}_{22} & \mathbf{P}_{23} \\ \mathbf{P}_{13}^T & \mathbf{P}_{23}^T & \mathbf{P}_{33} \end{bmatrix} \begin{bmatrix} \mathbf{A}_{11} & \mathbf{A}_{12} & \mathbf{A}_{13} \\ 0 & \mathbf{A}_{22} & \mathbf{A}_{23} \\ 0 & 0 & \mathbf{A}_{33} \end{bmatrix} = -\begin{bmatrix} \mathbf{Q}_{11} & \mathbf{Q}_{12} & \mathbf{Q}_{13} \\ \mathbf{Q}_{12}^T & \mathbf{Q}_{22} & \mathbf{Q}_{23} \\ \mathbf{Q}_{13}^T & \mathbf{Q}_{23}^T & \mathbf{Q}_{33} \end{bmatrix},$$

from which the following block equations can be written and solved in order:

$$\mathbf{A}_{11}^T\mathbf{P}_{11} + \mathbf{P}_{11}\mathbf{A}_{11} - -\mathbf{Q}_{11}$$

$$\mathbf{A}_{11}^T\mathbf{P}_{12} + \mathbf{P}_{12}\mathbf{A}_{22} = -\mathbf{Q}_{12} - \mathbf{P}_{11}\mathbf{A}_{12}$$

$$\mathbf{A}_{11}^T\mathbf{P}_{13} + \mathbf{P}_{13}\mathbf{A}_{33} = -\mathbf{Q}_{13} - \mathbf{P}_{11}\mathbf{A}_{13} - \mathbf{P}_{12}\mathbf{A}_{23}$$

$$\mathbf{A}_{22}^T\mathbf{P}_{22} + \mathbf{P}_{22}\mathbf{A}_{22} = -\mathbf{Q}_{22} - \mathbf{A}_{12}^T\mathbf{P}_{12} - \mathbf{P}_{12}^T\mathbf{A}_{12}$$

$$\mathbf{A}_{22}^T\mathbf{P}_{23} + \mathbf{P}_{23}\mathbf{A}_{33} = -\mathbf{Q}_{23} - \mathbf{A}_{12}^T\mathbf{P}_{13} - \mathbf{P}_{12}^T\mathbf{A}_{13} - \mathbf{P}_{22}\mathbf{A}_{23}$$

$$\mathbf{A}_{33}^T\mathbf{P}_{33} + \mathbf{P}_{33}\mathbf{A}_{33} = -\mathbf{Q}_{33} - \mathbf{A}_{13}^T\mathbf{P}_{13} - \mathbf{A}_{23}^T\mathbf{P}_{23} - \mathbf{P}_{13}^T\mathbf{A}_{13} - \mathbf{P}_{23}^T\mathbf{A}_{23}.$$

Thus $\mathbf{P}' - \mathbf{S}^T\mathbf{PS}$ can be constructed, and then $\mathbf{P}$ can be computed as $\mathbf{P} = (\mathbf{S}^T)^{-1}\mathbf{P}'\mathbf{S}^{-1}$.

11 Refer to Example 7 for the continuous-time case, and for the discrete-time case, refer to Example 15.

(a) Taking the Laplace transform gives

$$\hat{y}(s) = \frac{2}{(s+2)^2} = \frac{2}{s^2 + 4s + 4},$$

so a system which has this signal as free response $\mathbf{C}e^{t\mathbf{A}}\mathbf{X}(0)$ is given by

$$\mathbf{A} = \begin{bmatrix} 0 & 1 \\ -4 & -4 \end{bmatrix}, \quad \mathbf{X}(0) = \begin{bmatrix} 0 \\ 1 \end{bmatrix}, \quad \mathbf{C} = [2, 0].$$

The Lyapunov equation is

$$\mathbf{A}^T\mathbf{P} + \mathbf{PA} = \begin{bmatrix} 0 & -4 \\ 1 & -4 \end{bmatrix}\begin{bmatrix} p_{11} & p_{12} \\ p_{12} & p_{22} \end{bmatrix} + \begin{bmatrix} p_{11} & p_{12} \\ p_{12} & p_{22} \end{bmatrix}\begin{bmatrix} 0 & 1 \\ -4 & -4 \end{bmatrix}$$

$$= -\mathbf{C}^T\mathbf{C} = \begin{bmatrix} -4 & 0 \\ 0 & 0 \end{bmatrix}$$

and has solution

$$\mathbf{P} = \begin{bmatrix} 5/2 & 1/2 \\ 1/2 & 1/8 \end{bmatrix}$$

so the energy is

$$w = [0, 1]\begin{bmatrix} 5/2 & 1/2 \\ 1/2 & 1/8 \end{bmatrix}\begin{bmatrix} 0 \\ 1 \end{bmatrix} = 1/8.$$

(b) A system with $\{1/(2^k) + 1/(3^k)\}$ as free response sequence $\mathbf{C}\mathbf{A}^k\mathbf{X}(0)$ is given by

$$\mathbf{A} = \begin{bmatrix} 1/2 & 0 \\ 0 & 1/3 \end{bmatrix}, \quad \mathbf{X}(0) = \begin{bmatrix} 1 \\ 1 \end{bmatrix}, \quad \mathbf{C} = [1, 1].$$

The Lyapunov equation is

$$\mathbf{A}^T\mathbf{PA} - \mathbf{P} = \begin{bmatrix} 1/2 & 0 \\ 0 & 1/3 \end{bmatrix}\begin{bmatrix} p_{11} & p_{12} \\ p_{12} & p_{22} \end{bmatrix}\begin{bmatrix} 1/2 & 0 \\ 0 & 1/3 \end{bmatrix} - \begin{bmatrix} p_{11} & p_{12} \\ p_{12} & p_{22} \end{bmatrix}$$

$$= -\mathbf{C}^T\mathbf{C} = \begin{bmatrix} -1 & -1 \\ -1 & -1 \end{bmatrix}$$

and has solution

$$\mathbf{P} = \begin{bmatrix} 4/3 & 6/5 \\ 6/5 & 9/8 \end{bmatrix}$$

so the energy is

$$w = [1, 1]\begin{bmatrix} 4/3 & 6/5 \\ 6/5 & 9/8 \end{bmatrix}\begin{bmatrix} 1 \\ 1 \end{bmatrix} = 4/3 + 6/5 + 6/5 + 9/8 = 4.86.$$

1 Controllability will be tested, then observability, then a minimal externally equivalent system.

(a) i. The controllability matrix is

$$\mathscr{C} = \begin{bmatrix} 4 & -8 & 16 \\ -1 & 2 & -4 \\ 2 & -4 & 8 \end{bmatrix} \overset{\text{col}}{\sim} \begin{bmatrix} 4 & 0 & 0 \\ -1 & 0 & 0 \\ 2 & 0 & 0 \end{bmatrix},$$

which has rank $1 < n$, so the pair $(\mathbf{A}, \mathbf{B})$ is not controllable.

ii. The controllability matrix is

$$\mathscr{C} = \begin{bmatrix} 1 & 3 & -1 & -2 & -3 & 3 \\ 0 & 3 & 1 & -1 & -9 & -1 \end{bmatrix},$$

which has full row rank of 2 by inspection, so $(\mathbf{A}, \mathbf{B})$ is a controllable pair.

iii. The controllability matrix is

$$\mathscr{C} = \begin{bmatrix} 0 & 1 & 0 & -3 & 0 & 9 \\ 1 & 0 & -3 & 0 & 9 & 0 \\ 0 & 0 & 0 & 0 & 0 & 0 \end{bmatrix},$$

which has rank 2 by inspection, so the pair $(\mathbf{A}, \mathbf{B})$ is not controllable.

iv. The controllability matrix is

$$\mathscr{C} = \begin{bmatrix} 0 & 1 & 0 & -3 & 0 & 9 \\ 1 & 0 & -3 & 0 & 9 & 0 \\ 0 & 1 & 0 & -3 & 0 & 9 \end{bmatrix} \overset{\text{col}}{\sim} \begin{bmatrix} 1 & 0 & 0 & 0 & 0 & 0 \\ 0 & 1 & 0 & 0 & 0 & 0 \\ 1 & 0 & 0 & 0 & 0 & 0 \end{bmatrix},$$

which has rank 2, so the pair $(\mathbf{A}, \mathbf{B})$ is not controllable.

(b) i. The observability matrix is

$$\mathscr{O} = \begin{bmatrix} 1 & 2 & -1 \\ -3 & -4 & 4 \\ 9 & 12 & -12 \end{bmatrix} \overset{\text{row}}{\sim} \begin{bmatrix} 1 & 2 & -1 \\ 0 & 2 & 1 \\ 0 & 0 & 0 \end{bmatrix},$$

which has rank 2, so $(\mathbf{C}, \mathbf{A})$ is not an observable pair.

ii. The observability matrix is

$$\mathscr{O} = \begin{bmatrix} 1 & 3 \\ -1 & 3 \\ -5 & -5 \\ -1 & -7 \end{bmatrix},$$

which has full column rank $n = 2$ by inspection; therefore $(\mathbf{C}, \mathbf{A})$ is an observable pair.

iii. The observability matrix is

$$
\mathcal{O} = \begin{bmatrix} 0 & 1 & 0 \\ 0 & 0 & 1 \\ 0 & -3 & 0 \\ 0 & 0 & -3 \\ 0 & 9 & 0 \\ 0 & 0 & 9 \end{bmatrix},
$$

which by inspection has less than full column rank, so $(\mathbf{C}, \mathbf{A})$ is not an observable pair.

iv. The observability matrix is

$$
\mathcal{O} = \begin{bmatrix} 0 & 1 & 0 \\ 1 & 0 & 1 \\ 0 & -3 & 0 \\ -3 & 0 & -3 \\ 0 & 9 & 0 \\ 9 & 0 & 9 \end{bmatrix} \underset{\sim}{\text{row}} \begin{bmatrix} 1 & 0 & 1 \\ 0 & 1 & 0 \\ 0 & 0 & 0 \\ 0 & 0 & 0 \\ 0 & 0 & 0 \\ 0 & 0 & 0 \end{bmatrix},
$$

which has rank $2 < n$, so $(\mathbf{C}, \mathbf{A})$ are not an observable pair.

(c) i. From the controllability calculation, a similarity matrix can be written as

$$
\mathbf{S} = [\,\mathbf{V}, \mathbf{U}\,] = \begin{bmatrix} 1 & 0 & 4 \\ 0 & 1 & -1 \\ 0 & 0 & 2 \end{bmatrix},
$$

where the first two columns have been written by inspection to be independent of the third column, which is $\mathbf{U}$. The transformed system is then

$$
\mathbf{S}^{-1}\mathbf{AS} = \begin{bmatrix} -3 & -4 & 0 \\ 0 & 0 & 0 \\ -2 & -8 & -2 \end{bmatrix}, \quad \mathbf{S}^{-1}\mathbf{B} = \begin{bmatrix} 0 \\ 0 \\ 1 \end{bmatrix}, \quad \mathbf{CS} = [\,1 \quad 2 \quad 0\,],
$$

so that the controllable part of the system has matrices $\mathbf{A}_{22} = -2$, $\mathbf{B}_2 = 1$, $\mathbf{C}_2 = 0$, which is identically unobservable, so the smallest externally equivalent system for the given system has state dimension zero and the transfer matrix is $\mathbf{D} = 3$.

ii. The system is completely controllable and observable, and is therefore of minimal order as given. The transfer matrix is

$$
\hat{\mathbf{H}}(s) = 0 + \begin{bmatrix} 1 & 3 \\ -1 & 3 \end{bmatrix} \begin{bmatrix} s+2 & -1 \\ 1 & s+2 \end{bmatrix}^{-1} \begin{bmatrix} 1 & 3 & -1 \\ 0 & 3 & 1 \end{bmatrix}
$$

$$
= \begin{bmatrix} 1 & 3 \\ -1 & 3 \end{bmatrix} \frac{1}{s^2+4s+5} \begin{bmatrix} s+2 & 1 \\ -1 & s+2 \end{bmatrix} \begin{bmatrix} 1 & 3 & -1 \\ 0 & 3 & 1 \end{bmatrix}
$$

$$
= \frac{1}{s^2+4s+5} \begin{bmatrix} s-1 & 3s+7 \\ -s-5 & 3s+5 \end{bmatrix} \begin{bmatrix} 1 & 3 & -1 \\ 0 & 3 & 1 \end{bmatrix}
$$

$$= \frac{1}{s^2+4s+5} \begin{bmatrix} s-1 & 12s+18 & 2s+8 \\ -s-5 & 6s & 4s+10 \end{bmatrix}.$$

iii. The controllability matrix $\mathscr{C}$ has rank 2, and by inspection its first two columns are a basis for $\mathcal{R}(\mathscr{C})$. A similarity transformation matrix can be written as

$$\mathbf{S} = [\mathbf{V}, \mathbf{U}] = \begin{bmatrix} 0 & 0 & 1 \\ 0 & 1 & 0 \\ 1 & 0 & 0 \end{bmatrix},$$

so the transformed system is

$$\mathbf{S}^{-1}\mathbf{AS} = \begin{bmatrix} -3 & 0 & 0 \\ 0 & -3 & 0 \\ 0 & 0 & -3 \end{bmatrix}, \quad \mathbf{S}^{-1}\mathbf{B} = \begin{bmatrix} 0 & 0 \\ 1 & 0 \\ 0 & 1 \end{bmatrix}, \quad \mathbf{CS} = \begin{bmatrix} 0 & 1 & 0 \\ 1 & 0 & 0 \end{bmatrix}.$$

The controllable part of this system is

$$\mathbf{A}_{22} = \begin{bmatrix} -3 & 0 \\ 0 & -3 \end{bmatrix}, \quad \mathbf{B}_2 = \begin{bmatrix} 1 & 0 \\ 0 & 1 \end{bmatrix}, \quad \mathbf{C}_2 = \begin{bmatrix} 1 & 0 \\ 0 & 0 \end{bmatrix}.$$

Then, constructing the dual system and its controllability matrix gives

$$\mathscr{C} = \begin{bmatrix} 1 & 0 & -3 & 0 \\ 0 & 0 & 0 & 0 \end{bmatrix}.$$

The first column can be chosen as the basis $\mathbf{U}$, so that a similarity matrix can be written by inspection as $\mathbf{S} = [\mathbf{V}, \mathbf{U}] = \begin{bmatrix} 0 & 1 \\ 1 & 0 \end{bmatrix}$, giving the transformed system

$$\mathbf{A}'' = \begin{bmatrix} -3 & 0 \\ 0 & -3 \end{bmatrix}, \quad \mathbf{B}'' = \begin{bmatrix} 0 & 0 \\ 1 & 0 \end{bmatrix}, \quad \mathbf{C}'' = \begin{bmatrix} 0 & 1 \\ 1 & 0 \end{bmatrix}.$$

Taking the dual of the controllable part gives the final reduced system

$$\tilde{\mathbf{A}}_{22} = -3, \quad \tilde{\mathbf{B}}_2 = [1 \ \ 0], \quad \tilde{\mathbf{C}}_2 = \begin{bmatrix} 1 \\ 0 \end{bmatrix}.$$

The transfer matrix is

$$\hat{\mathbf{H}}(s) = 0 + \begin{bmatrix} 1 \\ 0 \end{bmatrix} (s+3)^{-1} [1 \ \ 0] = \begin{bmatrix} (s+3)^{-1} & 0 \\ 0 & 0 \end{bmatrix}.$$

iv. Constructing a basis $\mathbf{V}$ for $\mathcal{N}(\mathbf{U}^T)$ where $\mathbf{U}$ is a basis for the range of the controllability matrix yields

$$\begin{bmatrix} \mathbf{U}^T \\ \mathbf{I} \end{bmatrix} = \begin{bmatrix} 1 & 0 & 1 \\ 0 & 1 & 0 \\ 1 & 0 & 0 \\ 0 & 1 & 0 \\ 0 & 0 & 1 \end{bmatrix} \overset{\text{col}}{\sim} \begin{bmatrix} 1 & 0 & 0 \\ 0 & 1 & 0 \\ 1 & 0 & -1 \\ 0 & 1 & 0 \\ 0 & 0 & 1 \end{bmatrix} = \begin{bmatrix} \times & 0 \\ \times & \mathbf{V} \end{bmatrix},$$

so a suitable similarity matrix $[\mathbf{V}, \mathbf{U}]$ is

$$\mathbf{S} = \begin{bmatrix} -1 & 1 & 0 \\ 0 & 0 & 1 \\ 1 & 1 & 0 \end{bmatrix},$$

and the transformed system is

$$\mathbf{A}' = \mathbf{A}, \quad \mathbf{B}' = \begin{bmatrix} 0 & 0 \\ 0 & 1 \\ 1 & 0 \end{bmatrix}, \quad \mathbf{C}' = \begin{bmatrix} 0 & 0 & 1 \\ 0 & 2 & 0 \end{bmatrix},$$

of which the controllable part is

$$\mathbf{A}_{22} = \begin{bmatrix} -3 & 0 \\ 0 & -3 \end{bmatrix}, \quad \mathbf{B}_2 = \begin{bmatrix} 0 & 1 \\ 1 & 0 \end{bmatrix}, \quad \mathbf{C}_2 = \begin{bmatrix} 0 & 1 \\ 2 & 0 \end{bmatrix}.$$

Because $\text{rank}(\mathbf{C}_2) = 2$, this system is completely observable, with transfer matrix

$$\hat{\mathbf{H}}(s) = 0 + \begin{bmatrix} 0 & 1 \\ 2 & 0 \end{bmatrix} \frac{1}{s+3} \begin{bmatrix} 0 & 1 \\ 1 & 0 \end{bmatrix} = \frac{1}{s+3} \begin{bmatrix} 1 & 0 \\ 0 & 2 \end{bmatrix}.$$

---

3 If $\mathbf{A} = \text{diag}[\lambda_i]$ and $\mathbf{B}$ is written by rows as

$$\mathbf{B} = \begin{bmatrix} \mathbf{B}_1 \\ \vdots \\ \mathbf{B}_n \end{bmatrix},$$

then the i-th row of the controllability matrix $\mathscr{C}$ is $[\mathbf{B}_i, \ \lambda_i \mathbf{B}_i, \ \cdots \lambda_i^{n-1}\mathbf{B}_i]$. Then if any row $\mathbf{B}_i$ is zero, the i-th row of $\mathscr{C}$ is zero, implying that $\text{rank}(\mathscr{C}) < n$ and $(\mathbf{A}, \mathbf{B})$ is not a controllable pair.

Similarly, if

$$\mathbf{C} = [\mathbf{C}_1, \ \mathbf{C}_2, \ \cdots \mathbf{C}_n],$$

then the i-th column of $\mathscr{O}$ is $\begin{bmatrix} \mathbf{C}_i \\ \mathbf{C}_i \lambda_i \\ \vdots \\ \mathbf{C}_i \lambda_i^{n-1} \end{bmatrix}$, which is zero if $\mathbf{C}_i$ is zero, implying

that $\text{rank}(\mathscr{O}) < n$ and $(\mathbf{C}, \mathbf{A})$ is not an observable pair.

---

5 The Lyapunov equation is

$$\mathbf{A}\mathbf{W}_{\text{C}}\mathbf{A}^T - \mathbf{W}_{\text{C}} + \mathbf{B}\mathbf{B}^T =$$

$$\begin{bmatrix} -1 & 6 \\ 1 & 0 \end{bmatrix} \begin{bmatrix} w_{11} & w_{12} \\ w_{12} & w_{22} \end{bmatrix} \begin{bmatrix} -1 & 1 \\ 6 & 0 \end{bmatrix} - \begin{bmatrix} w_{11} & w_{12} \\ w_{12} & w_{22} \end{bmatrix} + \begin{bmatrix} 2 \\ 1 \end{bmatrix} [2 \ \ 1] = 0,$$

where the unknown $\mathbf{W}_{\text{C}}$ has been written as a symmetric matrix. One method of solving this linear equation is to equate each entry of the 2×2 left-hand side to 0, giving four equations, but only three independent equations in the three unknowns, written

$$\begin{bmatrix} 0 & -12 & 36 \\ -1 & 5 & 0 \\ 1 & 0 & -1 \end{bmatrix} \begin{bmatrix} w_{11} \\ w_{12} \\ w_{22} \end{bmatrix} = \begin{bmatrix} -4 \\ -2 \\ -1 \end{bmatrix}.$$

Solving, we get

$$\mathbf{W}_C = \begin{bmatrix} -4/3 & -2/3 \\ -2/3 & -1/3 \end{bmatrix},$$

which has zero determinant, so the system is not completely controllable.

Chapter 10   Solutions

1 In each case the elementary row and column operations are shown. The sequence of operations is not unique.

(a)

$$\mathbf{G}(s) = \begin{bmatrix} s+1 & 0 \\ s+2 & s+2 \end{bmatrix} \xrightarrow{H_{21}(-1)} \begin{bmatrix} s+1 & 0 \\ 1 & s+2 \end{bmatrix} \xrightarrow{H_{12}} \begin{bmatrix} 1 & s+2 \\ s+1 & 0 \end{bmatrix}$$

$$\xrightarrow{H_{21}(-(s+1))} \begin{bmatrix} 1 & s+2 \\ 0 & -(s+1)(s+2) \end{bmatrix}$$

$$\xrightarrow[K_2(\ \ 1)]{K_{21}(-(s+2))} \begin{bmatrix} 1 & 0 \\ 0 & (s+1)(s+2) \end{bmatrix}.$$

(b)

$$\mathbf{G}(s) = \begin{bmatrix} s+1 & 0 \\ 0 & s+2 \end{bmatrix} \xrightarrow{H_{12}(1)} \begin{bmatrix} s+1 & s+2 \\ 0 & s+2 \end{bmatrix} \xrightarrow[K_{12}]{K_{21}(-1)} \begin{bmatrix} 1 & s+1 \\ s+2 & 0 \end{bmatrix}$$

$$\xrightarrow[K_{21}(-(s+1))]{H_{21}(-(s+2))} \begin{bmatrix} 1 & 0 \\ 0 & -(s+2)(s+1) \end{bmatrix} \xrightarrow{K_2(-1)} \begin{bmatrix} 1 & 0 \\ 0 & (s+2)(s+1) \end{bmatrix}.$$

(c)

$$\mathbf{G}(s) = \begin{bmatrix} s^3+3s^2+5s+3 & s^3+3s^2+6s+4 \\ s^2+2s+1 & s^2+s \end{bmatrix}$$

$$\xrightarrow{H_{12}} \begin{bmatrix} s^2+2s+1 & s^2+s \\ s^3+3s^2+5s+3 & s^3+3s^2+6s+4 \end{bmatrix}$$

$$\xrightarrow[K_{21}(-1)]{H_{21}(-s-1)} \begin{bmatrix} s^2+2s+1 & -s-1 \\ 2s+2 & s^2+3s+2 \end{bmatrix} \xrightarrow{K_{12}} \begin{bmatrix} -s-1 & s^2+2s+1 \\ s^2+3s+2 & 2s+2 \end{bmatrix}$$

$$\xrightarrow[K_{21}(s+1)]{H_{21}(s+2)} \begin{bmatrix} -s-1 & 0 \\ 0 & s^3+4s^2+7s+4 \end{bmatrix} \xrightarrow{H_1(-1)} \begin{bmatrix} s+1 & 0 \\ 0 & s^3+4s^2+7s+4 \end{bmatrix}.$$

3 The poles and transmission zeros will be found from the Smith-McMillan form and from the system matrices.

(a) i. The transfer matrix is

$$\hat{\mathbf{H}}(s) = 2 + 8\frac{1}{s-3}1 = \frac{1}{s-3}(2s+2) = \frac{1}{\phi(s)}G(s).$$

The Smith form of $\mathbf{G}(s)$ is obtained as

$$\mathbf{G}(s) \xrightarrow{H_1(1/2)} s+1 = \mathbf{S}(s),$$

and the Smith-McMillan form is

$$\frac{1}{\phi(s)}\mathbf{S}(s) = \frac{s+1}{s-3} = \mathbf{M}(s).$$

Hence the pole is at 3 and the zero at -1.

ii. The poles are obtained from

$$\det\begin{bmatrix} \mathbf{A}-s\mathbf{I} & 0 \\ \mathbf{C} & -\mathbf{I} \end{bmatrix} = \det\begin{bmatrix} 3-s & 0 \\ 8 & -1 \end{bmatrix} = s-3;$$

therefore there is a pole at 3.

The zeros are obtained from

$$\det\begin{bmatrix} \mathbf{A}-s\mathbf{I} & \mathbf{B} \\ \mathbf{C} & \mathbf{D} \end{bmatrix} = \det\begin{bmatrix} 3-s & 1 \\ 8 & 2 \end{bmatrix} = -2(s+1);$$

therefore there is a zero at -1.

(b) i. The transfer matrix is

$$\hat{\mathbf{H}}(s) = 0 + \begin{bmatrix} 0 & 1 & 0 \\ 0 & 0 & 1 \end{bmatrix}\begin{bmatrix} s & 2 & 0 \\ -1 & s+3 & 0 \\ 0 & 0 & s+2 \end{bmatrix}^{-1}\begin{bmatrix} 0 & 0 \\ 1 & 0 \\ 0 & -2 \end{bmatrix}$$

$$= \begin{bmatrix} 0 & 1 & 0 \\ 0 & 0 & 1 \end{bmatrix}\frac{1}{(s+2)^2(s+1)}\begin{bmatrix} \times & \times & \times \\ \times & s(s+2) & 0 \\ \times & 0 & s^2+3s+2 \end{bmatrix}\begin{bmatrix} 0 & 0 \\ 1 & 0 \\ 0 & -2 \end{bmatrix}$$

$$= \frac{1}{(s+2)^2(s+1)}\begin{bmatrix} s^2+2s & 0 \\ 0 & -2(s^2+3s+2) \end{bmatrix} = \frac{1}{\phi(s)}G(s).$$

The Smith form of $\mathbf{G}(s)$ is obtained as

$$\mathbf{G}(s) \xrightarrow[\longrightarrow]{\substack{H_2(-1/2) \\ H_{12}(1)}} \begin{bmatrix} s^2+2s & s^2+3s+2 \\ 0 & s^2+3s+2 \end{bmatrix} \xrightarrow[\longrightarrow]{K_{21}(-1)} \begin{bmatrix} s^2+2s & s+2 \\ 0 & s^2+3s+2 \end{bmatrix}$$

$$\xrightarrow[\longrightarrow]{K_{12}} \begin{bmatrix} s+2 & s^2+2s \\ s^2+3s+2 & 0 \end{bmatrix} \xrightarrow[\longrightarrow]{H_{21}(-s-1)} \begin{bmatrix} s+2 & s^2+2s \\ 0 & -s^3-3s^2-2s \end{bmatrix}$$

$$\xrightarrow[\longrightarrow]{K_{21}(-s)} \begin{bmatrix} s+2 & 0 \\ 0 & -s^3-3s^2-2s \end{bmatrix} \xrightarrow[\longrightarrow]{H_2(-1)} \begin{bmatrix} s+2 & 0 \\ 0 & s(s+1)(s+2) \end{bmatrix}$$

$$= \mathbf{S}(s).$$

The Smith-McMillan form is

$$\frac{1}{\phi(s)}\mathbf{S}(s) = \begin{bmatrix} 1/((s+1)(s+2)) & 0 \\ 0 & s/(s+2) \end{bmatrix} = \mathbf{M}(s).$$

Therefore, there are poles at -1, -2, -2, and there is one zero at 0.

ii. Computing the determinant to obtain the pole polynomial gives

$$\det \begin{bmatrix} -s & -2 & 0 & 0 & 0 \\ 1 & -3-s & 0 & 0 & 0 \\ 0 & 0 & -2-s & 0 & 0 \\ 0 & 1 & 0 & -1 & 0 \\ 0 & 0 & 1 & 0 & -1 \end{bmatrix} = -(s+2)^2(s+1);$$

therefore the poles are at $-1, -2, -2$.
The zero polynomial is

$$\det \begin{bmatrix} -s & -2 & 0 & 0 & 0 \\ 1 & -3-s & 0 & 1 & 0 \\ 0 & 0 & -2-s & 0 & -2 \\ 0 & 1 & 0 & 0 & 0 \\ 0 & 0 & 1 & 0 & 0 \end{bmatrix} = 2s;$$

therefore there is one zero at 0.

(c) i. The transfer matrix is

$$\hat{\mathbf{H}}(s) = 0 + \begin{bmatrix} 0 & 1 & 0 \\ 0 & 0 & 1 \end{bmatrix} \begin{bmatrix} s & 4 & 0 \\ -1 & s+5 & 0 \\ 0 & 0 & s+4 \end{bmatrix}^{-1} \begin{bmatrix} 0 & 0 \\ 1 & 1 \\ -1 & 2 \end{bmatrix}$$

$$= \begin{bmatrix} 0 & 1 & 0 \\ 0 & 0 & 1 \end{bmatrix} \frac{1}{(s+1)(s+4)^2} \begin{bmatrix} \times & \times & \times \\ \times & s(s+4) & 0 \\ \times & 0 & s^2+5s+4 \end{bmatrix} \begin{bmatrix} 0 & 0 \\ 1 & 1 \\ -1 & 2 \end{bmatrix}$$

$$= \frac{1}{(s+4)^2(s+1)} \begin{bmatrix} s^2+4s & s^2+4s \\ -s^2-5s-4 & 2(s^2+5s+4) \end{bmatrix} = \frac{1}{\phi(s)}\mathbf{G}(s).$$

The Smith form of $\mathbf{G}(s)$ is obtained as

$$\mathbf{G}(s) \xrightarrow{H_{21}(1)} (s+4)\begin{bmatrix} s & s \\ -1 & 3s+2 \end{bmatrix} \xrightarrow{K_{21}(-1)} (s+4)\begin{bmatrix} s & 0 \\ -1 & 3s+3 \end{bmatrix}$$

$$\xrightarrow{H_{12}} (s+4)\begin{bmatrix} -1 & 3s+3 \\ s & 0 \end{bmatrix} \xrightarrow{H_1(-1)} (s+4)\begin{bmatrix} 1 & -3s-3 \\ s & 0 \end{bmatrix}$$

$$\xrightarrow{H_{21}(-s)} (s+4)\begin{bmatrix} 1 & -3(s+1) \\ 0 & 3s(s+1) \end{bmatrix} \xrightarrow[K_2(1/3)]{K_{21}(3(s+1))} (s+4)\begin{bmatrix} 1 & 0 \\ 0 & s(s+1) \end{bmatrix}$$

$$= \mathbf{S}(s).$$

The Smith-McMillan form is

$$\frac{1}{\phi(s)}\mathbf{S}(s) = \begin{bmatrix} 1/((s+1)(s+4)) & 0 \\ 0 & s/(s+4) \end{bmatrix} = \mathbf{M}(s).$$

Therefore, there are poles at $-1, -4, -4$, and there is one zero at 0.

ii. Computing the determinant to obtain the pole polynomial,

$$\det \begin{bmatrix} -s & -4 & 0 & 0 & 0 \\ 1 & -5-s & 0 & 0 & 0 \\ 0 & 0 & -4-s & 0 & 0 \\ 0 & 1 & 0 & -1 & 0 \\ 0 & 0 & 1 & 0 & -1 \end{bmatrix} = -(s+1)(s+4)^2;$$

therefore the poles are at $-1, -4, -4$.
The zero polynomial is

$$\det \begin{bmatrix} -s & -4 & 0 & 0 & 0 \\ 1 & -5-s & 0 & 1 & 1 \\ 0 & 0 & -4-s & -1 & -2 \\ 0 & 1 & 0 & 0 & 0 \\ 0 & 0 & 1 & 0 & 0 \end{bmatrix} = s;$$

therefore there is one zero at 0.

5 The required quantities are found as follows.

(a) The transfer matrix $\hat{\mathbf{H}}(s)$ is

$$\hat{\mathbf{H}}(s) = \mathbf{C}(s\mathbf{I}-\mathbf{A})^{-1}\mathbf{B} + \mathbf{D}$$

$$= \begin{bmatrix} 0 & 1 & 0 \\ 0 & 0 & 1 \end{bmatrix} \begin{bmatrix} \dfrac{1}{s^2+5s+4}\begin{bmatrix} s+5 & -4 \\ 1 & s \end{bmatrix} & 0 \\ 0 & 0 & \dfrac{1}{s+4} \end{bmatrix} \begin{bmatrix} 0 & 0 \\ 1 & 1 \\ -1 & 2 \end{bmatrix}$$

$$+ \begin{bmatrix} 0 & 0 \\ 0 & 0 \end{bmatrix}$$

$$= \begin{bmatrix} \dfrac{1}{(s+4)(s+1)} & \dfrac{s}{(s+4)(s+1)} & 0 \\ 0 & 0 & \dfrac{1}{s+4} \end{bmatrix} \begin{bmatrix} 0 & 0 \\ 1 & 1 \\ -1 & 2 \end{bmatrix}$$

$$= \begin{bmatrix} \dfrac{s}{(s+1)(s+4)} & \dfrac{s}{(s+1)(s+4)} \\ \dfrac{-1}{s+4} & \dfrac{2}{s+4} \end{bmatrix}$$

$$= \dfrac{1}{(s+1)(s+4)} \begin{bmatrix} s & s \\ -s-1 & 2(s+1) \end{bmatrix}.$$

(b) The Smith form of $\mathbf{G}(s)$ is obtained as

$$\begin{bmatrix} s & s \\ -s-1 & 2s+2 \end{bmatrix} \xrightarrow{H_{21}(1)} \begin{bmatrix} s & s \\ -1 & 3s+2 \end{bmatrix} \xrightarrow{\substack{K_{21}(-1) \\ H_{12}}} \begin{bmatrix} -1 & 3s+3 \\ s & 0 \end{bmatrix}$$

$$\xrightarrow{\substack{H_{21}(s) \\ K_{21}(3s+3)}} \begin{bmatrix} -1 & 0 \\ 0 & 3s^2+3s \end{bmatrix} \xrightarrow{\substack{H_1(-1) \\ H_2(1/3)}} \begin{bmatrix} 1 & 0 \\ 0 & s^2+s \end{bmatrix}.$$

Therefore the Smith-McMillan form is calculated as

$$\frac{1}{(s+1)(s+4)} \begin{bmatrix} 1 & 0 \\ 0 & s(s+1) \end{bmatrix} = \begin{bmatrix} \frac{1}{(s+1)(s+4)} & 0 \\ 0 & \frac{s}{s+4} \end{bmatrix}.$$

(c) From the Smith-McMillan form, there is one transmission zero at 0 and the transmission poles are at $-1, -4, -4$.

(d) The controllability matrix $\mathscr{C}$ and observability matrix $\mathscr{O}$ are

$$\mathscr{C} = [\mathbf{B}, \ \mathbf{AB}, \ \mathbf{A}^2\mathbf{B}] = \begin{bmatrix} 0 & 0 & -4 & -4 & 20 & 20 \\ 1 & 1 & -5 & -5 & 21 & 21 \\ -1 & 2 & 4 & -8 & -16 & 32 \end{bmatrix},$$

$$\mathscr{O} = \begin{bmatrix} \mathbf{C} \\ \mathbf{CA} \\ \mathbf{CA}^2 \end{bmatrix} = \begin{bmatrix} 0 & 1 & 0 \\ 0 & 0 & 1 \\ 1 & -5 & 0 \\ 0 & 0 & -4 \\ -5 & 21 & 0 \\ 0 & 0 & 16 \end{bmatrix},$$

both of which are of full rank 3. Therefore the transmission poles are the eigenvalues of $\mathbf{A}$, which are at $-1, -4, -4$, and the transmission zeros are the roots of

$$\det \begin{bmatrix} \mathbf{A} - s\mathbf{I} & \mathbf{B} \\ \mathbf{C} & \mathbf{D} \end{bmatrix} = \det \begin{bmatrix} -s & -4 & 0 & 0 & 0 \\ 1 & -s-5 & 0 & 1 & 1 \\ 0 & 0 & -s-4 & -1 & 2 \\ 0 & 1 & 0 & 0 & 0 \\ 0 & 0 & 1 & 0 & 0 \end{bmatrix} = -3s,$$

so there is one transmission zero at 0.

---

7   In each case, applying Equation (10.9), the matrices $\mathbf{D}(s)$, $\mathbf{N}(s)$ of the left-factorization $\mathbf{H}(s) = \mathbf{D}^{-1}(s)\,\mathbf{N}(s)$ are given from computation of the Smith-McMillan form by

$$\mathbf{D}(s) = \begin{bmatrix} \text{diag}\,[\psi_i(s)] & 0 \\ 0 & \mathbf{I}_{p-r} \end{bmatrix} \mathbf{P}(s), \quad \mathbf{N}(s) = \begin{bmatrix} \text{diag}\,[\epsilon_i(s)] & 0 \\ 0 & 0 \end{bmatrix} \mathbf{Q}^{-1}(s),$$

where $\mathbf{P}(s)\,\mathbf{H}(s)\,\mathbf{Q}(s) = \frac{1}{d(s)}\mathbf{P}(s)\,\mathbf{G}(s)\,\mathbf{Q}(s) = \frac{1}{d(s)}\mathbf{S}(s)$ and $\mathbf{S}(s)$ is the Smith form of $\mathbf{G}(s)$. The matrix $\mathbf{P}(s)$ is obtained from an identity matrix by doing row operations as they are performed on $\mathbf{G}(s)$ in the Smith-form computation, and $\mathbf{Q}(s)$ corresponds to the column operations. Because the sequence of row and column operations producing the Smith form is not unique, $\mathbf{D}(s)$ and $\mathbf{N}(s)$ are not unique. The matrices of Problem 3 will be investigated first.

(a) The Smith-McMillan computation gives

$$\mathbf{P}(s) = \mathbf{H}(1, 1/2) = 1/2, \quad \mathbf{Q}(s) = 1, \quad \text{diag}[\psi_i(s)] = s-3,$$
$$\text{diag}[\epsilon_i(s)] = s+1,$$

so that

$$\mathbf{D}(s) = (s-3)/2, \quad \mathbf{N}(s) = s+1.$$

(b) The computed matrices are as shown, using the same Smith-form computation as for Problem 3. Note that $r = 2$:

$$\phi(s) = s^3+5s^2+8s+4, \quad \mathbf{G}(s) = \begin{bmatrix} s^2+2s & 0 \\ 0 & -2s^2-6s-4 \end{bmatrix},$$

$$\mathbf{S}(s) = \begin{bmatrix} s+2 & 0 \\ 0 & s^3+3s^2+2s \end{bmatrix}, \quad \mathbf{P}(s) = \begin{bmatrix} 1 & -(1/2) \\ s+1 & -(1/2)s \end{bmatrix},$$

$$\mathbf{Q}(s) = \begin{bmatrix} -1 & s+1 \\ 1 & -s \end{bmatrix}, \quad \text{diag}[\epsilon_i(s)] = \begin{bmatrix} 1 & 0 \\ 0 & s \end{bmatrix},$$

$$\text{diag}[\psi_i(s)] = \begin{bmatrix} s^2+3s+2 & 0 \\ 0 & s+2 \end{bmatrix},$$

$$\mathbf{D}(s) = \begin{bmatrix} \text{diag}[\psi_i(s)] & 0 \\ 0 & \mathbf{I}_{p-r} \end{bmatrix} \mathbf{P}(s) = \begin{bmatrix} s^2+3s+2 & -(1/2)s^2-(3/2)s-1 \\ s^2+3s+2 & -(1/2)s^2-s \end{bmatrix},$$

$$\mathbf{N}(s) = \begin{bmatrix} \text{diag}[\epsilon_i(s)] & 0 \\ 0 & 0 \end{bmatrix} \mathbf{Q}^{-1}(s) = \begin{bmatrix} s & s+1 \\ s & s \end{bmatrix}.$$

(c) The computed matrices are as shown. Again $r = 2$:

$$\phi(s) = s^3+9s^2+24s+16, \quad \mathbf{G}(s) = \begin{bmatrix} s^2+4s & s^2+4s \\ -s^2-5s-4 & 2s^2+10s+8 \end{bmatrix},$$

$$\mathbf{S}(s) = \begin{bmatrix} s+4 & 0 \\ 0 & s^3+5s^2+4s \end{bmatrix}, \quad \mathbf{P}(s) = \begin{bmatrix} -1 & -1 \\ (1/3)s+(1/3) & (1/3)s \end{bmatrix},$$

$$\mathbf{Q}(s) = \begin{bmatrix} 1 & 3s+2 \\ 0 & 1 \end{bmatrix}, \quad \text{diag}[\epsilon_i(s)] = \begin{bmatrix} 1 & 0 \\ 0 & s \end{bmatrix},$$

$$\text{diag}[\psi_i(s)] = \begin{bmatrix} s^2+5s+4 & 0 \\ 0 & s+4 \end{bmatrix},$$

$$\mathbf{D}(s) = \begin{bmatrix} \text{diag}[\psi_i(s)] & 0 \\ 0 & \mathbf{I}_{p-r} \end{bmatrix} \mathbf{P}(s)$$
$$= \begin{bmatrix} -s^2-5s-4 & -s^2-5s-4 \\ (1/3)s^2+(5/3)s+(4/3) & (1/3)s^2+(4/3)s \end{bmatrix},$$

$$\mathbf{N}(s) = \begin{bmatrix} \text{diag}[\epsilon_i(s)] & 0 \\ 0 & 0 \end{bmatrix} \mathbf{Q}^{-1}(s) = \begin{bmatrix} 1 & -3s-2 \\ 0 & s \end{bmatrix}.$$

For the system of Problem 6 the matrices are as shown:

$$\phi(s) = s^3+7s^2+15s+9, \quad \mathbf{G}(s) = \begin{bmatrix} s^2+3s \\ -3s^2-12s-9 \end{bmatrix},$$

$$\mathbf{S}(s) = \begin{bmatrix} s+3 \\ 0 \end{bmatrix}, \quad \mathbf{Q}(s) = 1,$$

$$\mathbf{P}(s) = \mathbf{H}_1(-(1/3))\,\mathbf{H}_{21}((1/3)s)\,\mathbf{H}_{12}\,\mathbf{H}_{21}(3) = \begin{bmatrix} -1 & -(1/3) \\ s+1 & (1/3)s \end{bmatrix},$$

$$\begin{bmatrix} \text{diag}\,[\,\epsilon_i(s)\,] & 0 \\ 0 & 0 \end{bmatrix} = \begin{bmatrix} 1 \\ 0 \end{bmatrix}, \quad \begin{bmatrix} \text{diag}\,[\,\psi_i(s)\,] & 0 \\ 0 & \mathbf{I}_{p-r} \end{bmatrix} = \begin{bmatrix} s^2+4s+3 & 0 \\ 0 & 1 \end{bmatrix},$$

$$\mathbf{D}(s) = \begin{bmatrix} \text{diag}\,[\,\psi_i(s)\,] & 0 \\ 0 & \mathbf{I}_{p-r} \end{bmatrix} \mathbf{P}(s) = \begin{bmatrix} -s^2-4s-3 & -(1/3)s^2-(4/3)s-1 \\ s+1 & (1/3)s \end{bmatrix},$$

$$\mathbf{N}(s) = \begin{bmatrix} \text{diag}\,[\,\epsilon_i(s)\,] & 0 \\ 0 & 0 \end{bmatrix} \mathbf{Q}^{-1}(s) = \begin{bmatrix} 1 \\ 0 \end{bmatrix}.$$

Index